Electrosynthesis Faraday Discussion

John McIntyre Conference Centre, Edinburgh, United Kingdom and online

12 - 14 July 2023

FARADAY DISCUSSIONS
Volume 247, 2023

ROYAL SOCIETY
OF **CHEMISTRY**

The Faraday Community for Physical Chemistry of the Royal Society of Chemistry, previously the Faraday Society, was founded in 1903 to promote the study of sciences lying between chemistry, physics and biology.

Editorial Staff

Executive Editor
Michael A. Rowan

Deputy Editor
Vikki Pritchard

Development Editors
Bee Hockin, Andrea Carolina Ojeda Porras

Editorial Production Manager
Gisela Scott

Senior Publishing Editor
Robin Brabham

Publishing Editors
Kirsty McRoberts and Hugh Ryan

Editorial Assistant
Daphne Houston

Publishing Assistants
Julie Ann Roszkowski and Huw Hedges

Publisher
Jeanne Andres

Faraday Discussions (Print ISSN 1359-6640, Electronic ISSN 1364-5498) is published 8 times a year by the Royal Society of Chemistry, Thomas Graham House, Science Park, Milton Road, Cambridge, UK CB4 0WF.

Volume 247 ISBN 978-1-83767-094-9

2023 annual subscription price: print+electronic £1223 US $2154; electronic only £1165, US $2051. Customers in Canada will be subject to a surcharge to cover GST. Customers in the EU subscribing to the electronic version only will be charged VAT.

All orders, with cheques made payable to the Royal Society of Chemistry, should be sent to the Royal Society of Chemistry Order Department, Royal Society of Chemistry, Thomas Graham House, Science Park, Milton Road, Cambridge, CB4 0WF, UK Tel +44 (0)1223 432398; E-mail orders@rsc.org

If you take an institutional subscription to any Royal Society of Chemistry journal you are entitled to free, site-wide web access to that journal. You can arrange access via Internet Protocol (IP) address at www.rsc.org/ip

Customers should make payments by cheque in sterling payable on a UK clearing bank or in US dollars payable on a US clearing bank.

Faraday Discussions

Faraday Discussions are unique international discussion meetings that focus on rapidly developing areas of chemistry and its interfaces with other scientific disciplines.

Information for Authors

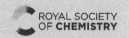

ROYAL SOCIETY
OF CHEMISTRY

MIX
Paper from responsible sources
FSC
FSC® C013604

Electrosynthesis

Faraday Discussions

www.rsc.org/faraday_d

A General Discussion on Electrosynthesis was held in Edinburgh, UK and online on the 12th, 13th and 14th of July 2023.

The Royal Society of Chemistry is the world's leading chemistry community. Through our high impact journals and publications we connect the world with the chemical sciences and invest the profits back into the chemistry community.

CONTENTS

ISSN 1359-6640; ISBN 978-1-83767-094-9

Faraday
Discussions
Volume: 247

**Electrosynthesis
Faraday Discussion**

ROYAL SOCIETY
OF CHEMISTRY

Cover
R. Price *et al.*, *Faraday Discuss.*, 2023, **247**, 268–288.

Whisky distillery co-products are electrochemically converted to hydrogen using a phosphomolybdic acid catalyst. A process overview through the Isle of Raasay Distillery's stillhouse window.

Image reproduced by permission of Robert Price, the University of Strathclyde and R&B Distillers Ltd. from R. Price *et al.*, *Faraday Discuss.*, 2023, **247**, 268–288.

INTRODUCTORY LECTURE

PAPERS AND DISCUSSIONS

WILEY

CONCLUDING REMARKS

ADDITIONAL INFORMATION

Note Added After First Publication

We regret that for the first publication of this volume on the 1st of November 2023, the placement of the General Discussion sections in the order of the papers was incorrect. The order of these discussion sections has now been corrected and page numbers shown in this volume are correct.

The Royal Society of Chemistry apologises for these errors and any consequent inconvenience to authors and readers.

Faraday Discussions

ROYAL SOCIETY
OF CHEMISTRY

PAPER

Spiers Memorial Lecture: Old but new organic electrosynthesis: history and recent remarkable developments

Toshio Fuchigami (iD)

Received 27th June 2023, Accepted 7th July 2023

DOI: 10.1039/d3fd00129f

Organic electrosynthesis has a long history. However, this chemistry is still new. Recently, we have seen its second renaissance with organic electrosynthesis being considered a typical green chemistry process. Therefore, a number of novel electrosynthetic methodologies have recently been developed. However, there are still many problems to be solved from a green and sustainable viewpoint. After an explanation of the historical survey of organic electrosynthesis, this paper focuses on recent remarkable developments in new electrosynthetic methodologies, such as novel electrodes, recyclable nonvolatile electrolytic solvents and recyclable supporting electrolytes, as well as new types of electrolytic flow cells. Furthermore, novel types of organic electrosynthetic reactions will be mentioned.

1. Introduction

Organic electrosynthesis has a long history, but it is still new, and much attention has been paid to organic electrosynthesis because it is a typical green synthetic method. In this paper, the historical background of organic electrosynthesis is described and then the first and second renaissances of electrosynthesis are illustrated.

The Italian physicist Volta invented the so-called Voltaic pile in 1800.[1] Three years later, Petrov in Russia published a paper on the electrolysis of alcohols and aliphatic oils.[2] One year later, Grotthuss in Lithuania, who proposed the ionic conducting mechanism, found that a diluted solution of indigo white could be easily electrochemically oxidized to indigo blue.[3] In 1828, Wöhler successfully prepared urea for the first time from ammonium cyanate.[4] In 1833, Faraday discovered Faraday's Law,[5] and one year later he found that hydrocarbons were formed by the electrolysis of an aqueous solution of potassium acetate. Unfortunately, he could not identify the products. In 1849, Wöhler's disciple, Kolbe, discovered the electrochemical oxidation of isovaleric acid salt forming mainly an isobutyl radical and CO_2.[6] Actually, the main product was found to be a dimer of

Department of Electronic Chemistry, Tokyo Institute of Technology, Japan. E-mail: fuchi@echem.titech.ac.jp

an isobutyl radical. This type of reaction is well known as Kolbe electrolysis. Consequently, Faraday and Kolbe are pioneers in the study of organic electro-chemistry. From the end of the 19th century to the first half of the 20th century, various electrochemical processes, exemplified by the reduction of nitrobenzenes to aniline derivatives, have been intensively investigated. Organic electrochem-istry was also developed along with the discovery of new electroanalytical tech-niques, such as polarography, which was developed by Heyrovský and Tachi in 1922.[7]

After World War II, the development of cyclic voltammetry and aprotic polar solvents such as DMF (that have a relatively wide potential window and good solubility of poorly soluble organic compounds) enabled electrochemical measurements of various organic compounds, as exemplified by the detection of unstable intermediates, such as radical cations and anions. According to Heyr-ovsky's disciple, Zuman, polarography was among the top five most powerful analytical methods between 1950 and 1975, and contributed greatly to the development of physical organic chemistry.[8]

2.   The first renaissance of organic electrolysis

In 1964, Baizer developed a highly useful industrial process, the electrochemical hydrodimerization of acrylonitrile, which is useful for the production of adipo-nitrile.[9] Since then, the development of organic electrosynthesis has been markedly advanced by incorporating new types of organic reactions, modern organic synthesis, and physical organic chemistry. In addition, cyclic voltammetry and related electroanalytical techniques, including spectroscopic analysis, assis-ted in the understanding of the kinetics and mechanisms of organic electrode processes in the last decade of the 20th century. This period is the first renais-sance of organic electrochemistry. Interestingly, Nobel laureates Haber, Corey, Noyori and Suzuki used to work on organic electrosynthesis, such as cathodic reduction of nitrobenzene, Kolbe electrolysis of optically active carboxylic acid,[10] anodic glycoxylation,[11] and anodic oxidation of organoboron compounds,[12] respectively. These facts clearly indicate that organic electrosynthesis is highly attractive even for such great chemists.

With the development of synthetic organic chemistry since the 1980s, organic electrochemists have focused their efforts on the precise control of organic electrochemical reactions, such as regioselectivity, chemoselectivity, stereo-selectivity, and asymmetric synthesis. By the end of the 20th century, a number of new methodologies for organic electrosynthesis had been developed in order to overcome the drawbacks of electrosynthesis as follows.

2.1.   Expansion of the electrochemical reaction field (from two-dimensional reactions to three-dimensional reactions)

An electrochemical reaction is a typical heterogeneous reaction; therefore, the reaction field is limited. In order to expand the reaction dimension, a "fluidized bed electrode"[13] and "mediators" were developed. The former is mainly based on chemical engineering; therefore, the details are not explained here.

   2.1.1   Mediator. In order to enlarge the reaction field and avoid electrode passivation, and also to decrease the overpotentials of electron transfer, various inorganic mediators, such as multivalent metal ions, and halide ions, such as Cl⁻, Br⁻ and I⁻, have been used for a long time. Quite recently, halide ions have been shown to be a powerful mediator for the electrosynthesis of protected dehydroamino acids.[14] Different from halide ion mediators such as Cl^-, Br^- and I^-, the "fluoride ion mediator" is not a redox catalyst, but it greatly enhances the anodic substitution of organosulfur and selenium compounds. For example, sulfides do not undergo anodic α-alkoxylation in general unless they have a strongly electron-withdrawing group (EWG), such as CN and CF_3 groups, at their α-position. However, it was found that fluoride ions markedly promoted the anodic methoxylation of sulfides devoid of an α-EWG.[15]

As shown in Scheme 1, methoxylation proceeds *via* a fluorosulfonium ion intermediate (**A**) similarly to a Pummerer-type reaction. This mediator can be widely applied to various open-chain and sulfur-containing heterocycles and selenides.[16] Moreover, fluoride ions markedly promote anodic cyclization reactions, as shown in Scheme 2.[17] Notably, ultrasonication enhances the cyclization of intermediate **B**.

Moreover, a fluoride ion mediator enables the following intramolecular cyclization of trifluoromethylated sulfides (Schemes 3 and 4).[18,19]

Triarylamines, which are well known as outer-sphere electron transfer reagents, were used as a mediators for the first time for the oxidative deprotection of ketone dithioacetals by Steckhan.[20] Fullerene C_{60} undergoes reversible multiple electron transfer of up to 6 electrons depending on the measurement conditions. It was demonstrated to be a "power variable mediator" for the electrocatalytic dehalogenation of various *vic*-dihalo compounds.[21]

Franke and Little developed a new mediator, phenanthro[9,10-*d*]imidazoles for the effective anodic oxidation of benzyl alcohols to aldehydes. Notably, making

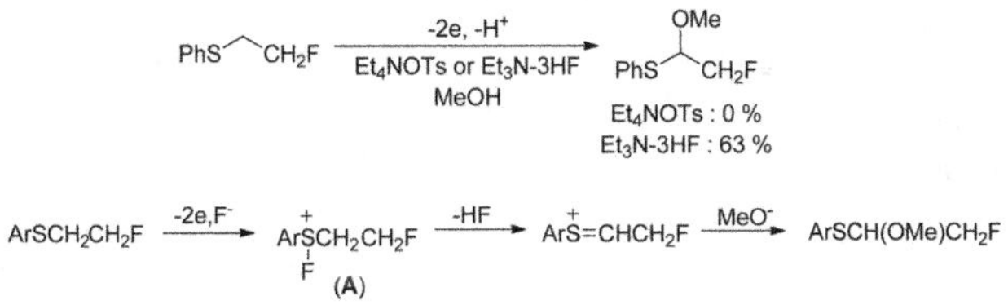

Scheme 1 Fluoride-ion-mediated anodic methoxylation of sulfides and reaction mechanism.

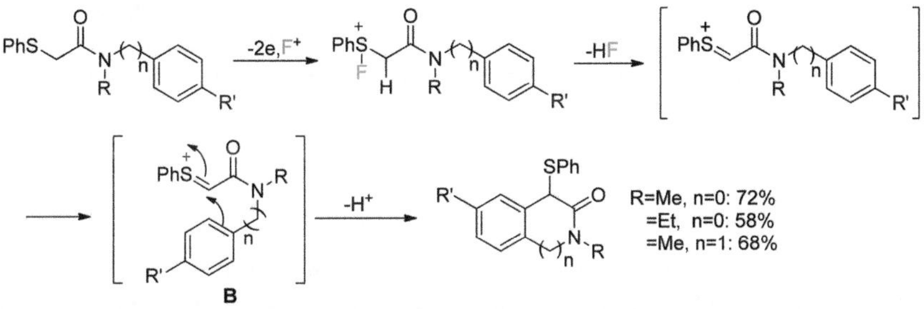

Scheme 2 Fluoride ion-mediated anodic intramolecular cyclization to form oxindole and 3-oxotetrahydroisoquinoline derivatives.

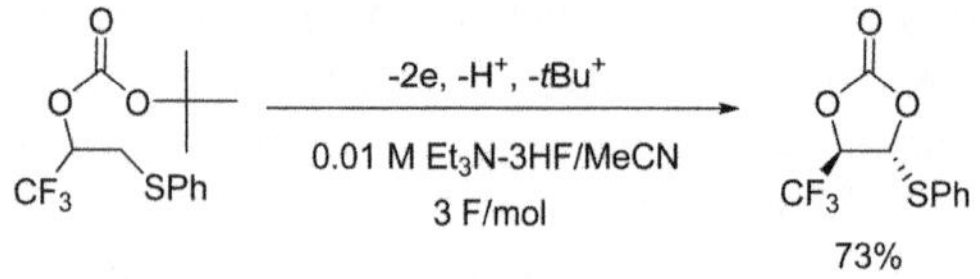
-2e, -H+, -tBu+
0.01 M Et3N-3HF/MeCN
3 F/mol
CF3
SPh
73%

Scheme 3 Fluoride-ion-mediated anodic intramolecular cyclization to form trifluoromethylated ethylene carbonate derivative.

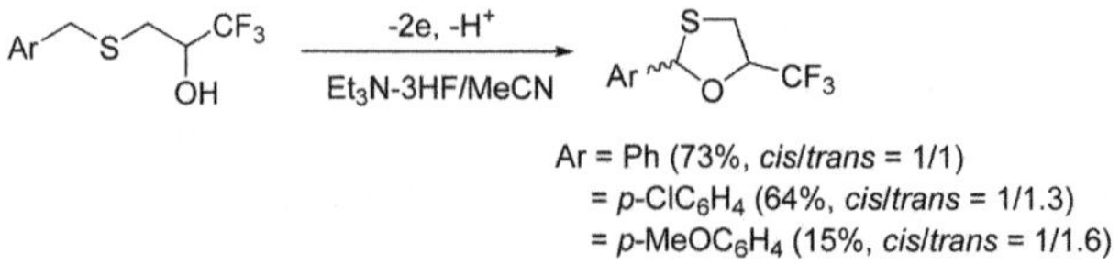
Ar
S
CF3
OH
-2e, -H+
Et3N-3HF/MeCN
Ar = Ph (73%, cis/trans = 1/1)
= p-ClC6H4 (64%, cis/trans = 1/1.3)
= p-MeOC6H4 (15%, cis/trans = 1/1.6)

Scheme 4 Fluoride-ion mediated anodic intramolecular cyclization to form CF_3-containing 1,3-oxathiolanes.

a fused framework at C-4 and C-5 of the substrates significantly changes their redox properties.[22] Baran *et al.* reported scalable, anodic oxidation of unactivated C–H bonds using a quinuclidine mediator.[23] They also developed *N*-ammonium ylide mediators for site-specific and chemoselective C(sp^3)–H oxidation.[24]

2.1.2 Reactive electrode (sacrificial electrode). These electrodes not only act as a sacrificial anode avoiding anodic degradation of products once formed at a cathode, but also form reactive organometallic reagents, such as the Reformatsky reagent, *in situ* from a Zn anode.

2.2. Photocatalysis using a semi-conductor (wireless electrolysis)

Fox and Ohtani intensively studied photocatalytic synthesis utilizing semiconductors.[25] A typical example is shown in the following figure (Fig. 1).[26]

Oxidation and reduction occur simultaneously at two sites on the small particle. Since the oxidation and reduction sites are very close to each other, various unique organic synthetic reactions have been developed. Photocatalytic reactions can be considered as wireless electrolytic reactions. Another example of "wireless electrolysis" is the bipolar electrode reaction, as explained in Section 4.1.9.

2.3. SPE (solid polymer electrolyte) electrolysis

Ogumi and Yoshizawa developed this system, which has big advantages, such as being supporting-electrolyte-free and having easy product separation. They applied this system to the cathodic reduction of olefins.[27] However, due to the short lifetime of SPEs and the contamination of SPE membranes, electrosynthesis using SPEs has not been explored.

2.4. Electrogenerated base (EGB) and acid (EGA)

EGBs and EGAs are utilized for syntheses that are difficult to achieve with other ordinary bases and acids. For example, the α-CF_3 enolate is unstable and will undergo defluorination, but when an EGB is used, a stable enolate can be generated, as shown in Scheme 5.[29]

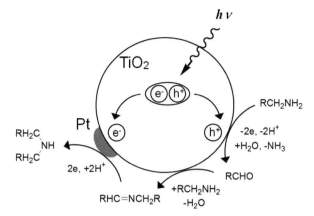

Fig. 1 Examples of photocatalytic organic reactions. Image reproduced with permission from Wiley.[28]

2.5. Paired electrosynthesis

A typical inorganic electrosynthesis of an aqueous solution of sodium chloride to produce sodium hydroxide and chlorine gas utilizes both cathodic and anodic reactions. In contrast, organic electrosynthesis usually uses either an anodic or cathodic reaction, and products at the counter electrode are wasted. Baizer first demonstrated paired electrosynthesis using both anodic and cathodic reactions to produce useful products.[30] At most, 200% current efficiency could be expected in the ideal case of paired electrosynthesis. An example of industrialized paired electrosynthesis is shown in Section 3 (Scheme 9).

2.6. Modified electrodes

2.6.1 Electrode modified with chiral molecules and chiral macromolecules.
Miller obtained 48% asymmetric yield in the cathodic reduction of 2-acetylpyridine to the corresponding alcohol in the presence of 0.5 mM of strychnine salt.[31] The asymmetric induction seems to be attributable to the chiral reaction fields constructed by the physical adsorption of optically active alkaloids on the cathode surface. However, Miller's results are not always reproducible. Later on, polymer-

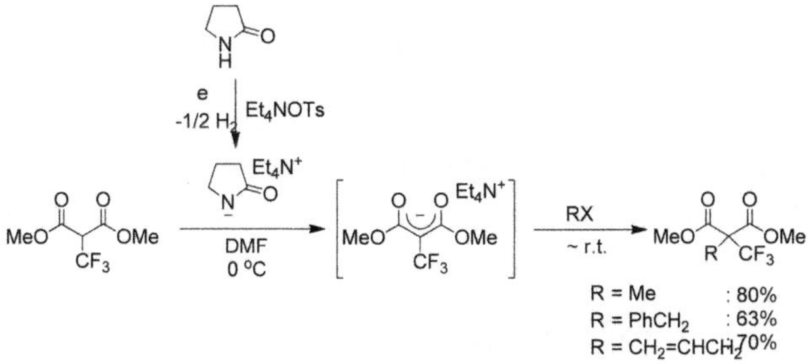

R = Me   : 80%
R = PhCH₂ : 63%
R = CH₂=CHCH₂ : 70%

Scheme 5 Generation of the α-CF₃ enolate using an EGB.

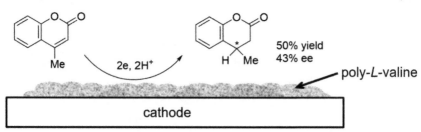

Fig. 2   Asymmetric reduction using a chiral polymer-modified electrode.

modified electrodes were developed in order to immobilize various catalysts. In order to solve problems such as the instability of electrodes modified with chiral adsorbates and the low density of the modified chiral compounds, chiral polymer-modified electrodes were developed. For instance, the authors prepared various electrodes coated with optically active poly(amino acid)s and applied these electrodes to the asymmetric reduction of olefins. Thus, 43% ee was obtained in the reduction of 4-methylcumarin at a cathode coated with poly(L-valine), as shown in Fig. 2.[32]

2.6.2 Hydrophobic electrodes

(a) An electrode coated with LB-films of a quaternary ammonium salt and a PTFE composite-plated electrode. Most electrodes are hydrophilic, but hydrophobic electrodes can be prepared as follows. An electrode coated with LB-films of a quaternary ammonium salt is hydrophobic, but its durability is not high enough.[33] Therefore, a PTFE composite-plated electrode was developed. This hydrophobic electrode can be prepared by composite-plating electrodes, such as Ni, Zn, Pb, *etc.*, in the corresponding metal-salt–poly(tetrafluoroethylene) (PTFE) dispersion plating bath. In general, hydrogen and oxygen evolution occur competitively at these electrodes during electrolytic reactions of organic compounds in an aqueous solution. Therefore, it is not easy to cathodically reduce carbonyl compounds in an aqueous acidic solution and anodically oxidize alcohols in an aqueous alkaline solution. However, nickel/PTFE composite-plated Ni electrodes are used for the reduction and oxidation, as mentioned above, to provide the corresponding alcohols and carbonyl compounds, respectively, with much higher current efficiency compared to unplated Ni electrodes (Scheme 6).[34,35] This is not because of the higher hydrogen- and oxygen-overpotentials of the composite-plated electrode suppressing hydrogen and oxygen evolution, but because of the substrate-collecting effect as a result of the strong hydrophobic interaction between the hydrophobic electrode surface and the hydrophobic organic substrates.

(b) PTFE-fiber-coated electrodes. PTFE-fibre-coated electrodes, which are prepared by wrapping the electrode with PTFE strings, are hydrophobic. It has been demonstrated that anodic oxidation of hydroquinones at this electrode in the presence of dienes in $NaClO_4/MeNO_2$ generated quinones, which underwent a Diels–Alder reaction with the dienes to provide cyclic products in excellent yields.[36] Use of an uncoated glassy carbon anode also generates quinones from hydroquinones, but the coexisting dienes are easily oxidized and decompose prior to the reaction with the quinones. However, when a PTFE-fibre-coated glassy carbon electrode is used, the easily oxidizable diene is maintained on the PTFE-

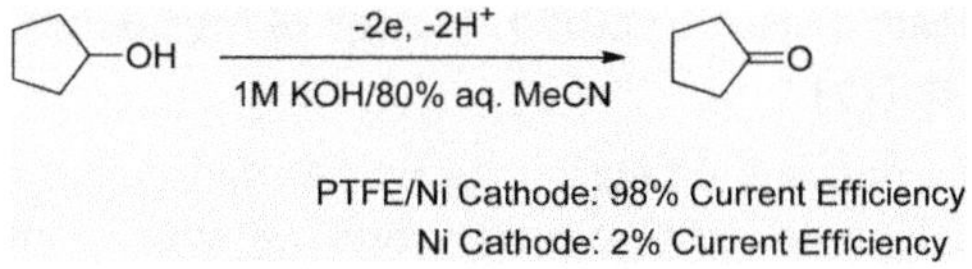

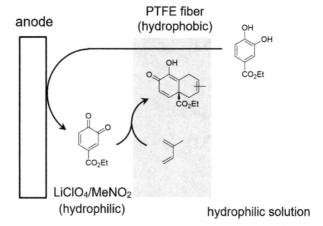

Scheme 6   Anodic oxidation of cyclopentanol at a PTFE composite-plated anode.

Fig. 3   Diels–Alder reaction of anodically generated quinone derivatives with diene on an electrode surface modified with PTFE fibres.

fibres and only the polar hydroquinones can reach the electrode through the hydrophobic fibres, as shown in Fig. 3. Next, the anodically generated hydrophobic quinones immediately react with the dienes adsorbed on the PTFE fibres, and the resulting hydrophobic products are also maintained on the fibres to avoid further oxidation of the products.

Foreign metal adatom-modified electrodes are known to be effective for the selective cathodic reduction of dinitrobenzene, and can also be applied to the catalyst for fuel cells.[37] Furthermore, biologically modified electrodes are useful for clarification of bioelectrochemical systems.[38]

2.7.   Electrosynthesis under specific environments

2.7.1   Under high pressure. In order to increase the efficiency of the cathodic reduction of CO_2 at a tin-based cathode, it has been carried out under high pressure. Increasing the CO_2 pressure yields a high current efficiency of formic acid and formate up to 90%.[39]

2.7.2   Under a high magnetic field. A long time ago, Mogi investigated the redox behavior of a p-toluenesulfonate-doped polypyrrole (PPy/TsO⁻) film electrochemically polymerized in magnetic fields. It was found that magnetic fields higher than 1 T drastically changed the redox potential to the more negative side.[40]

2.7.3   In a supercritical liquid. Tokuda et al. achieved efficient cathodic fixation of CO_2 in supercritical CO_2-DMF media, while Atobe et al. prepared a conducting polymer of pyrrole and thiophene in supercritical CHF_3 without any polar solvent additives.[41,42]

2.8. Electroauxiliary-assisted synthesis (design of organic reactions from molecular-level orbitals)

A silyl group is not only a good leaving group as a cation, but also decreases the oxidation potentials of heteroatom compounds owing to an interaction between orbitals of a heteroatom and a C–Si σ-bond (β-effect).[43] Recently, a similar effect was found in the case of a heteroatom and allylic compounds bearing a BF₃ (or *in situ*-formed BF₃) moiety at the α-position as follows.[44]

$$PhSCH_2BF_3^- \; Bu_4N^+ \; (0.88 \; V \; \textit{vs. SCE}) \; \textit{vs.} \; PhSCH_3 \; (1.41 \; V \; \textit{vs. SCE})$$

$$PhOCH_2BF_3^- \; Bu_4N^+ \; (1.28 \; V \; \textit{vs. SCE}) \; \textit{vs.} \; PhOCH_3 \; (1.81 \; V \; \textit{vs. SCE})$$

3. Progress of industrialization

Soon after the development of the Monsanto adiponitrile process, the Asahi Chemical Company (now the Asahi Kasei Corporation) in Japan developed an emulsion-based cathodic hydrocoupling of acrylonitrile using the thermomorphicity of acrylonitrile and water containing 10% quaternary ammonium salt, and started operation of the process in 1971.[45] Both processes were operated in a divided cell equipped with a separator-like cation exchange membrane during the early stage of the processes. Later on, both companies improved their processes, and in the mid-1980s, they established processes with undivided cells in order to reduce the higher cell voltage caused by the separator resistance. The Simons fluorination process has been running since 1949, but quite recently, the process has ceased due to the PEFAS (per- and polyfluoroalkyl substances) problem.

The Dow Chemical Company has commercialized the regioselective cathodic dechlorination of 3,4,5,6-tetrachloropicolinic acid in aqueous solution to 3,6-dichloropicolinic acid, which is useful as a precursor for agrochemicals, as shown in Scheme 7.[46] The high regioselectivity is attributable to the controlled orientation of the substrate at the cathode surface owing to the dipole moment of the molecule.

The electrochemical production of cysteine from cystine has been running for a long time in the world. Since the separation of cysteine from cystine is quite difficult, the electrolysis has to be continued until all the starting cystine has been

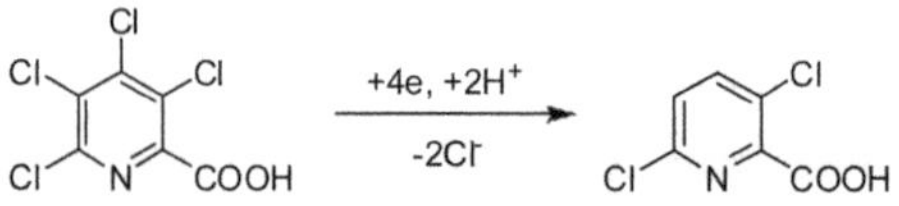

Scheme 7 Regioselective electrochemical reduction of 3,4,5,6-tetrachloropicolinic acid.

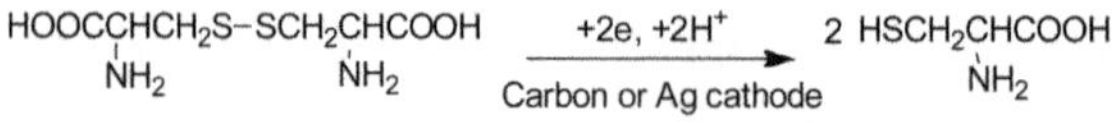

Scheme 8 Electrochemical synthesis of cysteine from cystine.

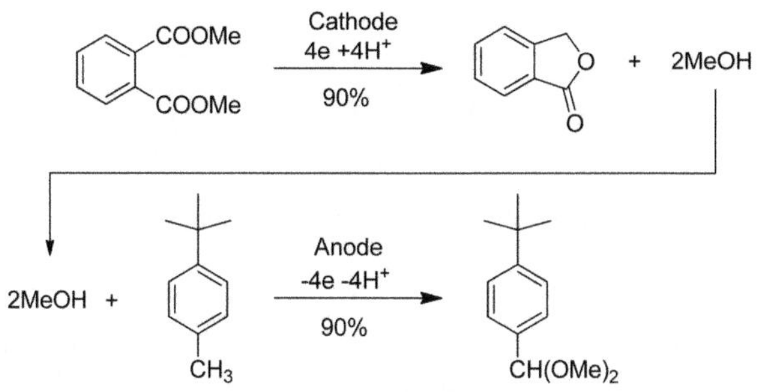

Scheme 9 BASF paired electrosynthesis.

completely consumed, which decreases current efficiency owing to predominant hydrogen evolution. This problem has to be overcome (Scheme 8).[47]

At the end of the last century, a novel paired electrosynthesis of phthalide and p-t-butyl benzaldehyde was developed and industrialized by BASF in Germany (Scheme 9).[48]

4. The second renaissance of organic electrosynthesis

The 21st century is recognized as the ecological century. Organic electrosynthesis is expected to be a typical green chemistry process because it does not require any hazardous or toxic reagents, and produces less waste than conventional chemical synthesis. As mentioned above, BASF stated that electrosynthesis would be the most promising green synthetic process based on the industrialized paired electrosynthesis. These facts have prompted both organic electrochemists and synthetic organic chemists to make great efforts to develop new types of electrosynthetic reactions and novel electrolytic systems in order to achieve green and sustainable chemistry. In fact, a number of organic electrolytic systems toward green sustainable chemistry have been developed to date as follows.

4.1. New green electrolytic systems

4.1.1 Cation pool method and its application to parallel combinatorial synthesis. Yoshida and Suga developed a cation pool method.[49] Carbocations are anodically generated and accumulated in relatively high concentration in the absence of nucleophiles. Then, carbocations are allowed to react with various nucleophiles. This method enables parallel combinatorial synthesis.

4.1.2 Recyclable supporting electrolyte/electrolyte-free system. In the past, supporting electrolytes have never been recovered and reused. From an atom-economy point of view, recyclable supporting electrolytes that are supported on silica gel or polymer beads have been developed (Scheme 10).[50,51]

4.1.3 New electrolytic solvents: ionic liquids (VOC-free electrolysis). Ionic liquids consist of cations and anions without any solvent, and they are in a liquid state around room temperature. Since ionic liquids have unique properties, such as non-flammability, thermal stability, non-volatility and reusability, they have been intensively studied as a green solvent from both basic and applied aspects.[52]

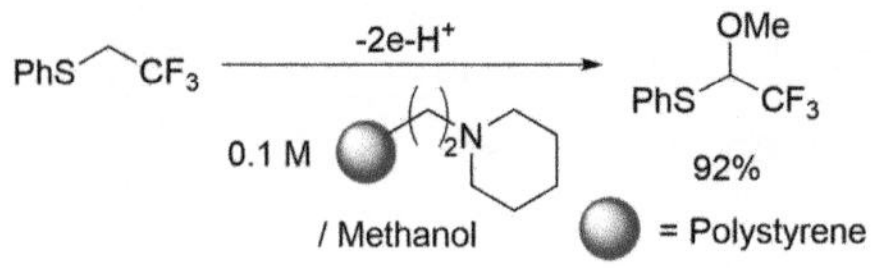

Scheme 10   Electrochemical methoxylation using a recyclable supporting electrolyte.

However, there have still been only a limited number of papers dealing with organic electrolytic reactions and electrosynthesis.[53]

CO_2 well dissolves into ionic liquids. Electrochemical synthesis of carbamates was also achieved by the electrolysis of a solution of CO_2 and an amine in [bmim]BF_4 followed by the addition of an alkylating agent, as shown in Scheme 11.[54]

Electroreductive dehalogenation of *vic*-dihalides using a Co(ii)salen complex in [bmim]BF_4 has also been achieved, as shown in Scheme 12.[55] The product isolation is much easier compared to the similar dehalogenation in ordinary molecular solvents since the Co(ii)salen complex remains in the ionic liquid phase during product extraction with non-polar organic solvents such as diethyl ether. Furthermore, the catalyst/ionic liquid system can be recycled to some extent.

Mattes' group and the author's group independently achieved electrooxidative polymerization of pyrrole, thiophene, and aniline in moisture-stable imidazolium ionic liquids.[56–58] The authors employed [EMIM][OTf] for the electro-polymerization. It was found that the polymerization of pyrrole in the ionic liquid proceeds much faster than that in conventional media, such as aqueous and acetonitrile solutions containing 0.1 M [EMIM][OTf] as a supporting electrolyte.[57,58] It is known that in the radical–radical coupling, further oxidation of oligomers and polymer deposition in the electrooxidative polymerization are favorably affected because the reaction products are accumulated in the vicinity of the electrode surface under slow diffusion conditions, and consequently the polymerization rate is increased. It is reasonable to assume that the

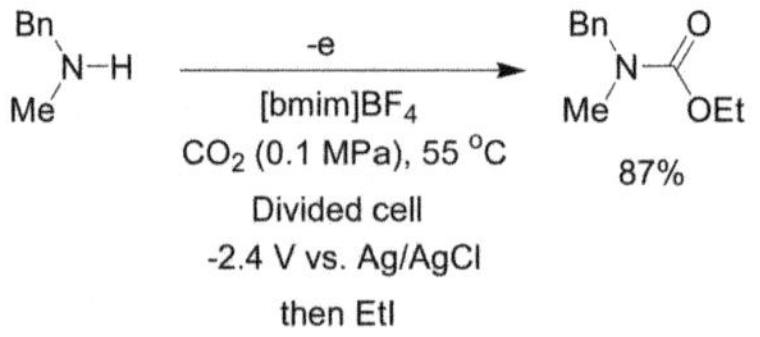

Scheme 11   Electrochemical synthesis of carbamates in an ionic liquid.

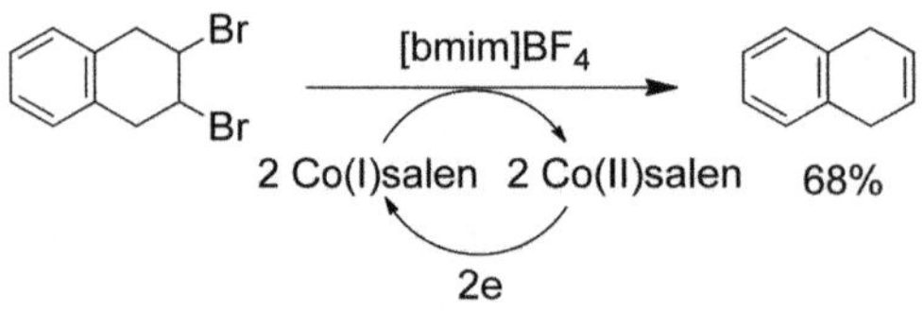

Scheme 12   Electroreductive dehalogenation of *vic*-dihalides using a Co(ii)salen complex.

polymerization rate in [EMIM][OTf] is higher than that in the conventional media, since neat [EMIM][OTf] (viscosity: 42.7 cP) has a higher viscosity than the others. Electropolymerization of thiophene is also accelerated in this ionic liquid. It is notable that the surface of the polypyrrole film prepared in neat [EMIM][OTf] is so smooth that no grains are observed. Both the electrochemical capacity and electroconductivity are markedly increased when the polypyrrole and poly-thiophene films are prepared in the ionic liquid. This may be attributable to the extremely high concentration of anions as the dopants, which results in a much higher doping level.

4.1.4 New electrolytic cells.[59,60] Batch electrosynthesis requires generally high concentrations of supporting electrolytes. In order to solve this problem, the following new type of cells have been developed.

(a) Micro-flow cell and thin-layer flow cell. Yoshida and Suga *et al.* also developed micro-flow cells for oxidative generation of unstable organic cations at low temperatures.[61] Marken *et al.* developed a "thin-layer flow cell" that enables electrolysis to be carried out even without any supporting electrolyte.[62] Later, Atobe *et al.* successfully used a parallel laminar micro-flow system for anodic substitution reactions with readily oxidizable nucleophiles such as allylsilane, as exemplified in Fig. 4.[63]

(b) Proton-exchange membrane (PEM)-type and anion exchange membrane (AEM)-type reactors. In order to overcome the drawback of the SPE system, Atobe *et al.* applied a membrane electrode assembly (MEA) containing a PEM to electrosynthesis reactions, such as hydrogenation of toluene to methylcyclohexane (a so-called hydrogen carrier).[64,65] On the other hand, the use of an AEM reactor enabled preparation of aldehydes from alcohols.[66]

4.1.5 New electrodes

(a) Boron-doped diamond (BDD) electrode.[67,68] Compared to cathode materials, anode materials are severely limited. However, boron-doped diamond electrodes have been developed. This electrode has many advantages, such as a wide potential window and high durability. During the early stages of BDD utilization, it was applied to the degradation of organic hazardous waste. Compton and

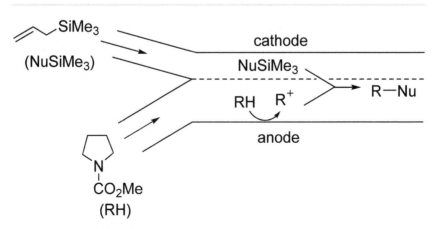

Fig. 4 Anodic substitution of *N*-(methoxycarbonyl)pyrrolidine with allylsilane using parallel laminar flow in the reactor.

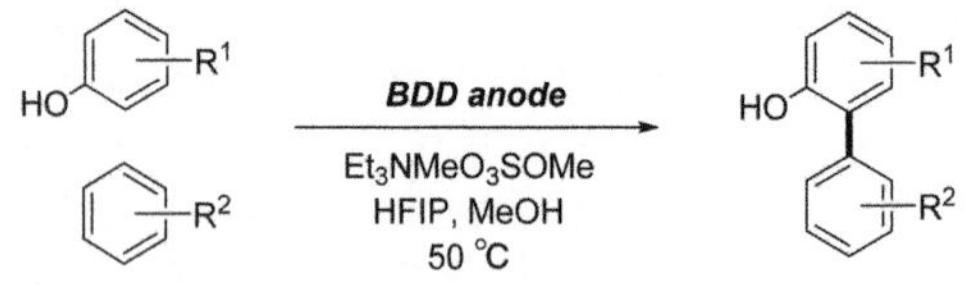

Scheme 13 Anodic homo- and cross-coupling of arenes.

Marken *et al.* used a BDD anode for the Kolbe reaction. Notably, Waldvogel *et al.* successfully used a BDD anode for anodic homo- and cross-coupling of arenes in hexafluoroisopropanol (HFIP) (Scheme 13).[69]

A Pt anode has been used for anodic cyanation so far. However, quite recently, it has been shown that a BDD anode is effective for anodic cyanation of an alkylamine and *N*-methylpyrrole and can also be used for anodic fluorination.[70]

(b) *12CaO·7Al$_2$O$_3$ electride.* This novel electrode has a small work function and is therefore expected to be very effective for the cathodic reduction of organic compounds. Conventional cathodic carboxylation of olefins with CO$_2$ forms dicarboxylic acid products. In contrast, the use of a 12CaO·7Al$_2$O$_3$ electride cathode in a divided cell mainly produces monocarboxylic products.[71] This electrode is also effective for transformation of organoboron compounds to the corresponding alcohols.[72] This is due to the highly efficient generation of superoxide ions at this novel electrode (Scheme 14).

4.1.6 Recyclable mediatory system

(a) *Mediators bearing an ionic tag.* For a long time, mediators were not recycled. However, recyclable mediatory systems, such as a solid (silica gel and polymer)-support mediator and mediators bearing an ionic tag, have been developed from an atom-economy perspective. Thus, a novel indirect electrochemical fluorination system was developed that uses an iodoarene mediator bearing an ionic tag in poly(HF) salt ionic liquids (Scheme 15).[73]

(b) *Carborane mediator.* With the electron-deficient nature of a carborane cage, LUMOs of carboranes are generally low-lying and reversible reduction waves are observed in the cyclic voltammograms. Diaryl-substituted carboranes have been applied to the electron-mediatory system for organic electrosynthesis. Only 1 mol% of carborane mediator worked in the electrocatalytic cycle of the debromination reaction of 1,2-dibromo-1,2-diphenylethane to give stilbene under a potentiostatic condition.[74] The introduction of an ammonium moiety at the *para*-position of the aryl substituent made the carborane mediator immobilized in a polar solvent, such as DMF. Since the olefin products were extracted with non-

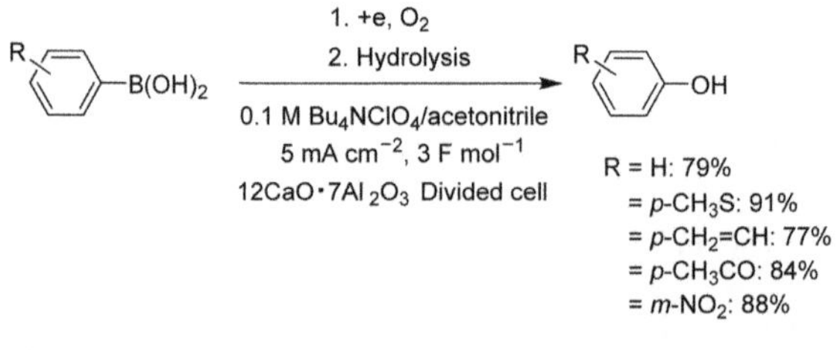

Scheme 14 Cathodic reduction of arylboronic acids in the presence of oxygen.

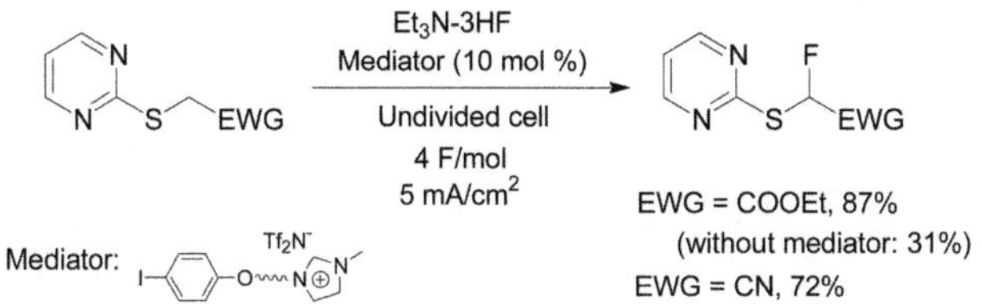

Scheme 15 Anodic fluorination using an iodoarene mediator bearing an ionic tag.

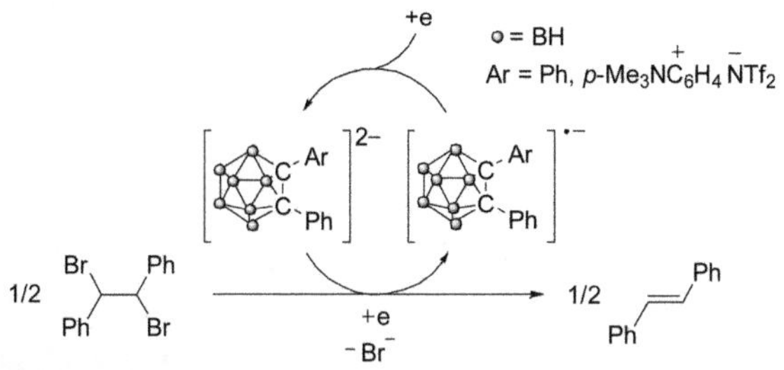

Scheme 16 Cathodic mediatory reduction using carborane derivatives.

polar solvents, such as hexane, the recyclability of the carborane mediator in a DMF electrolytic solution was successfully demonstrated by the ionic-tag approach (Scheme 16).[75]

4.1.7 Electrochemical polymer reaction (solid-phase reaction). Electron transfer between a conducting polymer and an electrode generates its doped state, followed by a chemical reaction in the presence of appropriate reactants. Inagi *et al.* developed the electrochemical polymer reaction as a novel method for post-functionalization of conducting polymer films. Thus, fluorodesulfurization and halogenation can be achieved. In this method, the resulting pure polymer products on the anode can be isolated without purification (Fig. 5).[76]

4.1.8 Paired electrochemical polymer reaction. The paired electrochemical polymer reaction is also realized as shown in the following scheme (Scheme 17).[77]

4.1.9 Bipolar electrode reaction. The bipolar electrode has both anodic and cathodic surfaces simultaneously, and is applied to the paired electrosynthesis method developed by BASF as mentioned above. Inagi *et al.* successfully fabricated gradient conducting polymer films and also demonstrated the patterning application (Fig. 6).[78,79]

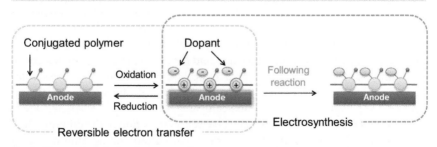

Fig. 5 Electrochemical polymer reactions.

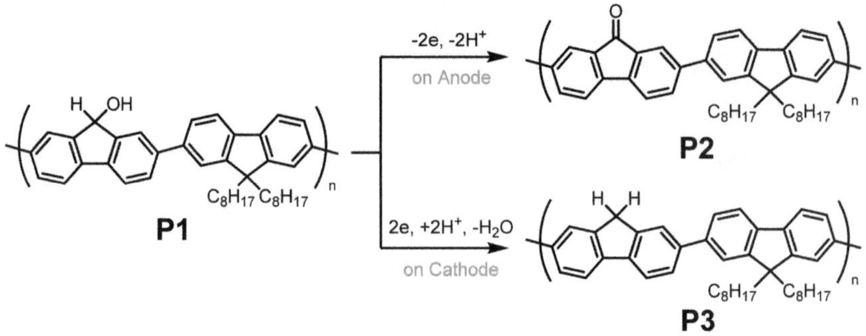

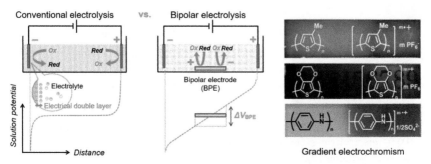

Scheme 17 Paired electrochemical polymer reaction.

Fig. 6 Bipolar electrolysis and preparation of gradient conducting polymer films.

4.1.10 Electropolymerization without an electric power supply. Quite recently, anodic polymerization of pyrrole without an electric power supply has been realized under the flow of electrolyte in a microchannel without an electric power supply.[80]

4.2. Novel green electrosynthetic reactions

Recently, a variety of new efficient electrosynthetic reactions have been developed as reviewed by Baran *et al.*[81]

4.2.1 Template-directed method. A regioselective intermolecular coupling reaction is generally difficult compared to an intramolecular one. Template-directed reactions *via* intramolecular reactions enable selective coupling reactions that are otherwise difficult to achieve. As shown in Scheme 18, regioselective anodic coupling of phenols can be achieved using a boron template.[82]

4.2.2 C–N and N–N coupling *via* N-centered radicals. Radical chemistry has been markedly developed owing to the rapid development of photoredox chemistry and electrosynthesis. N-centered radicals are useful reactive species for nitrogen-containing organic compounds. However, their anodic generation and synthetic application are less studied compared to other reactive species. Previously, Tokuda *et al.* reported that the one-electron oxidation of the lithium salt of an unsaturated amine derivative generates the corresponding aminyl radical intermediate, which undergoes intramolecular cyclization to provide the

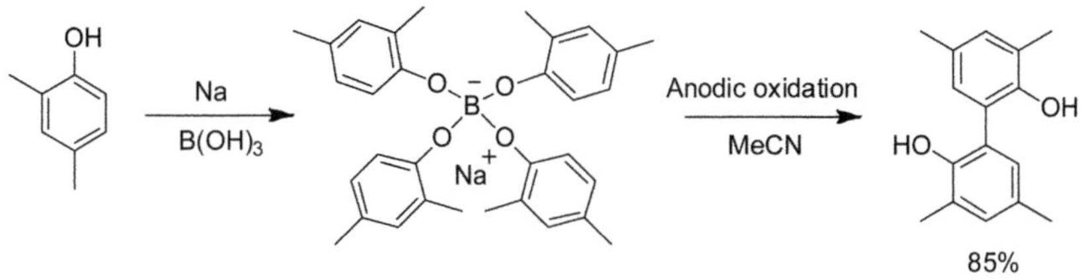

Scheme 18 Regioselective anodic coupling of phenols using a boron template.

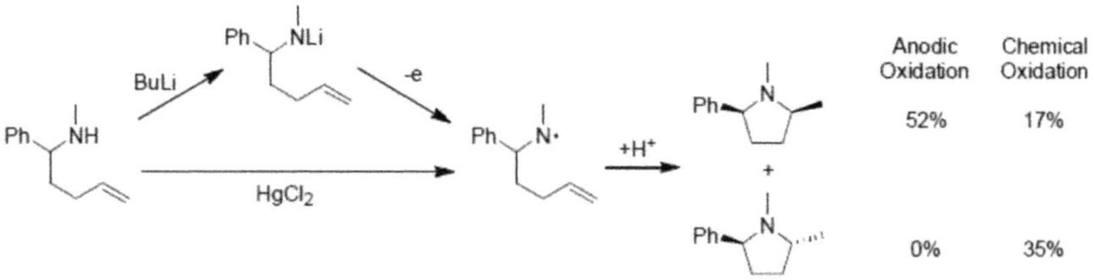

Scheme 19 Stereoselective anodic cyclization.

corresponding pyrrolidine derivative (Scheme 19).[83] Notably, the *cis*-form product is formed exclusively.

Around the same time, the authors found by chance that the amination of tetrahydrofuran took place by the anodic oxidation of R_2NLi prepared in tetrahydrofuran (Scheme 20).[84] Even the free amine underwent similar amination *via* N-centered radicals.

Xu and Moeller achieved anodic intramolecular cyclization of the anodically generated aminyl radical to afford proline derivatives (Scheme 21).[85]

Recently, electrosynthesis using anodically generated N-centered radicals has been intensively studied as reviewed by Xiong and Xu.[86] Baran *et al.* achieved anodic *N,N*-coupling of xiamycine A to provide dixiamycine.[87] This paper stimulated many organic synthetic chemists and electrochemists in the world to work on organic electrosynthesis (Scheme 22).

4.2.3 Electrosynthesis of π-expanded molecules. Recently, the electrosynthesis of π-expanded molecules has been extensively studied by intramolecular C–N, C–O, C–S, and C–P bond formation. Fused benzimidazole derivatives have

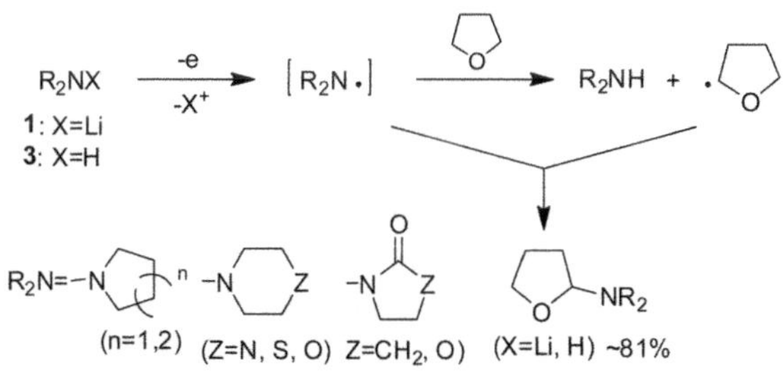

Scheme 20 Anodic amination of tetrahydrofuran.

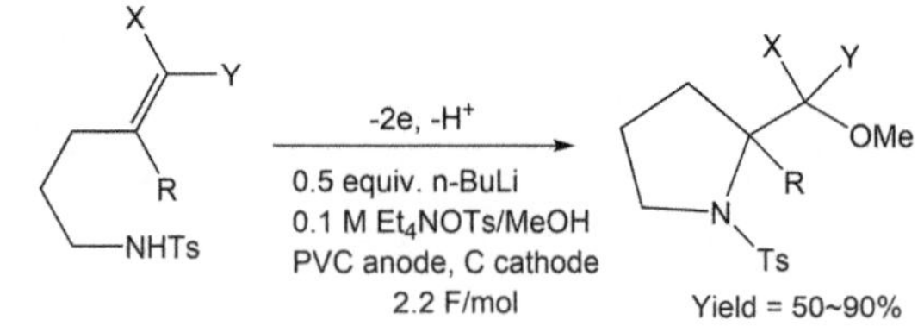

Scheme 21 Intramolecular cyclization of anodically generated aminyl radical.

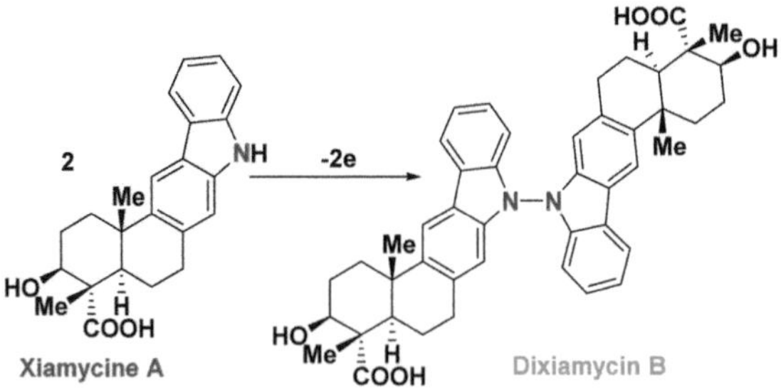

Scheme 22 Anodic N,N-coupling of xiamycine A.

been prepared by utilizing anodically generated aminyl radicals.[88] Xu and Wirth *et al.* reported the preparation of thiazoles by anodic oxidation of thioamides using an undivided batch cell. The use of a flow cell enables gram-scale synthesis.[89] On the other hand, Mitsudo *et al.* achieved Br ion-mediated cyclization forming π-expanded thienoacenes.[90]

4.2.4   Electrocatalytic cross-coupling and cross-metathesis. Chiba *et al.* found that electrocatalytic cross coupling between different olefins forms cyclobutane derivatives, as shown in Scheme 23.[91]

They also achieved electrochemical enol ether/olefin cross-metathesis in a lithium perchlorate/nitromethane electrolyte solution (Scheme 24).[92]

4.2.5   Electrophotocatalytic reactions. Since photosynthesis and electrosynthesis are green processes, the marriage of photochemistry and electrochemistry has been shown to be a powerful synthetic method.[93] For example, Ackerman *et al.* developed the first electrophotochemical C–H trifluoromethylation reaction enabled by utilizing both electrochemical and photochemical reactions.[94]

4.2.6   Selective electrochemical fluorination. Organofluorine compounds are key materials applied in daily life because of their versatile utility as functional materials, pharmaceuticals, and agrochemicals. The development of the selective fluorination of organic molecules under safe conditions is therefore one of the

Scheme 23 Electrocatalytic cross coupling.

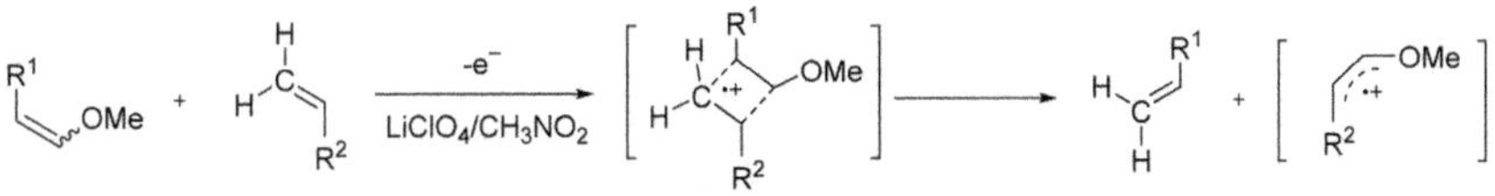

Scheme 24 Electrocatalytic cross-metathesis.

most important subjects in modern synthetic organofluorine chemistry.[95] Different from well-established electrochemical perfluorination, selective electrochemical fluorination is a rather new field and methodology. It has not been well developed because of low reaction selectivity, the low nucleophilicity of fluoride ions, and competitive anode passivation. Therefore, the selective electrochemical fluorination of only limited substrates, such as aromatics, olefins, benzyl ketones, and α-phenylacetates, was reported before the early 1990s. Fluorination can be generally achieved in aprotic solvents (acetonitrile, dichloromethane, dimethoxyethane (DME), nitromethane, sulfolane, *etc.*) containing fluoride ions to provide mostly mono- and/or difluorinated products.[96]

(a) Electrochemical fluorination of heteroatom and heterocyclic compounds.[97,98] In the 1990, the author's group and Laurent's group independently achieved anodic α-fluorination of organosulfur compounds such as α-(phenylthio)acetate in Et$_3$N–3HF/MeCN (Scheme 25: Y = Ph, R = COOEt).[99] These are the first reports of selective electrochemical fluorination of heteroatom compounds. The fluorination proceeds in a manner similar to a Pummerer-type mechanism *via* the fluorosulfonium cation (**C**), as shown in Scheme 26.[100]

However, anode passivation takes place during anodic fluorination. Therefore, pulse electrolysis is necessary to avoid the anode passivation. Although anodic passivation takes place during electrochemical fluorination, particularly in Et$_3$N–3HF/MeCN, electrochemical fluorination is widely applicable, as exemplified in the following schemes (Schemes 27 and 28).[101,102]

Interestingly, dehydrodimers readily derived from benzothiazines undergo electrochemical fluorination accompanied by C–C double bond cleavage to provide *gem*-difluorinated benzothiazine derivatives, as shown in Scheme 29.[103]

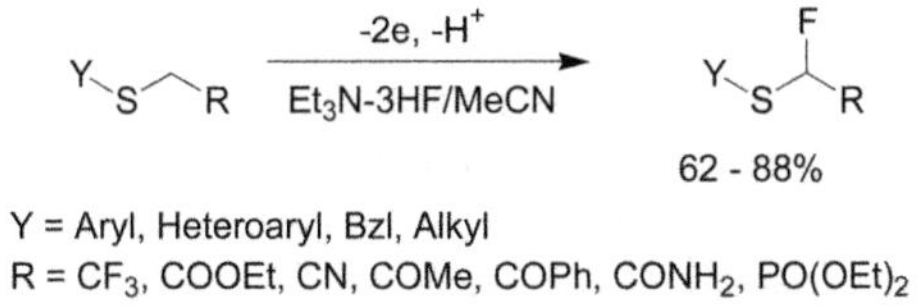

Y = Aryl, Heteroaryl, Bzl, Alkyl
R = CF$_3$, COOEt, CN, COMe, COPh, CONH$_2$, PO(OEt)$_2$

Scheme 25 Electrochemical fluorination of various sulfides.

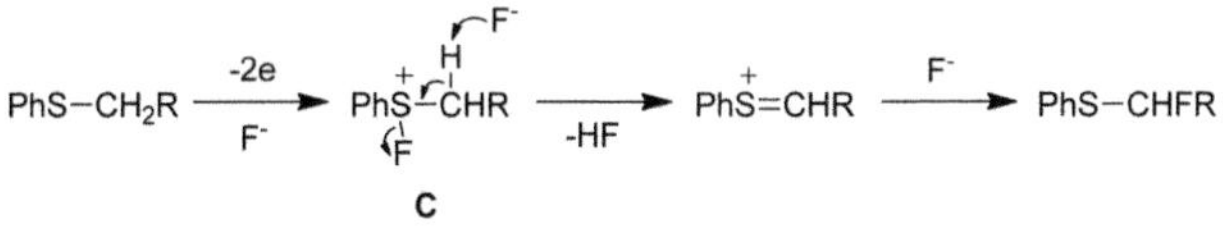

Scheme 26 Reaction mechanism for anodic fluorination.

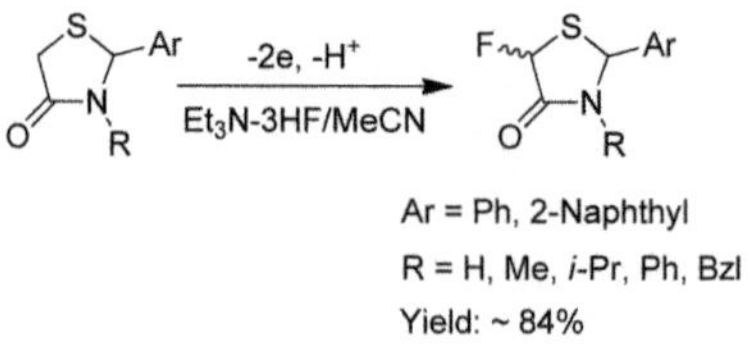
-2e, -H+
Et3N-3HF/MeCN
Ar = Ph, 2-Naphthyl
R = H, Me, i-Pr, Ph, Bzl
Yield: ~ 84%

Scheme 27 Regioselective electrochemical fluorination.

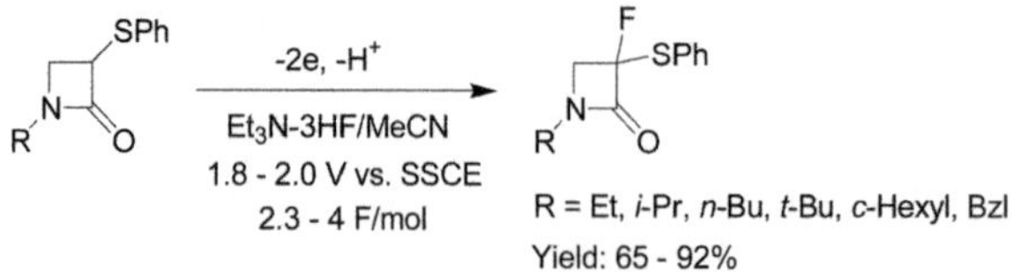
-2e, -H+
Et3N-3HF/MeCN
1.8 - 2.0 V vs. SSCE
2.3 - 4 F/mol
R = Et, i-Pr, n-Bu, t-Bu, c-Hexyl, Bzl
Yield: 65 - 92%

Scheme 28 Electrochemical fluorination of β-lactams.

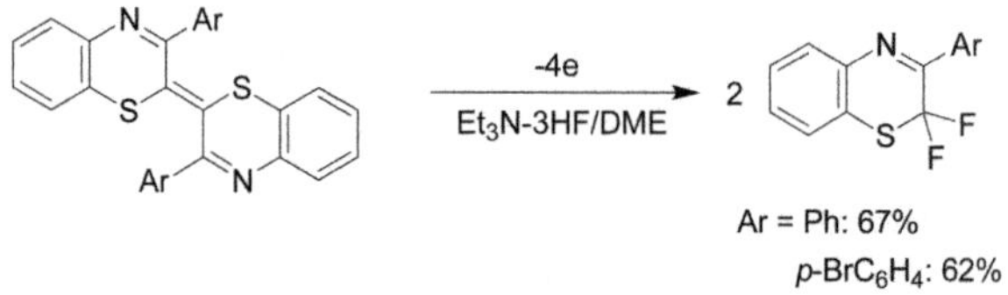
-4e
Et3N-3HF/DME
Ar = Ph: 67%
p-BrC6H4: 62%

Scheme 29 Electrochemical fluorination accompanied by C–C double bond cleavage.

Notably, the fluorination product selectivity is markedly affected by electrolytic solvents, as shown in Scheme 30.[104] This is the first example of solvent effects on fluorinated product selectivity.

Moreover, the authors have also found marked solvent effects on anodic fluorination. Ethereal solvents such as ethoxyethane (DME) are much more suitable than MeCN for the electrochemical fluorination of various heterocyclic sulfides since DME enhances the nucleophilicity of the fluoride ions and also suppresses anode passivation (Scheme 31).[105]

Mediators are also effective in avoiding anodic passivation. It was demonstrated that *in situ*-formed hypervalent difluoroiodoarene mediated *gem*-difluoro-desulfurization of dithioacetals, as shown in Fig. 7.[106,107] This is the first example of the catalytic use of a hypervalent iodo compound.

(b) Electrochemical fluorination in poly(HF) salt ionic liquids. Solvent-free electrochemical fluorination is an alternative method for preventing anode

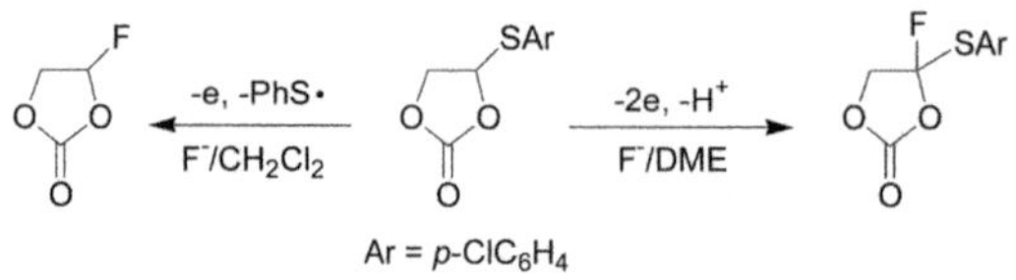
-e, -PhS•
F-/CH2Cl2
-2e, -H+
F-/DME
Ar = p-ClC6H4

Scheme 30 Solvent effects on fluorination product selectivity.

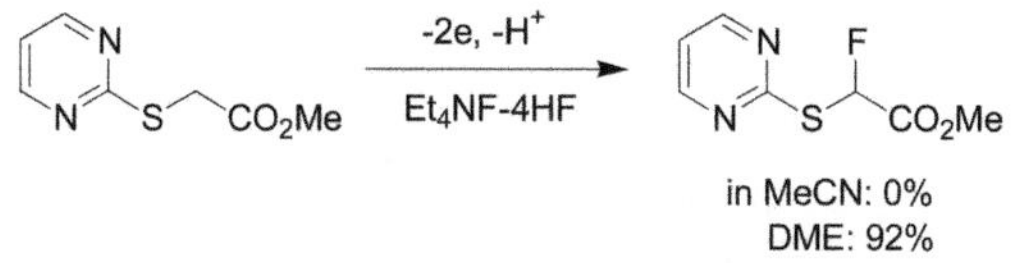

Scheme 31　Solvent effects on electrochemical fluorination.

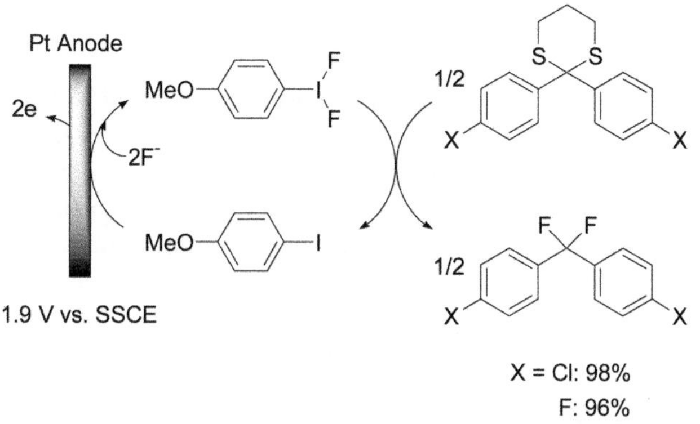

Fig. 7 Hypervalent difluoroarene-mediated *gem*-difluoro-desulfurization of dithioacetals.

passivation and acetoamidation. Anodic difluorination of α-(phenylthio)acetate **1** in Et$_3$N·3HF/MeCN is difficult because of severe anode passivation. However, anodic difluorination of **1** can be achieved in the same ionic liquid under ultrasonication, as shown in Scheme 32.[108]

Thus, the current efficiency for the anodic difluorination of **1** was greatly improved compared to that without ultrasonication.

Since phthalide is hardly oxidized (E_p^{ox} = 2.81 V *vs*. SCE), anodic fluorination does not proceed in either an organic solvent or solvent-free systems. However, fluorination is highly efficient in a double ionic liquid system consisting of [emim]OTf and Et$_3$N–5HF, as shown in Scheme 33.[109] In this double ionic liquid system, the cationic intermediate generated from the phthalide was expected to have a TfO$^-$ counter anion (activated cation **B** in Scheme 33), which readily reacted with F$^-$ to provide the fluorinated phthalide in good to excellent yields.

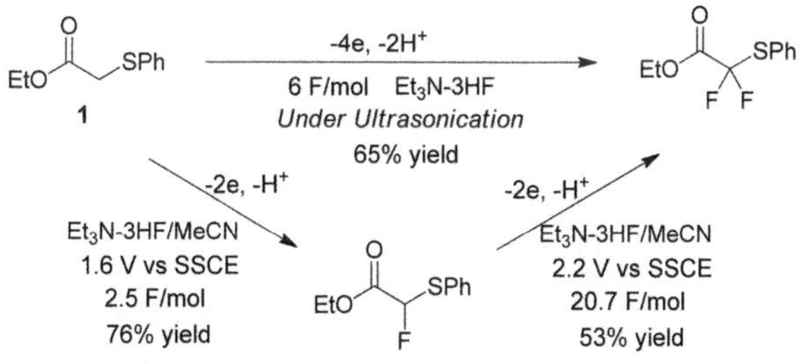

Scheme 32　Electrochemical fluorination in Et$_3$N−3HF ionic liquid under ultrasonication.

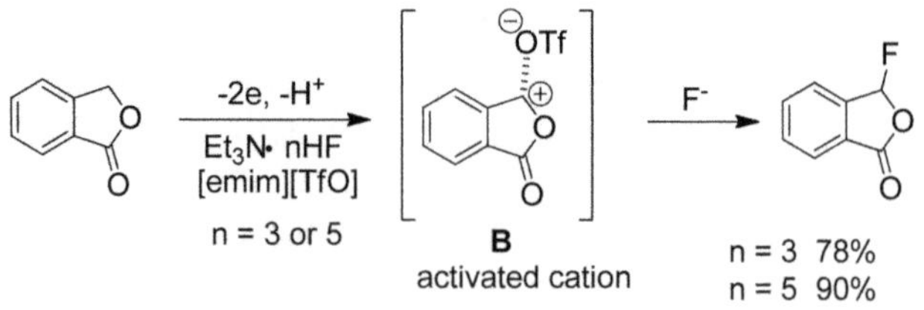

Scheme 33   Electrochemical fluorination of phthalide in a double ionic liquid system.

(c)   Selective electrochemical fluorination using inorganic fluoride salts. In most of the electrochemical fluorination reactions described above, HF-containing salts or poly(HF) salt ionic liquids are used as the fluorine source. Although HF salts are safer than anhydrous HF, it is desirable to search for safer and cheaper fluorine sources, such as alkali metal salts and alkaline earth metal salts. The authors reported the first successful case of electrochemical fluorination utilizing alkali metal fluorides solubilized in organic solvents. KF or CsF can be dissolved in acetonitrile to a sufficient concentration ($\sim$0.05 M) by complexation with PEG, and can be used as a supporting electrolyte and fluorine source. Here, the ether oxygen of PEG encapsulates the alkali metal cation, and the OH group at the end of PEG interacts with the fluoride ion to achieve efficient dissolution. By using this electrolyte, C–H fluorination of triphenylmethane, fluorodesulfurization at the benzylic position, and difluorination of triphenylphosphine and triphenylantimony have been achieved (Scheme 34).[110]

Dissolving CsF and KF at high concentrations in fluorinated alcohols or their mixed solvents with acetonitrile has also been successfully achieved. Fluorinated alcohols such as 1,1,1,3,3-hexafluoroisopropanol (HFIP) and TFE are widely used as solvents in electrochemical oxidation because of their high oxidation tolerance. By using the 0.3 M CsF solution of MeCN/HFIP (8/2) as an electrolyte for electrochemical fluorination, the benzylic C–H fluorination of various compounds was realized (Scheme 35).[111]

(d)   Anodic [18F] fluorination. ^{18}F-Labeling of bioactive compounds is useful for positron emission tomography (PET), which detects cancer, *etc.* Anodic [^{18}F] fluorination of phenylalanine and electrosynthesis of a model compound of [^{18}F]-DOPA have been reported.[112]

(e)   Electrochemical alkyl C–H fluorination. It is generally difficult to carry out anodic fluorination of an unactivated alkyl C–H bond. However, Hara *et al.* has also successfully carried out electrochemical fluorination of adamantanes in

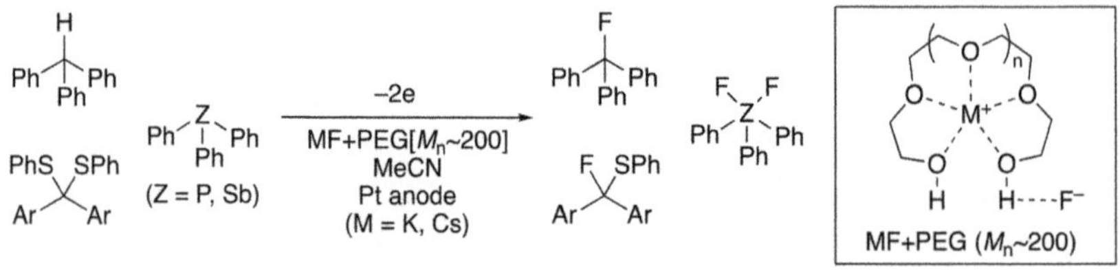

Scheme 34   Electrochemical fluorination using inorganic fluoride salts in the presence of PEG.

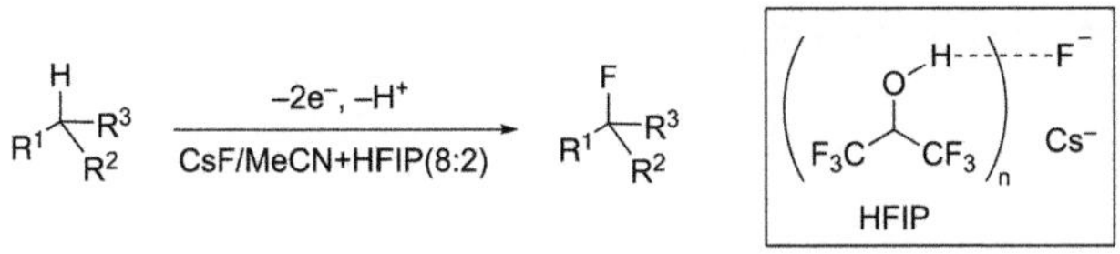

Scheme 35 Electrochemical fluorination in CsF/MeCN–HFIP.

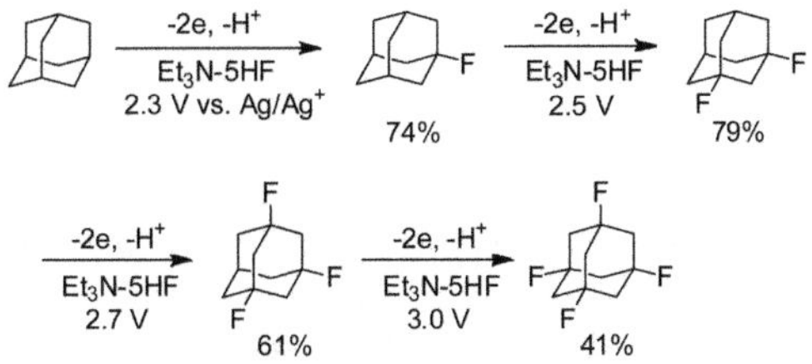

Scheme 36 Electrochemical fluorination of adamantane in Et_3N–5HF ionic liquid.

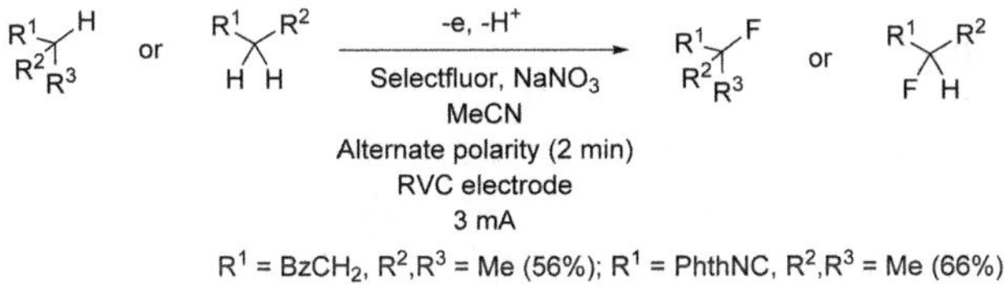

$R^1 = BzCH_2$, R^2,$R^3 = Me$ (56%); $R^1 = PhthNC$, R^2,$R^3 = Me$ (66%)

Scheme 37 Electrochemical fluorination using a nitrate mediator and Selectfluor.

Et_3N–5HF.[113] Mono-, di-, tri-, and tetrafluoroadamantanes were selectively prepared from adamantanes by controlling the oxidation potentials, and the fluorine atoms were introduced selectively at the tertiary carbons, as shown in Scheme 36.

Baran et al. achieved electrochemical alkyl C–H fluorination using a nitrate mediator, Selectfluor as a fluorine source, and carbon-based electrodes.[114] A tertiary carbon is preferentially fluorinated more than a secondary carbon since an anodically generated NO_3 radical abstracts a hydrogen radical from the tertiary carbon preferentially, followed by a reaction with Selectfluor to provide the fluorinated product (Scheme 37).

5. Future perspective of organic electrosynthesis

As explained above, we are currently experiencing the second renaissance of organic electrochemistry developed as an integrated field including not only organic electrosynthesis, but also materials chemistry, as well as organic electronic devices, catalysis chemistry, biochemistry, medicinal chemistry, and environmental chemistry. I believe that such hybridization will promote the sustainable development of attractive novel organic electrosynthesis. Most importantly, we need to develop an entirely new organic synthetic methodology to manufacture organic substances for our daily lives.

Conflicts of interest

There are no conflicts to declare.

Acknowledgements

The author thanks Prof. Shinsuke Inagi for his valuable suggestions for preparing this manuscript.

References

1 G. A. Volta, *Nat. Philos. Chem. Arts*, 1800, **4**, 179.
2 *Academician V. V. Petrov*, ed. S. I. Vavilov, Academy of Sciences of the U.S.S.R., Moscow-Leningrad, 1940.
3 C. T. J. T. Grotthuss, *Ann. Chim.*, 1806, **58**, 54.
4 F. Wöhler, *Ann. Phys. Chem.*, 1828, **88**, 253.
5 M. Faraday, *Ann. Phys. Leipzig*, 1834, **47**, 438.
6 H. J. Kolbe, *J. Prakt. Chem.*, 1847, **41**, 137.
7 J. Heyrovsky, *Chem. Listy*, 1922, **16**, 256.
8 P. Zuman, *Chem. Rec.*, 2012, **12**, 46.
9 M. M. Baizer, *CHEMTECH*, 1980, 161.
10 E. J. Corey and R. R. Sauers, *J. Am. Chem. Soc.*, 1959, **81**, 1739–1743.
11 R. Noyori and I. Kurihara, *J. Org. Chem.*, 1986, **51**, 4320.
12 T. Taguchi, Y. Takahashi, M. Itoh and A. Suzuki, *Chem. Lett.*, 1974, **3**, 1021.
13 R. Backhurst, J. M. Coulson, F. Goodridge, R. E. Plinley and M. Fleischmonn, *J. Electrochem. Soc.*, 1969, **116**, 1600.
14 M. Gausmann, N. Kreidt and M. Christmann, *Org. Lett.*, 2023, **25**(13), 2228.
15 T. Fuchigami, H. Yano and A. Konno, *J. Org. Chem.*, 1991, **56**, 6731.
16 T. Fuchigami and S. Inagi, *Curr. Opin. Electrochem.*, 2020, **24**, 24.
17 Y. Shen, M. Atobe and T. Fuchigami, *Org. Lett.*, 2004, **6**, 2441.
18 M. Isokawa, M. Sano, K. Kubota, K. Suzuki, S. Inagi and T. Fuchigami, *J. Electrochem. Soc.*, 2017, **164**, G121.
19 M. Sano, S. Inagi and T. Fuchigami, *J. Electrochem. Soc.*, 2018, **165**, G171.
20 E. Steckhan, *Angew. Chem., Int. Ed. Engl.*, 1986, **25**, 683.
21 T. Fuchigami, M. Kasuga and A. Konno, *J. Electroanal. Chem.*, 1996, **411**, 115.
22 R. Franke and R. D. Little, *J. Am. Chem. Soc.*, 2014, **136**, 427.
23 F. Baran, *et al.*, *J. Am. Chem. Soc.*, 2017, **139**, 7448.
24 F. Baran, *et al.*, *J. Am. Chem. Soc.*, 2021, **143**, 7859.
25 M. A. Fox and M. T. Dulay, *Chem. Rev.*, 1993, **93**, 341.
26 S. Nishimoto, B. Ohtani, T. Yoshikawa and T. Kagiya, *J. Am. Chem. Soc.*, 1983, **105**, 7180.
27 Z. Ogumi, K. Nishio and S. Yoshizawa, *Electrochim. Acta*, 1981, **26**, 1779.
28 T. Fuchigami, M. Atobe and S. Inagi, *Fundamental and Applications of Organic Electrochemistry: Synthesis, Materials, Devices*, Wiley, 2014.
29 T. Fuchigami and Y. Nakagawa, *J. Org. Chem.*, 1987, **52**, 5276.
30 M. M. Baizer, *Tetrahedron*, 1984, **40**, 935.
31 J. Kopilov, E. Kariv and L. L. Miller, *J. Am. Chem. Soc.*, 1977, **99**, 3450.
32 S. Abe, T. Nonaka and T. Fuchigami, *J. Am. Chem. Soc.*, 1983, **105**, 3630.
33 Y. Kunugi, T. Fuchigami, H.-J. Tien and T. Nonaka, *Chem. Lett.*, 1989, **18**, 757.

34 Y. Kunugi, T. Fuchigami and T. Nonaka, *J. Electroanal. Chem.*, 1990, **287**, 385.

35 Y. Kunugi, T. Nonaka, Y.-B. Chong and N. Watanabe, *Electrochim. Acta*, 1992, **37**, 353.

36 K. Chiba, M. Fukuda, S. Kim, Y. Kitano and M. Tada, *J. Org. Chem.*, 1999, **64**, 7654.

37 A. D. Jannakoudakis and G. Kokkinidis, *Electrochim. Acta*, 1982, **27**, 1199.

38 (*a*) L. Gorton, *Electroanalysis*, 1995, **7**, 23; (*b*) I. Taniguchi, *Anal. Sci. Supp.*, 2001, **17**, i355.

39 M. Ramdin, A. R. T. Morrison, M. Groen, R. Haperen, R. Kler, L. J. P. Broeke, J. P. M. Trusler, W. Jong and T. J. H. Vlugt, *Ind. Eng. Chem. Res.*, 2019, **58**, 1834.

40 I. Mogi, *Chem. Lett.*, 1996, **25**, 419.

41 M. Tokuda, *J. Nat. Gas Chem.*, 2006, **15**, 275.

42 M. Atobe, H. Ohsuka and T. Fuchigami, *Chem. Lett.*, 2004, **33**, 618.

43 J. Yoshida, T. Maekawa, T. Murata, S. Matsunaga and S. Isoe, *J. Am. Chem. Soc.*, 1990, **112**, 1962.

44 S. Inagi and T. Fuchigami, *Curr. Opin. Electrochem.*, 2017, **2**, 32.

45 (*a*) M. M. Baizer, *J. Electrochem. Soc.*, 1964, **111**, 215; (*b*) H. Putter, in *Organic Electrocghemistry*, ed. H. Lund and O. Hammerich, Marcel Dekker, Inc., New York, 4th edn, 2001, ch. 31.

46 J. Chausaard, M. Troupel and Y. Robin, *J. Appl. Electrochem.*, 1989, **19**, 345.

47 J. D. Genders, N. L. Weinberg and C. Zauodzinski, *Extended Abstracts, The Electrochem. Soc. Spring Meeting*, Montreal, 1990, vol. 90-1, p. 890.

48 H. Pütter, in *Organic Electrochemistry*, ed. H. Lund and O. Hammerich, Marcel Dekker, 4th edn, 2001, ch. 31.

49 M. Okajima, S. Suga, K. Itami and J. Yoshida, *J. Am. Chem. Soc.*, 1999, **121**, 9546.

50 T. Tajima and T. Fuchigami, *J. Am. Chem. Soc.*, 2005, **127**, 2848.

51 T. Tajima and T. Fuchigami, *Angew. Chem., Int. Ed.*, 2005, **44**, 4760.

52 P. Hapiot and C. Lagrost, *Chem. Rev.*, 2008, **108**, 2238.

53 J. Yoshida, K. Kataoka, R. Horcajada and A. Nagaki, *Chem. Rev.*, 2008, **108**, 2265.

54 M. Feroci, M. Orsini, L. Rossi, G. Sotgiu and A. Inesi, *J. Org. Chem.*, 2007, **72**, 200.

55 Y. Shen, M. Atobe, T. Tajima and T. Fuchigami, *Electrochemistry*, 2004, **72**, 849.

56 W. Lu, A. G. Fadeev, B. Qi, E. Smela, B. R. Mattes, J. Ding, G. M. Spinks, J. Mazurkiewicz, D. Zhou, G. G. Wallace, D. R. MacFarlane, S. A. Forsyth and M. Forsyth, *Science*, 2002, **297**, 983.

57 K. Sekiguchi, M. Atobe and T. Fuchigami, *Electrochem. Commun.*, 2002, **4**, 881.

58 K. Sekiguchi, M. Atobe and T. Fuchigami, *J. Electroanal. Chem.*, 2003, **557**, 1.

59 M. Atobe, H. Tateno and Y. Matsumura, *Chem. Rev.*, 2018, **118**, 4541.

60 M. Elsherbini and T. Wirth, *Acc. Chem. Res.*, 2019, **52**, 3287.

61 S. Suga, M. Okajima, K. Fujiwara and J. Yoshida, *J. Am. Chem. Soc.*, 2001, **123**, 7941.

62 C. A. Paddon, M. Atobe, T. Fuchigami, P. He, P. Watts, S. J. Haswell, G. J. Pritchard, S. D. Bull and J. Marken, *J. Appl. Electrochem.*, 2006, **36**, 617.

63 D. Horii, T. Fuchigami and M. Atbe, *J. Am. Chem. Soc.*, 2007, **129**, 11692.

64 A. Fukazawa, K. Takano, Y. Matsumura, K. Nagasawa, S. Mitsushima and M. Atobe, *Bull. Chem. Soc. Jpn.*, 2018, **91**, 897.

65 K. Takano, H. Tateno, Y. Matsumura, A. Fukasawa, T. Kashiwagi, K. Nakabayashi, K. Nagasawa, S. MIshima and M. Atobe, *Bull. Chem. Soc. Jpn.*, 2016, **89**, 1178.

66 Y. Ido, A. Fukasawa, Y. Furutani, Y. Sato, N. Shida and M. Atobe, *ChemSusChem*, 2021, **14**, 5405.

67 S. Lips and S. R. Waldvoge, *ChemElectroChem*, 2019, **6**, 1649.

68 D. M. Heard and A. J. J. Lennox, *Angew. Chem., Int. Ed.*, 2020, **59**, 18866.

69 (*a*) A. Kirste, B. Elsler, G. Schnakenburg and S. Waldvogel, *J. Am. Chem. Soc.*, 2012, **134**, 3571; (*b*) S. Lips, D. Schollmeyer, R. Franke and S. Waldvogel, *Angew. Chem., Int. Ed.*, 2018, **57**, 13325.

70 H. Nojima, T. Shimoda, S. Inagi and T. Fuchigami, *Isr. J. Chem.*, 2023, e202300025.

71 J. Li, S. Inagi, T. Fuchigami, H. B. Yin and S. Ito, *Electrochem. Commun.*, 2014, **44**, 45.

72 J. Li, B. Yin, T. Fuchigami, S. Inagi, H. Hosono and S. Ito, *Electrochem. Commun.*, 2012, **17**, 52.

73 T. Sawamura, S. Inagi and T. Fuchigami, *Org. Lett.*, 2010, **12**, 644.

74 K. Hosoi, S. Inagi, T. Kubo and T. Fuchigami, *Chem. Commun.*, 2011, **47**, 8632.

75 S. Inagi, K. Hosoi, T. Kubo, N. Shida and T. Fuchigami, *Electrochemistry*, 2013, **81**, 368.

76 S. Inagi, S. Hayashi and T. Fuchigami, *Chem. Commun.*, 2009, 1718.

77 S. Inagi, H. Nagai, I. Tomita and T. Fuchigami, *Angew. Chem., Int. Ed.*, 2013, **52**, 6616.

78 Y. Ishiguro, S. Inagi and T. Fuchigami, *J. Am. Chem. Soc.*, 2012, **134**, 4034.

79 Y. Koizumi, N. Shida, M. Ohira, N. Nishiyama, T. Tomita and S. Inagi, *Nat. Commun.*, 2016, **7**, 10404.

80 S. Iwai, T. Suzuki, H. Sakagami, K. Miyamoto, Z. o. Chen, M. Konishi, E. Villani, N. Shida, I. Tomita and S. Inagi, *Commun. Chem.*, 2022, **5**, 66.

81 M. Yan, Y. Kawamata and P. S. Baran, *Chem. Rev.*, 2017, **117**, 13230.

82 I. M. Malkowsky, C. E. Rommel, R. Fröhlich, U. Griesbach, H. Pütter and S. R. Waldvogel, *Chem.–Eur. J.*, 2006, **12**, 7482.

83 M. Tokuda, Y. Yamada, T. Takagi, H. Suginome and A. Furusaki, *Tetrahedron Lett.*, 1985, **26**, 6085.

84 T. Fuchigami, T. Sato and T. Nonaka, *J. Org. Chem.*, 1986, **51**, 366.

85 H.-C. Xu and K. D. Moeller, *J. Am. Chem. Soc.*, 2008, **130**(41), 13542.

86 P. Xiong and H.-C. Xu, *Acc. Chem. Res.*, 2019, **52**, 3339.

87 B. R. Rosen, E. W. Werner, A. G. O'Brien and P. S. Baran, *J. Am. Chem. Soc.*, 2014, **136**, 5571.

88 H.-B. Zhao, Z.-W. Hou, Z.-J. Liu, Z.-F. Zhou, J. Song and H. Xu, *Angew. Chem., Int. Ed.*, 2017, **56**, 587.

89 A. A. Folgueiras-Amador, X.-Y. Qian, H.-C. Xu and T. Wirth, *Chem.–Eur. J.*, 2018, **24**, 487.

90 K. Mitsudo, R. Matsuo, T. Yonezawa, H. Inoue, H. Mandai and S. Suga, *Angew. Chem., Int. Ed.*, 2020, **59**, 7803.

91 K. Chiba, T. Miura, S. Kim, Y. Kitano and M. Tada, *J. Am. Chem. Soc.*, 2001, **123**, 11314.

92 T. Miura, S. Kim, Y. Kitano, M. Tada and K. Chiba, *Angew. Chem., Int. Ed.*, 2006, **45**, 1461.

93 Y. Yu, P. Guo, J.-S. Zhong, Y. Yuan and K.-Y. Ye, *Org. Chem. Front.*, 2020, 7, 131.

94 Y. Qiu, A. Scheremetjew, L. H. Finger and L. Ackermann, *Chem.–Eur. J.*, 2020, **26**, 3241.

95 M. Inoue, Y. Sumi and N. Shibata, *ACS Omega*, 2020, **5**, 10633.

96 T. Fuchigami and S. Inagi, *Electrochemical Fluorination PATAI'S Chemistry of Functional Groups, Organofluorine Compounds*, Wiley, 2022.

97 T. Fuchigami and S. Inagi, *Chem. Commun.*, 2011, **47**, 10211.

98 T. Fuchigami and S. Inagi, *Acc. Chem. Res.*, 2020, **53**, 322.

99 T. Fuchigami, M. Shimojo, A. Konno and K. Nakagawa, *J. Org. Chem.*, 1990, **55**, 6074–6075.

100 (*a*) A. Konno, K. Nakagawa and T. Fuchigami, *J. Chem. Soc., Chem. Commun.*, 1991, 1027; (*b*) T. Fuchigami, A. Konno, Y. Nakagawa and M. Shimojo, *J. Org. Chem.*, 1994, **59**, 5937.

101 T. Fuchigami, S. Narizuka and A. Konno, *J. Org. Chem.*, 1992, **57**, 3755.

102 S. Narizuka and T. Fuchigami, *J. Org. Chem.*, 1993, **58**, 4200–4201.

103 R. M. Shaaban, S. Inagi and T. Fuchigami, *Electrochim. Acta*, 2009, **54**, 2635.

104 H. Ishii, N. Yamada and T. Fuchigami, *Chem. Commun.*, 2000, 1617.

105 Y. Hou and T. Fuchigami, *J. Electrochem. Soc.*, 2000, **147**, 4567.

106 T. Fuchigami and T. Fujita, *J. Org. Chem.*, 1994, **59**, 1269.

107 T. Fujita and T. Fuchigami, *Tetrahedron Lett.*, 1996, **37**, 4725.

108 T. Sunaga, M. Atobe, S. Inagi and T. Fuchigami, *Chem. Commun.*, 2009, 956.

109 M. Hasegawa, H. Ishii and T. Fuchigami, *Green Chem.*, 2003, **5**, 512.

110 T. Sawamura, K. Takahashi, S. Inagi and T. Fuchigami, *Angew. Chem., Int. Ed.*, 2012, **51**, 4413.

111 N. Shida, H. Takenaka, A. Gotou, T. Isogai, A. Yamauchi, Y. Kishikawa, Y. Nagata, I. Tomita, T. Fuchigami and S. Inagi, *J. Org. Chem.*, 2021, **86**, 16128.

112 (*a*) G. J. Kienzle and G. Reischl, *J. Labelled Compd. Radiopharm.*, 2005, **48**, 259; (*b*) Q. He, Y. Wang, I. Alfeazi and S. Sadeghi, *Appl. Radiat. Isot.*, 2014, **92**, 52.

113 M. Aoyama, T. Fukuhara and S. Hara, *J. Org. Chem.*, 2008, **73**, 4186.

114 Y. Takahira, M. Chen, Y. Kawamata, P. Mykhailiuk, H. Nakamura, B. K. Peters, S. H. Reisberg, C. Li, L. Chen, T. Hoshikawa, T. Shibuguchi and P. S. Baran, *Synlett*, 2019, **30**, 1178.

Faraday Discussions

PAPER

Magnetic field-enhanced redox chemistry on-the-fly for enantioselective synthesis†

Gerardo Salinas,‡[a] Serena Arnaboldi, ‡[b] Patrick Garrigue,[a] Giorgia Bonetti,[c] Roberto Cirilli,[d] Tiziana Benincori [c] and Alexander Kuhn *[a]

Received 15th February 2023, Accepted 22nd March 2023

DOI: 10.1039/d3fd00041a

Chemistry on-the-fly is an interesting concept, extensively studied in recent years due to its potential use for recognition, quantification and conversion of chemical species in solution. In this context, chemistry on-the-fly for asymmetric synthesis is a promising field of investigation, since it can help to overcome mass transport limitations, present for example in conventional organic electrosynthesis. Herein, the synergy between a magnetic field-enhanced self-electrophoretic propulsion mechanism and enantioselective redox chemistry on-the-fly is proposed as an efficient method to boost stereoselective conversion. We employ Janus swimmers as redox-active elements, exhibiting a well-controlled clockwise or anticlockwise motion with a speed that can be increased by one order of magnitude in the presence of an external magnetic field. While moving, these bifunctional objects convert spontaneously on-the-fly a prochiral molecule into a specific enantiomer with high enantiomeric excess. The magnetic field-enhanced self-mixing of the swimmers, based on the formation of local magnetohydrodynamic vortices, leads to a significant improvement of the reaction yield and the conversion rate.

1 Introduction

Chemistry on-the-fly is a concept defined as the localized chemical conversion of a substance into a desired product *via* a specifically functionalized mobile platform.[1] In recent years, this approach has gained considerable attention in the field of micro- and nano-swimmers, since the intrinsically high active surface area and continuous motion enables a localized mixing of the reactants.[1,2] The efficiency of these self-propelled miniaturized reactors has been extensively exploited

[a]Univ. Bordeaux, CNRS, Bordeaux INP, ISM UMR 5255, 33607 Pessac, France. E-mail: kuhn@enscbp.fr

[b]Dip. Di Chimica, Univ. degli Studi di Milano, 20133 Milan, Italy

[c]Dip. di Scienza e Alta Tecnologia, Univ. degli Studi dell'Insubria, 22100 Como, Italy

[d]Istituto Superiore di Sanità, Centro Nazionale per il Controllo e la Valutazione dei Farmaci, 00161 Rome, Italy

† Electronic supplementary information (ESI) available. See DOI: https://doi.org/10.1039/d3fd00041a

‡ These authors contributed equally.

for a wide range of physico-chemical systems, from organic pollutant degradation[3-7] to metal ion removal[8-10] and biorecognition.[11-13] In this context, the possible enantiodiscrimination, quantification and synthesis of chiral analytes *via* chemistry-on-the-fly approaches is an interesting challenge. Recently, different enantioselective self-propelled devices have been designed by functionalizing the swimmer surface with different chiral recognition elements, *e.g.* cyclodextrins, D- or L-amino acid oxidase and also inherently chiral oligomers.[14-16] In particular the latter approach presents outstanding enantioselectivity, related to diastereomeric interactions between the chiral oligomer surface and the analyte antipodes in solution, resulting in thermodynamic differences in terms of redox potentials, which allow the preferential conversion of only one of the analyte enantiomers.[17,18] This energetic differentiation has been exploited to design unconventional approaches for the electrochemical recognition of chiral probes.[19-21] Furthermore, these oligomeric systems were used for the enantioselective conversion of molecules with a prostereogenic carbon–oxygen double bond into the corresponding chiral hydroxyl derivatives *via* chemistry-on-the-fly.[22] In particular, self-propelled Zn particles, functionalized with oligomers of 2,2-bis[2-(5,2-bithienyl)]-3,3-bithianaphthene (BT_2T_4), were employed as mobile microreactors for the enantioselective synthesis of the antipodes of phenylethanol and mandelic acid. This approach is based on the coupling of the spontaneous oxidation of Zn in acidic media, acting as a source of electrons, and the enantioselective reduction of the prochiral starting compound on the surface of the inherently chiral oligomer. The dynamic behaviour of such Zn/BT_2T_4 hybrid objects induces a continuous mixing of the solution, allowing a more efficient renewal of the prochiral starting compound at the swimmer interface, overcoming the characteristic mass-transport limitations of conventional electroorganic synthesis.[23-26] However, the motion is triggered by an asymmetric formation/release of bubbles, which partially block the active surface area of the device, and thus might decrease the global yield of the stereoselective conversion.

An interesting concept to enhance the mixing of the solution in a more controlled and efficient way is to induce a local Lorentz force on the ion flux present around the self-electrophoretic device. In general, the presence of a magnetic field, orthogonal to an electrode surface, generates a Lorentz force which triggers the formation of a magnetohydrodynamic (MHD) flow along the edges of the electrode.[27,28] This phenomenon has been extensively explored for improving overall reaction kinetics, *e.g.* in electrocatalysis and electrodeposition of metals and polymers.[29-31] Furthermore, due to the possibility to control the fluid flow around the electrode surface, this concept has been also extended to redox magnetohydrodynamic microfluidics and self-propulsion of active matter.[31-34] Recently, the synergy between the spontaneous ion flux produced by self-electrophoretic swimmers and an external magnetic field was proposed as an interesting alternative to boost their propulsion speed by up to 2 orders of magnitude.[35] These Lorentz force-driven Janus swimmers exhibit a predictable clockwise or anticlockwise rotational motion as a function of the magnetic field orientation. The concept is complementary to already well-studied magnetic field-driven swimmers, where motion is triggered either by a pulling mechanism or rotating/undulating magnetic fields.[36-44] In a first order approximation, the above mentioned Zn/BT_2T_4 swimmers behave as self-electrophoretic systems, with a flux of cations from the anodic to the cathodic sites of the Janus object (or *vice versa* for anions), induced by the spontaneous redox reactions. Thus, we propose in the present contribution the use of this ion flux, in combination with an external magnetic field, to design swimmers able to perform redox chemistry on-the-fly for enantioselective synthesis. The synergy between these two main ingredients provides a substantial boost of the dynamic behaviour of the redox swimmers, reflected by an improved efficiency of the stereoselective conversion.

2 Experimental

2.1 Electrosynthesis of enantiopure oligo-(R)-BT$_2$T$_4$ or oligo-(S)-BT$_2$T$_4$

Zn wires modified with a thin Pt layer were obtained by spontaneous deposition of Pt when dipping a zinc wire (GoodFellow, 99.99%, diameter (d) = 250 μm) in a 20 mM H$_2$PtCl$_6$ solution under constant stirring for 2 minutes. Electro-generation of the enantiopure oligo-(R)-BT$_2$T$_4$ or oligo-(S)-BT$_2$T$_4$ (Fig. 1a) was carried out on these core/shell wires in a vial containing 5 cm^3 of a 0.1 M solution of lithium perchlorate (LiClO$_4$; Sigma-Aldrich) in acetonitrile (MeCN; Sigma-Aldrich) and 5 mM (R)- or (S)-enantiopure monomers. The three-electrode system was composed of the Zn/Pt wire, a Pt wire and a Ag wire, acting as working, counter and pseudo-reference electrodes, respectively. The potentiody-namic synthesis of the oligomers was performed by applying a potential sweep from 0 V to 1.2 V, with a constant scan rate (100 mV s^{-1}) for 36 cycles. All elec-trochemical experiments were carried out with a PalmSense potentiostat con-nected to a personal computer.

2.2 Characterization

The morphology and composition of the Zn/Pt/oligo-(R)-BT$_2$T$_4$ or (S)-BT$_2$T$_4$ hybrid device was characterized by scanning electron microscopy (SEM) and energy dispersive X-ray (EDX) spectroscopy, using a Vega3 Tescan 20.0 kV microscope.

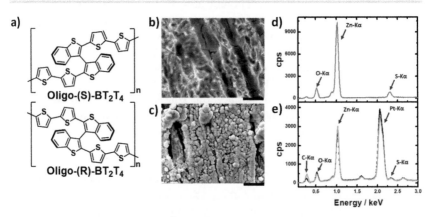

Fig. 1 (a) Chemical structures of oligo-(S)- and oligo-(R)-2,2-bis[2-(5,2-bithienyl)]-3,3-bithianaphthene (oligo-(S)- and oligo-(R)-BT$_2$T$_4$). SEM images of the surface of (b) pristine and (c) Pt/oligo-BT$_2$T$_4$ modified Zn. Scale bar: 5 μm. EDX signals around the emission peaks of carbon, oxygen, zinc, platinum and sulphur of (d) a pristine Zn wire and (e) a Zn/Pt (black line) and Zn/Pt/oligo-BT$_2$T$_4$ (orange line) hybrid device.

2.3 Enantioselective synthesis with the redox chemistry-on-the-fly swimmers

After depositing the respective oligo-BT_2T_4 enantiomers, the hybrid wire was cut into short pieces, leading to swimmers where bare Zn is exposed at both cross sections (average length of 1.1 ± 0.2 mm and diameter of 250 μm). In order to break the symmetry of the devices, one Zn face was carefully covered with varnish (Scheme S1†). For the enantioselective synthesis, six of these Janus swimmers were positioned at the air/water interface of an aqueous 0.1 M H_2SO_4 solution containing 50 mM acetophenone (AP) for 1 hour. For the redox synthesis in the presence of a magnetic field, a rectangular FeNdB magnet (magnetic field $B = 200$ mT, area = 98 cm^2) was placed below the reaction chamber. After one hour, the products were extracted with heptane and the solution was analyzed on a Jasco high-performance liquid chromatography (HPLC) machine (LC-4000) equipped with a photodiode array (PDA) and circular dichroism (CD) detectors at wavelengths of 210 nm and 260 nm, respectively. HPLC analysis was carried out using a chiral column Chiralpak IB N-5 (250 × 4.6 mm, 5 μm) and an n-heptane/2-propanol 92 : 8 (v/v) mixture as the mobile phase (flow rate = 0.5 mL min^{-1}). The dynamic behaviour of the swimmers was monitored by using a CCD camera (Canon EOS 70D, Objective Canon Macro Lens 100 mm 1 : 2.8). Video processing and tracking was performed with ImageJ software.

3   Results and discussion

The hybrid swimmers were designed by following a two-step approach. First, the macroscopic Zn wire was modified with a thin Pt shell (7 ± 1 μm) by the spontaneous reduction of $PtCl_6^{2-}$, to guarantee a good electrical contact between the Zn core and the corresponding oligomer. In the second step, the inherently chiral oligomer was generated by the potentiodynamic oligomerization of (R)-BT_2T_4 and (S)-BT_2T_4, respectively (Fig. 1a). In order to study the composition and evaluate the changes in morphology, SEM analysis (Fig. 1b and c), coupled with EDX spectroscopy (Fig. 1d and e), was carried out with a 1 cm long Zn/Pt/oligo-BT_2T_4 wire. Three different regions have been examined: the pristine Zn wire, a Zn/Pt region and a hybrid Zn/Pt/oligo-BT_2T_4 section. As can be seen from the SEM images, the hybrid region presents a porous globular morphology in comparison to pristine Zn (Fig. 1b and c). The EDX signal for the bare Zn extremity shows the presence of mainly zinc (Zn-Kα), oxygen (O-Kα) and sulphur (S-Kα), probably due to the formation of zinc sulphate during the cleaning of the wire (Fig. 1d). The Zn/Pt part of the wire presents characteristic signals of platinum (Pt-Kα) (Fig. 1e, black line), which decrease for the hybrid part of the device, accompanied by an increase in the carbon signal (C-Kα) (Fig. 1e, orange line), evidencing the presence of the conjugated oligomer on top of the platinum surface.

   After this first characterization of the composition of the Zn/Pt/oligo-BT_2T_4 objects, the propulsion mechanism, driven by the induced Lorentz-force, was examined. Since the use of inherently chiral molecules allows the control of the stereochemical outcome of a reaction, these hybrid materials are an interesting ingredient for asymmetric organic conversions. In this context, we have chosen as a model reaction, the reduction of the ketone moiety of a prochiral precursor, acetophenone, leading to the corresponding enantiomers of 1-phenylethanol (1-PE). Theoretically, in an acidic solution, the oxidation of Zn and the reduction of

protons on Pt occur spontaneously with a standard redox potential difference of
0.76 V. Under these conditions, and in the presence of the inherently chiral surface,
a certain fraction of the produced electrons can be used to trigger the asymmetric
reduction of the prochiral molecule to only one of the enantiomers (Fig. 2a). The
intrinsic electron flux from the anodic to the cathodic part of the device is
accompanied by a movement of ions towards or away from the outer surface at each
extremity of the Janus object (Fig. 2a, pink arrows). As stated above, in the presence
of a magnetic field, orthogonal to the air/water interface, the resulting Lorentz force
induces the formation of a MHD flow at each extremity of the device (Fig. 2a, green
circles). In order to visualize both MHD vortices in the vicinity of the swimmer,
a macro-Zn/Pt/oligo-BT$_2$T$_4$ object (length (l) $\approx$ 1 cm) was immobilized below the
air/water interface of a 100 mM H$_2$SO$_4$ and 50 mM AP solution. The set-up was
placed at the center of a rectangular FeNdB magnet ($B \approx$ 200 mT, $A = 98$ cm^2) with
the north pole facing upwards. A glassy carbon bead ($d = 500$–1000 μm), acting as
a tracer particle, was positioned at the air/water interface in two different regions of
the device. Two circular hydrodynamic flow patterns, with a specular clockwise and
anticlockwise orientation, are formed above the wire (Fig. 2b and Video S1†). The
circular motion exhibits an oscillating speed, which can be correlated with the
relative position of the bead moving above the hybrid wire (Fig. S1†). Maximum
speed values of 0.8 and 1.5 mm s^{-1} (above the Zn cross section and the Pt/oligo-
BT$_2$T$_4$ region, respectively), are recorded when the tracer bead is closest to the
middle part of the device. The differences in size of the circular hydrodynamic flow
patterns and the maximum speed values depend on the size of the active surface
where each reaction is taking place. Thus, since the oxidation of Zn occurs only at
the cross section of the wire where the bare metal is exposed, a rather focused MHD
vortex is produced. In contrast, as the reduction of prochiral molecules and
consumption of protons takes place everywhere along the Pt/oligo-BT$_2$T$_4$ surface,
this causes the formation of stronger MHD flow with a larger vortex diameter. This
continuous mixing of the solution, induced by the two cooperating MHD vortices,

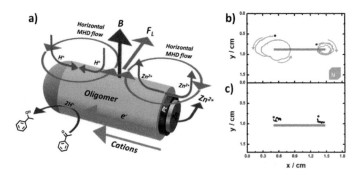

Fig. 2   (a) Schematic illustration of the formation of the MHD flow on the surface of a self-
propelled swimmer with a representation of the associated chemical reactions, the
spontaneous ionic currents, the magnetic field B, the induced Lorentz force F_L and the
resulting MHD convection. (b) Tracking of the trajectory of a tracer particle (glassy carbon
bead, $d = 500$–1000 μm) moving on the air/water interface (0.1 M H$_2$SO$_4$ and 50 mM AP)
above a macroscopic Zn/Pt/oligo-BT$_2$T$_4$ device in the presence of a magnetic field (north
pole up). (c) Same experiment as in (b), but in the absence of the magnetic field. The grey
dots indicate the initial position of the tracer particle.

should increase the mass transport of the prochiral compound towards the catalytic surface, thus improving the overall production rate. In strong contrast to these experiments, only random motion of the tracer particle is observed in the absence of the magnetic field (Fig. 2c and Video S1†).

After this characterization of the MHD flow at the macroscopic scale, the influence of the orientation of the magnetic field on the displacement of miniaturized Janus swimmers was studied during the redox conversion of AP to 1-PE. Six Janus $Zn/Pt/oligo\text{-}(R)\text{-}BT_2T_4$ swimmers, with an average length of 1.1 ± 0.2 mm, were positioned at the air/water interface of a 100 mM H_2SO_4 and 50 mM AP solution, allowing them to move freely for 1 hour. The set-up was placed at the center of a rectangular FeNdB magnet ($B \approx 200$ mT, area $(A) = 98$ cm^2) with either the north or south pole facing upwards. The MHD vortices, observed at the macroscopic scale (*vide supra*), should also provide at these smaller scales a significant driving force to continuously propel the swimmers.

However, as these objects are no longer immobilized, like in the previous experiments, an additional Lorentz force, which acts on the ion flux parallel to the

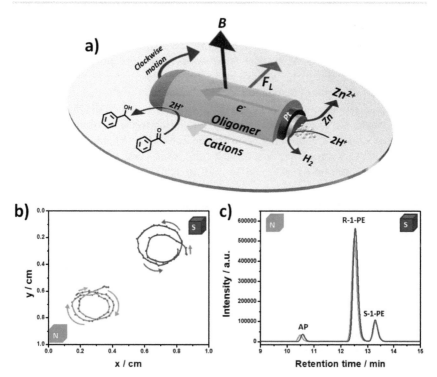

Fig. 3 (a) Schematic illustration of the clockwise motion of a self-propelled Janus swimmer with a representation of the associated chemical reactions, the spontaneous ionic currents, the magnetic field B and the resulting Lorentz force F_L. (b) Tracking plots of two individual $Zn/Pt/oligo\text{-}(R)\text{-}BT_2T_4$ swimmers moving at the air/water interface of a 0.1 M H_2SO_4 and 50 mM AP solution, as a function of the magnetic field orientation (indicated in the figure). (c) Chromatograms of the product mixtures obtained after 1 hour of asymmetric synthesis carried out by the continuous displacement of 6 $Zn/Pt/oligo\text{-}(R)\text{-}BT_2T_4$ swimmers placed at the air/water interface of a 0.1 M H_2SO_4 and 50 mM AP solution, as a function of the magnetic field orientation (indicated in the figure).

main axis of the object, has to be taken into account (Fig. 3a, yellow arrow).[35] This contributes to the overall displacement of the swimmer, by causing a torque force that translates into a predictable clockwise or anticlockwise motion, as a function of the orientation of the magnetic field.

Consequently, the Janus swimmers present a clockwise motion when the north pole of the magnet is orientated upwards, whereas an anticlockwise rotation is observed by inverting the orientation of the magnetic field (south pole upwards) (Fig. 3b and Video S2† for single swimmers and Video S3† for an ensemble of swimmers). After one hour of self-mixing, the products of the stereoselective reaction were extracted with heptane and analyzed by enantioselective HPLC. As mentioned previously, the spontaneous oxidation of Zn is the driving force of the enantioselective reduction of the prochiral compound, since a fraction of the liberated electrons is shuttled to the inherently chiral surface. Thus, by functionalizing the surface of the swimmers with only one enantiomer of oligo-BT_2T_4, in this case the (R)-oligomer, the enantioselective reduction of AP to R-1-PE is expected. From the analysis of the HPLC chromatograms it was possible to confirm the transformation of AP into R-1-PE with a yield of $\approx 80 \pm 4\%$ and a %ee $\approx 70 \pm 1\%$, independent of the magnetic field orientation (Fig. 3c, red curve for north pole up and blue curve for south pole up). The high stereoselectivity was confirmed by chromatograms with online CD detection, where the negative and positive signals can be attributed to the synthesis of the (R)- and (S)-enantiomer, respectively (Fig. S2†). Again, the enantioselectivity does not depend on the orientation of the magnetic field, showing that the stereoselective reaction is not influenced by any spin polarization effect.[45–47]

Using the same experimental conditions, we have evaluated the magnetic field enhancement of the redox conversion by comparing the yield of the reaction in the absence and in the presence of an orthogonal magnetic field. Six Janus Zn/Pt/oligo-(S)-BT_2T_4 swimmers ($l = 1.1 \pm 0.2$ mm), were positioned at the air/water interface of a 100 mM H_2SO_4 and 50 mM AP solution, allowing them to move freely for 1 hour. The set-up was placed at the center of a rectangular FeNdB magnet ($B \approx 200$ mT, $A = 98$ cm^2) with the north pole facing upwards. Under exactly the same conditions, an independent experiment was carried out with six freshly modified Zn/Pt/oligo-(S)-BT_2T_4 swimmers without a magnet. In the presence of the magnetic field, the self-propelled devices present a clockwise motion with an average speed of 0.7 ± 0.1 mm s^{-1}, which is roughly one order of magnitude faster than that measured for the random motion in the absence of the magnet (0.1 ± 0.03 mm s^{-1}) (Fig. 4a and Video S4†). The stereoselectivity of the reaction is preserved, since both redox conversions present a similar %ee value ($\approx 70\%$); however, the enantioselective synthesis in the presence of the magnetic field leads to higher yields ($\approx 80\%$) in comparison with the low value measured for the experiment without a magnet ($\approx 10\%$), as illustrated in Fig. 4b. This is due to the self-mixing capabilities of the swimmers, based on the magnetic field-induced motion.

In a control experiment, the dynamic behaviour and the yield of products generated by the spontaneous motion of six achiral Janus Zn/Pt swimmers were tested. For comparison, these objects were positioned at the air/water interface of a 100 mM H_2SO_4 and 50 mM AP solution, over a magnet with the north pole facing upwards, for 1 h. The Janus swimmers present a clockwise motion with an average speed of 0.4 ± 0.1 mm s^{-1} and a very low conversion ($\approx 0.5\%$) (Fig. S3 and Video

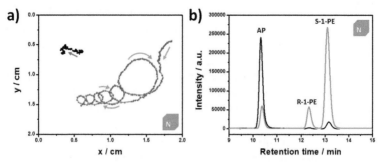

Fig. 4 (a) Tracking plots of two individual Zn/Pt/oligo-(S)-BT$_2$T$_4$ swimmers moving at the air/water interface of a 0.1 M H$_2$SO$_4$ and 50 mM AP solution, in the absence (black line) and in the presence (red line) of an orthogonal magnetic field (north pole upwards). (b) Chromatograms of the product mixtures obtained after 1 hour of asymmetric synthesis carried out by 6 Zn/Pt/oligo-(S)-BT$_2$T$_4$ swimmers, placed at the air/water interface of a 0.1 M H$_2$SO$_4$ and 50 mM AP solution, in the absence (black line) and in the presence (red line) of an orthogonal magnetic field (north pole upward).

S5†). These results indicate that in this case, the large majority of the electrons liberated by the dissolution of zinc is used for the reduction of protons, as this reaction is thermodynamically and kinetically favoured on platinum, compared to the reduction of the prochiral molecule. There is also a significant difference in speed in comparison with the swimmers modified with the inherently chiral oligomer, since the continuous production of H$_2$ bubbles everywhere on the Pt shell does not allow such an efficient directional propulsion of the Zn/Pt objects. As expected, in the case of achiral swimmers, the %ee is zero, within the error bars. Finally, in order to compare the conversion efficiency, the reaction rate of AP was calculated for achiral and chiral swimmers (see the ESI for details†). In the presence of the orthogonal magnetic field, values of 0.001 and 0.30 mmol h^{-1} mm^{-1} were obtained for the Zn/Pt and the Zn/Pt/oligo-(S)-BT$_2$T$_4$ swimmers, respectively. Once again, this provides evidence that the presence of the inherently chiral oligomer not only induces stereoselectivity, but also greatly promotes the global reactivity by more than two orders of magnitude. Furthermore, the efficiency of the redox chemistry-on-the fly is enhanced in the presence of the magnetic field by up to one order of magnitude, in comparison with the conversion in the absence of the magnet (0.03 mmol h^{-1} mm^{-1}), due to an improved mixing of the solution caused by the MHD effect.

4   Conclusions

We have demonstrated the possibility of boosting the propulsion of Zn/Pt/oligo-BT$_2$T$_4$ hybrid swimmers, using the effect of magnetohydrodynamic (MHD) convection. The chiral Janus objects are functionalized with the enantiomers of an inherently chiral oligomer. When they are placed in acidic solutions containing a prochiral starting compound, the spontaneous redox reactions that occur on their surfaces trigger the stereospecific conversion of the prochiral molecule into the desired enantiomer. The inherently helical molecular structure of the polymer provides a chiral environment which is the basis for quite important thermodynamic differences when comparing its interactions with the two antipodes of a given chiral molecule. This naturally also affects the outcome of the reduction of a prochiral ketone into a chiral alcohol if it occurs in such a chiral environment. In the transition state, the intermediate species is more likely to evolve towards a final enantiomer with energetically more favorable stereochemistry, as we have already demonstrated in one of our previous studies.[22]

The magnetic field enhancement of the redox conversion has been evaluated by comparing the yield of an asymmetric model reaction, the reduction of acetophenone into the enantiomers of phenylethanol, in the absence and in the presence of an orthogonal magnetic field. In the presence of a magnet, the Janus swimmers exhibit well-defined clockwise or anticlockwise motion, with an up to one order of magnitude higher speed. This leads to a significant improvement of the reaction yield and of the conversion rate. Without a magnet, the reaction is still stereospecific, but proceeds with a one order of magnitude lower conversion rate.

We can conclude that this better performance is based on the fact that the MHD effect, generated by the presence of a magnet, induces controlled active micro-mixing with a more efficient renewal of the prochiral starting compound at the oligomer surface. In addition, the MHD effect facilitates the detachment of bubbles from the swimmer surface, thus partially liberating the active surface area, which otherwise would be blocked by the presence of gas. This leads to a considerable increase in the global yield of the stereoselective conversion. Moreover, in contrast to other concepts of magnetically driven macro-, micro- and nano-swimmers described in the literature, where the presence of a ferromagnetic component is mandatory to trigger motion,[48] the approach presented here allows, in an unconventional way, the synergy between the local electric and global magnetic fields to be taken advantage of, without the need for ferromagnetic constituents in the swimmer architecture. This offers a higher degree of freedom in the design of motile functional devices, for which the presence of an external magnet will lead to a more efficient propulsion. With respect to future applications, these chiral Janus swimmers might be potentially used for the scale-up of asymmetric reduction processes. Furthermore, the magnetic field-enhanced conversion efficiency could also be of interest for other chemistry-on-the fly applications, as long as redox processes are involved in the reaction mechanism.

Author contributions

AK: conceptualization, funding acquisition, project administration, writing – review & editing; GS and SA: investigation, data curation, writing – review & editing; PG, RC, GB: investigation; TB: investigation, writing – review & editing.

Conflicts of interest

There are no conflicts to declare.

Acknowledgements

This work has been funded by the European Research Council (ERC) under the European Union's Horizon 2020 research and innovation program (grant agreement no. 741251, ERC Advanced grant ELECTRA) and under the HORIZON-ERC-2021 work program (grant agreement no. 101040798, ERC Starting grant CHEIR).

References

1 E. Karshalev, B. Esteban-Fernandez de Avila and J. Wang, *J. Am. Chem. Soc.*, 2018, **140**, 3810–3820.

2 Y. Hu, W. Liu and Y. Sun, *Adv. Funct. Mater.*, 2022, **32**, 2109181.

3 C. Zheng, X. Song, Q. Gan and J. Lin, *J. Colloid Interface Sci.*, 2023, **630**, 121.

4 M. Bayraktaroglu, B. Jurado-anchez and M. Uygun, *J. Hazard. Mater.*, 2021, **418**, 126268.

5 Y. Hu, W. Liu and Y. Sun, *ACS Appl. Mater. Interfaces*, 2020, **12**, 41495–41505.

6 M. Ren, W. Guo, H. Guo and X. Ren, *ACS Appl. Mater. Interfaces*, 2019, **11**, 22761–22767.

7 C. C. Mayorga-Martinez, J. Vyskocil, F. Novotny and M. Pumera, *J. Mater. Chem. A*, 2021, **9**, 14904–14910.

8 D. A. Uygun, B. Jurado-Sanchez, M. Uygun and J. Wang, *Environ. Sci.: Nano*, 2016, **3**, 559–566.

9 W. Yang, J. Li, Y. Lyu, X. Yan, P. Yang and M. Zuo, *J. Cleaner Prod.*, 2021, **309**, 127294.

10 T. Hou, S. Yu, M. Zhou, M. Wu, J. Liu, X. Zheng, J. Li, J. Wang and X. Wang, *Nanoscale*, 2020, **12**, 5227–5232.

11 K. Yuan, C. Cuntin-Abal, B. Jurado-Sanchez and A. Escarpa, *Anal. Chem.*, 2021, **93**, 16385–16392.

12 R. Maria-Hormigos, A. Molinero-Fernandez, M. A. Lopez, B. Jurado-Sanchez and A. Escarpa, *Anal. Chem.*, 2022, **94**, 5575–5582.

13 A. Molinero-Fernandez, M. Moreno-Guzman, L. Arruza, M. A. Lopez and A. Escarpa, *ACS Sens.*, 2020, **5**, 1336–1344.

14 J. Muñoz, M. Urso and M. Pumera, *Angew. Chem., Int. Ed.*, 2022, **134**, e202116090.

15 L. Garcia-Carmona, M. Moreno-Guzman, M. C. Gonzalez and A. Escarpa, *Biosens. Bioelectron.*, 2017, **96**, 275–280.

16 S. Arnaboldi, G. Salinas, A. Karajic, P. Garrigue, T. Benincori, G. Bonetti, R. Cirilli, S. Bichon, S. Gounel, N. Mano and A. Kuhn, *Nat. Chem.*, 2021, **13**, 1241–1247.

17 S. Arnaboldi, T. Benincori, R. Cirilli, W. Kutner, M. Magni, P. R. Mussini, K. Noworyta and F. Sannicolo, *Chem. Sci.*, 2015, **6**, 1706–1711.

18 G. Salinas, M. Niamlaem, A. Kuhn and S. Arnaboldi, *Curr. Opin. Colloid Interface Sci.*, 2022, **61**, 101626.

19 S. Arnaboldi, G. Salinas, G. Bonetti, R. Cirilli, T. Benincori and A. Kuhn, *ACS Meas. Sci. Au*, 2021, **1**, 110–116.

20 G. Salinas, S. Arnaboldi, G. Bonetti, R. Cirilli, T. Benincori and A. Kuhn, *Chirality*, 2021, **33**, 875–882.

21 S. Arnaboldi, G. Salinas, G. Bonetti, R. Cirilli, T. Benincori and A. Kuhn, *Biosens. Bioelectron.*, 2022, **218**, 114740.

22 S. Arnaboldi, G. Salinas, G. Bonetti, P. Garrigue, R. Cirilli, T. Benincori and A. Kuhn, *Angew. Chem., Int. Ed.*, 2022, **61**, e202209098.

23 S. B. Beil, D. Pollok and S. R. Waldvogel, *Angew. Chem., Int. Ed.*, 2021, **60**, 14750–14759.

24 S. Mohle, M. Zirbes, E. Rodrigo, T. Gieshoff, A. Wibe and S. R. Waldvogel, *Angew. Chem., Int. Ed.*, 2018, **57**, 6018–6041.

25 R. D. Little, *J. Org. Chem.*, 2020, **85**, 13375–13390.

26 G. Hilt, *ChemElectroChem*, 2020, **7**, 395–405.

27 V. Gatard, J. Deseure and M. Chatenet, *Curr. Opin. Electrochem.*, 2020, **23**, 96–105.

28 S. Luo, K. Elouarzaki and Z. J. Xu, *Angew. Chem., Int. Ed.*, 2022, **61**, e202203564.

29 Y. Zhang, C. Liang, J. Wu, L. Han, B. Zhang, Z. Jiang, S. Li and P. Xu, *ACS Appl. Energy Mater.*, 2020, **3**, 10303–10316.

30 I. Mogi, R. Morimoto and R. Aogaki, *Curr. Opin. Electrochem.*, 2018, **7**, 1–6.

31 G. Salinas, C. Lozon and A. Kuhn, *Curr. Opin. Electrochem.*, 2023, **38**, 101220.

32 F. Z. Khan, J. A. Hutcheson, C. J. Hunter, A. J. Powless, D. Benson, I. Fritsch and T. J. Muldoon, *Anal. Chem.*, 2018, **90**, 7862–7870.

33 J. C. Sikes, K. Wonner, A. Nicholson, P. Cignoni, I. Fritsch and K. Tschulik, *ACS Phys. Chem. Au*, 2022, **2**, 289–298.

34 A. Zharov, V. Fierro and A. Celzard, *Appl. Phys. Lett.*, 2020, **117**, 104101.

35 G. Salinas, K. Tieriekhov, P. Garrigue, N. Sojic, L. Bouffier and A. Kuhn, *J. Am. Chem. Soc.*, 2021, **143**, 12708–12714.

36 V. M. Kadiri, C. Bussi, A. W. Holle, K. Son, H. Kwon, G. Schütz, M. G. Gutierrez and P. Fischer, *Adv. Mater.*, 2020, **32**, 2001114.

37 X. Z. Chen, J. H. Liu, M. Dong, L. Müller, G. Chatzipirpiridis, C. Hu, A. Terzopoulou, H. Torlakcik, X. Wang, F. Mushtaq, J. Puigmarti-Luis, Q. D. Shen, B. J. Nelson and S. Pane, *Mater. Horiz.*, 2019, **6**, 1512–1516.

38 S. Sanchez, A. A. Solovev, S. M. Harazim and O. G. Schmidt, *J. Am. Chem. Soc.*, 2011, **133**, 701–703.

39 T. Qui, T. C. Lee, A. G. Mark, K. I. Morozov, R. Münster, O. Mierka, S. Turek, A. M. Leshansky and P. Fischer, *Nat. Commun.*, 2014, **5**, 5119.

40 L. Baraban, D. Makarov, R. Streubel, I. Mönch, D. Grimm, S. Sanchez and O. G. Schmidt, *ACS Nano*, 2012, **6**, 3383–3389.

41 A. Ghosh and P. Fischer, *Nano Lett.*, 2009, **9**, 2243–2245.

42 R. Dreyfus, J. Baudry, M. L. Roper, M. Fermigier, H. A. Stone and J. Bibette, *Nature*, 2005, **437**, 862–865.

43 K. E. Peyer, L. Zhang and B. J. Nelson, *Nanoscale*, 2013, **5**, 1259–1272.

44 L. Zhang, J. J. Abbott, L. Dong, B. E. Kratochvil, D. Bell and B. J. Nelson, *Appl. Phys. Lett.*, 2009, **94**, 064107.

45 R. Naaman and D. H. Waldeck, *J. Phys. Chem. Lett.*, 2012, **3**, 2178–2187.

46 R. Naaman, Y. Paltiel and D. H. Waldeck, *Annu. Rev. Biophys.*, 2022, **51**, 99–114.

47 Y. Wolf, Y. Liu, J. Xiao, N. Park and B. Yan, *ACS Nano*, 2022, **16**, 18601–18607.

48 H. Zhou, C. C. Mayorga-Martinez, S. Pane, L. Zhang and M. Pumera, *Chem. Rev.*, 2021, **121**(8), 4999–5041.

Faraday Discussions

ROYAL SOCIETY
OF CHEMISTRY

PAPER

Electrochemical hydrogen isotope exchange of amines controlled by alternating current frequency†

Nibedita Behera, [ID] ‡ Disni Gunasekera, [ID] ‡ Jyoti P. Mahajan, [ID] Joseph Frimpong, Zhen-Fei Liu [ID] and Long Luo [ID] *

Received 18th February 2023, Accepted 13th March 2023

DOI: 10.1039/d3fd00044c

Here, we report an electrochemical protocol for hydrogen isotope exchange (HIE) at α-C(sp^3)–H amine sites. Tetrahydroisoquinoline and pyrrolidine are selected as two model substrates because of their different proton transfer (PT) and hydrogen atom transfer (HAT) kinetics at the α-C(sp^3)–H amine sites, which are utilized to control the HIE reaction outcome at different applied alternating current (AC) frequencies. We found the highest deuterium incorporation for tetrahydroisoquinolines at 0 Hz (i.e., under direct current (DC) electrolysis conditions) and pyrrolidines at 0.5 Hz. Analysis of the product distribution and D isotope incorporation at different frequencies reveals that the HIE of tetrahydroisoquinolines is limited by its slow HAT, whereas the HIE of pyrrolidines is limited by the overoxidation of its α-amino radical intermediates. The AC-frequency-dependent HIE of amines can be potentially used to achieve selective labeling of α-amine sites in one drug molecule, which will significantly impact the pharmaceutical industry.

Introduction

Hydrogen isotope (D or T)-labeled organic molecules are essential to support drug discovery and development in the pharmaceutical industry. For example, deuterated compounds are used as stable isotope-labeled internal standards for quantification purposes in all "omics" fields, taking advantage of the mass shift arising from the substitution of H by D in a given molecule.[1-4] Furthermore, due to the primary kinetic isotope effect, D incorporation at specific positions of a bioactive molecule can potentially decrease its metabolism rate and/or prevent the formation of toxic metabolites, leading to the emergence of deuterated drugs.[5,6] The labeling efficiency and selectivity requirements vary with the application.[7] In the first application discussed above, multiple incorporations of

---

Department of Chemistry, Wayne State University, Detroit, Michigan 48202, USA. E-mail: long.luo@wayne.edu

† Electronic supplementary information (ESI) available. See DOI: https://doi.org/10.1039/d3fd00044c

‡ Equally contributing (first) authors.

hydrogen isotopes are often required. In the last one, selective hydrogen labeling at a specific position is more critical than multiple-site labeling.

Among all hydrogen isotope labeling methods, hydrogen isotope exchange (HIE) is most appealing because it allows fast and direct incorporation of D atoms at a late or the final stage of synthesizing the active pharmaceutical ingredient by replacing H with D, reducing the costs and time spent preparing intermediates or precursors for *de novo* syntheses. The state-of-the-art HIE strategies include: (i) homogeneous catalysis using transition metal complexes, such as [Ir],[8,9] [Co],[10] [Ni],[11] and [Fe];[12] (ii) heterogeneous catalysis using transition metals and their nanoparticles, such as Ru,[13] Rh,[7] and Pt;[14] and (iii) photoredox catalysis using molecular photocatalysts coupled with hydrogen atom transfer (HAT) catalysts (typically, thiols).[15,16] Transition-metal-complex-catalyzed HIE reactions typically target aromatic $C(sp^2)$–H sites. Heterogeneous catalysts such as Pd, Ru, Rh, Ir, and Pt can catalyze both $C(sp^2)$–H and $C(sp^3)$–H activation processes.[17] Photoredox protocols can efficiently and selectively install D at α-amino $C(sp^3)$–H bonds in a single step,[8,9] which is powerful because it can achieve high D incorporation. However, the current photocatalytic protocols are limited in labeling α- and β-$C(sp^3)$–H sites in tertiary amines.

The substrate scope of all existing HIE approaches is typically restricted to certain chemotypes and they cannot selectively label sites with similar chemical reactivity (Scheme 1A). The isotope-incorporation level also varies. Taking photocatalytic HIE of amines as an example:[15] photoexcitation of the photocatalyst generates an excited state to oxidize the amine to an amine radical cation, which then undergoes facile deprotonation at the α-$C(sp^3)$–H position to give an α-amino radical. At the same time, a deuterated thiol is generated *in situ* from the thiol HAT catalyst *via* H/D exchange with D_2O. Then, HAT between the deuterated thiol and α-amino radical results in a D-substituted amine and thiol radical. Electron transfer between the reduced photocatalyst and thiol radical regenerates the photocatalyst and deuterated thiol through protonation of the thiol anion. However, this protocol does not differentiate various α-amino $C(sp^3)$–H bonds (*e.g.*, all the α-amino sites of clomipramine are activated simultaneously, resulting in a high D incorporation of 7.22 D per molecule), so they are not suited for applications that require selective hydrogen labeling at a specific position. Therefore, methods that can label substrates with site-selective D incorporation (particularly among sites with similar chemical reactivity) are highly desirable.

Alternating current (AC) electrolysis provides an alternative way to sequentially perform redox-opposite reactions, like in a photoredox catalytic cycle, by periodically reversing the voltage polarity.[18-22] Unlike photoredox catalysis, AC electrolysis offers a facile way to tune the redox potentials and time duration of two redox events *via* voltage amplitude and AC frequency (*f*), enabling selective activation of sites with similar chemical reactivities. For example, we have previously shown that the electrochemical oxidation of tertiary amines to α-amino radicals takes place at an optimal substrate-dependent *f*.[23] At high *f*, the interconversion between an amine and its cationic radical is faster than the deprotonation of the cationic radical, so the formation of α-amino radicals is suppressed. However, at low *f*, the amine radical cation can be further oxidized to an iminium cation after deprotonation. Therefore, an optimal *f* is required to oxidize a tertiary amine to form its α-amino radical. Furthermore, because the deprotonation kinetics for amine substrates varies significantly, the optimal *f* differs by over two orders of

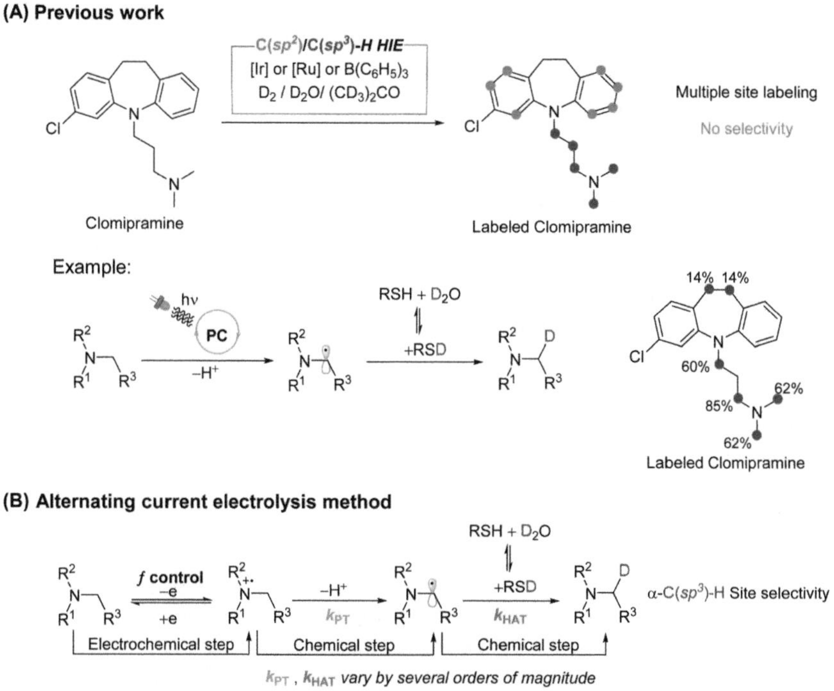

(A) Previous work

Clomipramine → Labeled Clomipramine

C(sp^2)/C(sp^3)-H HIE
[Ir] or [Ru] or B(C_6H_5)$_3$
D_2 / D_2O/ $(CD_3)_2CO$

Multiple site labeling

No selectivity

Example:

Labeled Clomipramine

(B) Alternating current electrolysis method

k_{PT}, k_{HAT} vary by several orders of magnitude

Scheme 1 Schematic illustration of (A) previously reported approaches for hydrogen isotope exchange (HIE) at C(sp^2)–H and C(sp^3)–H sites, highlighting the D-labelling of α-amine sites of clomipramine using a photocatalyst (PC) and HAT thiol catalyst as an example. (B) AC electrolysis method for deuteration at α-C(sp^3)–H sites that can potentially achieve α-amine site selectivity through AC-frequency (f)-controlled electrochemical and chemical (including PT and HAT) steps.

magnitude, providing a convenient handle to selectively activate α-amino C–H bonds in a molecule. AC electrolysis has also been found to affect the product selectivity of nickel-catalyzed cross-coupling reactions,[24] influence the chemo-selectivity of carbonyl-compound and (hetero)arene reduction,[18,25] decrease the undesired reductive precipitation of metals,[26,27] and promote cross-coupling reactions such as formal C–O/O–H cross-metathesis.[28–30]

Inspired by these previous works, we hypothesize that selective HIE of amine sites can be achieved by their selective activation using an optimal AC frequency. Scheme 1B depicts our site-selective electrochemical HIE idea. The electro-chemical HIE of amines includes three steps: (1) an amine substrate is electro-chemically oxidized to its cation radical form, (2) the radical cation undergoes deprotonation to generate an amino radical in the presence of a base, and (3) HAT between the amino radical and deuterated thiol produces the D-labelled mole-cule. Because the rates of the last two chemical steps (*i.e.*, proton transfer (PT) and HAT) can vary by several orders of magnitude and depend on the bond dissoci-ation energy (BDE) of the α-C(sp^3)–H bonds of the amine, we can control the time available for these two chemical steps by supplying different f values. More specifically, for amine sites with slow PT and HAT, the amine cation radical does not have time to be deprotonated and undergo HAT at high f (or with a short anodic pulse) before it gets reduced back to its neutral molecular form. In contrast, amine sites with fast PT and HAT can finish the HIE even at high f, allowing site-selective HIE between amine sites with different PT and HAT kinetics using the AC frequency.

Results and discussion

In this study, we chose two tertiary amines with different α-amino sites, *N*-benzyl-tetrahydroisoquinoline (**1**) and *N*-(4-methoxyphenyl)pyrrolidine (**2**), as the model substrates to test our proposed site-selectivity idea in Scheme 1B. If our proposed idea is feasible, then the HIE of **1** and **2** should occur at two different f values. Fig. 1A illustrates our electrochemical HIE reaction design. We used methyl thioglycolate (MTG) as the HAT catalyst and precursor of the base for the deprotonation step and D_2O as the deuterium source. The amine substrate is first oxidized to generate an α-amino radical in the presence of the base during the positive pulse of the AC waveform. Then, the formed α-amino radical undergoes HAT with a deuterated thiol (RSD), which is *in situ* generated from the thiol and D_2O, to afford the deuterated amine and thiyl radical (RS˙). Finally, RS˙ is reduced back to the thiolate (RS⁻) during the following negative voltage pulse, which acts as a base or is protonated to regenerate RSD, completing the HAT catalytic cycle. According to the reaction mechanism, oxidation of the amine and reduction of the thiol are required to drive the electrochemical HIE reaction of amines. We measured the redox potentials of **1**, **2**, and MTG using cyclic voltammetry. The cyclic voltammograms of **1** and **2** (Fig. 1B) show anodic peaks at 0.6 V and 0.2 V *vs.* Ag/Ag⁺, respectively, assigned to single-electron oxidation of these amines to the

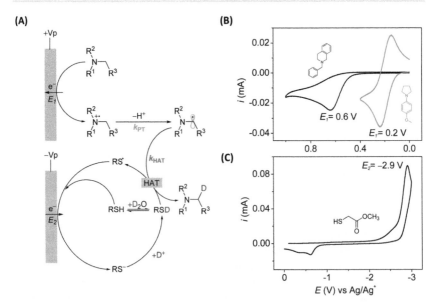

Fig. 1 (A) Proposed mechanism for AC-driven HIE of amines. Cyclic voltammograms of (B) **1** (black) and **2** (red) and (C) methyl thioglycolate (MTG) in *N,N′*-dimethylacetamide (DMA) containing 0.1 M LiClO₄. Scan rate: 0.1 V s⁻¹.

corresponding cation radicals.[23] The voltammogram of MTG (Fig. 1C) shows that electrochemical reduction of the thiol occurs at -2.9 V vs. Ag/Ag$^+$. Thus, we set the amplitudes of the AC waveform to be 3.5 V and 3.1 V after iR correction (see the ESI†) to drive the HIE reactions of 1 and 2, respectively.

As previously discussed, the electrochemical HIE of amines comprises three steps: amine oxidation, PT, and HAT. We first measured the relative deprotonation kinetics (k_{PT}) using a voltammetric method (see the ESI†). Briefly, we varied the scan rate and compared the cyclic voltammograms of the amine with and without the base. At a low scan rate, the deprotonation step can be completed and consequently lead to overoxidation of the amine, resulting in a higher peak than the reference voltammogram without the base ($i_1 > i_0$). As the scan rate is increased, the available deprotonation time decreases, and the difference between i_0 and i_1 shrinks. Therefore, we can estimate the deprotonation time from the voltammograms where $i_1 \approx i_0$. For 1, the deprotonation time is $\sim$5 ms; for 2, it is $\sim$100 ms, giving a 20:1 ratio for the relative deprotonation kinetics. Note that the deprotonation only occurs at site a of 1 because of its weaker C–H bond than sites b and c.

Next, we estimated the relative HAT kinetics (k_{HAT}) from the calculated C–H BDE values for 1 and 2. The BDE values for sites a, b, c, and d are 77, 82.3, 93, and 89.2 kcal mol^{-1}, respectively. Thus, the corresponding relative k_{HAT} ratios between sites a, b, and c are estimated to be $1:90:7 \times 10^5$ and $1:3 \times 10^4$, following the Evans–Polanyi correlation (see the ESI for calculation details†).[31] The k_{PT} and k_{HAT} values for 1 and 2 are summarized in Table 1. From the relative HAT and PT kinetics, we predict that the HIE at site a of 1 requires a much lower (by a few orders of magnitude) f than that at site d of 2 due to its small k_{HAT}.

To test our prediction, we conducted the HIE reactions of 1 and 2 using a sine waveform with different f values from 0 to 50 Hz (note: 0 Hz means DC electrolysis). For $f = 0$ Hz, we used two experimental setups: one home-built cell with an electrode–electrode distance (d) of 1 mm and one commercial IKA setup with $d = 5$ mm (see details in the Methods section). For other f values, we only used our home-built

Table 1 BDEs, oxidation potentials, and relative HAT and PT kinetics (k_{HAT} and k_{PT}) of 1 and 2. BDE values were calculated using density functional theory (see Computational methods). The relative k_{HAT} was calculated from the BDE values.[31] The oxidation potential was obtained from the cyclic voltammograms in Fig. 1, and the relative k_{PT} (*for site a of 1) was obtained from our previous report,[23] as discussed in the ESI†

	1	2
BDE (kcal mol^{-1})	$a = 77.0$, $b = 82.3$, $c = 93.0$	$d = 89.2$
Relative HAT kinetics (k_{HAT})	$a:b:c = 1:90:7 \times 10^5$	$a:d = 1:3 \times 10^4$
Oxidation potential (V vs. Ag/Ag$^+$)	0.6	0.2
Relative deprotonation kinetics (k_{PT})	20*	1.0

setup. Table 2 shows the f-dependent D incorporation for **1** and **2**. At $f = 0$ Hz, we observed the highest D incorporation of 60% for **1** with the IKA setup and reduced D incorporation with our setup (40%). With increasing f, D incorporation dramatically decreases and then falls below the quantification limit of NMR ($\sim$7%), in agreement with the prediction that the slow HAT of **1** requires a long time or low f to be completed. In contrast, the highest D incorporation of 89% for **2** was observed at $f = 0.5$ Hz. However, at $f = 0$ Hz, D incorporation plummets to only <8% and 16% using the IKA setup and our setup, respectively. D incorporation slightly decreased to $\sim$80% at $f > 0.5$ Hz. The high D incorporation observed in **2** at high f is also consistent with our prediction. However, the low D incorporation at $f = 0$ Hz is not predicted by the proposed reaction scheme in Scheme 1B because the fast chemical steps (PT + HAT) should not prevent HIE from happening at low f.

The different f-dependent D-incorporation values for **1** and **2** encouraged us to conduct a detailed analysis of the products to understand the electrochemical HIE reaction mechanism. Fig. 2A and C show the product distribution in terms of starting material (SM) conversion during the electrochemical HIE reactions of **1** and **2**, respectively, at different f values. Note that here we assigned an equivalent f for the DC electrolysis condition, considering that when the substrate or intermediates diffuse between the anode and cathode, they will experience an opposite redox environment, just like during the AC electrolysis. Therefore, we calculated the equivalent AC frequency from the diffusion time between the two electrodes in the electrolytic cell (for example, 3.5×10^{-5} Hz for the IKA setup and 8.7×10^{-4} Hz for our homebuilt setup for **1**; see the ESI for calculation details†). Also, the deuterated SM (*i.e.*, the desired product) and SM cannot be separated, so they were plotted as a group in the product distribution analysis. For the HIE of **1**, we observed significantly increased formation of dimers of **1** (from 0 to 40%) accompanied by a dramatic decrease in the D incorporation (from $\sim$60% to <10%, as shown in Fig. 2B) as f increased from $\sim$10^{-5} to $\sim$10^1 Hz. The formation of dimers indicates the successful deprotonation of the amine cation radicals to α-amino radicals, and the low D incorporation indicates that the HAT step in this

---

Table 2 AC frequency-dependent D incorporation during HIE of **1** and **2**. The applied voltage amplitudes for **1** (0.3 equiv. MTG) and **2** (0.8 equiv. MTG) are 3.5 and 3.1 V, respectively

f (Hz)	1a	2a
0.00	60% (IKA), 40% (our setup)	<8% (IKA), 16% (our setup)
0.05	13%	78%
0.50	<6%	89%
10.0	<7%	78%
50.0	<7%	80%

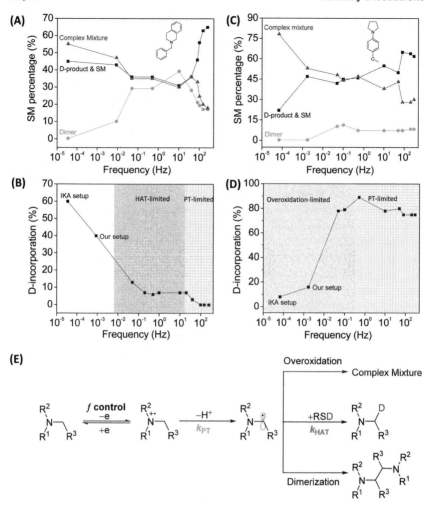

Fig. 2 f-dependence of product distribution in terms of starting material (SM) conversion and D incorporation for **1** (A and B) and **2** (C and D). (E) Illustration of the f-controlled pathways towards forming the D-substituted amine, dimer, and overoxidized complex mixture.

frequency range limits the HIE reaction. More quantitatively, the D incorporation reaches a plateau value of ~7% at $f = $ ~0.1 Hz, suggesting that the HAT between the deuterated thiol and α-amino radical typically requires a time (t_{HAT}) longer than ~10 s. As f increases beyond 20 Hz, the D incorporation decreases to zero, the dimer formation is suppressed, and the SM is recovered (~60%), suggesting that the deprotonation step is now the limiting factor due to the conversion of the amine cation radical back to the SM. Therefore, we can estimate the deprotonation time (t_{PT}) to be on the order of ~0.05 s.

In comparison, for the HIE of **2**, the highest D incorporation of 89% was achieved at 0.5 Hz (Fig. 2D). The D incorporation decreases drastically at frequencies <0.5 Hz and under DC conditions, accompanied by low SM and deuterated-product yields and significant formation of a complex mixture due to overoxidation. For example, DC electrolysis using an IKA setup resulted in ~10% D incorporation, ~20% yield of the SM and deuterated product, and >75% yield of overoxidized products, indicating that the HIE is limited by the overoxidation reactions. At $f >$ 0.5 Hz, the D incorporation slightly decreases, but the yield of the SM and deuterated product increases, similar to what happened for the HIE of **1** at $f > \sim$20 Hz (Fig. 2B). Such opposite correlation between D incorporation and SM/D-product yield is an indicator of a PT-limited process. Similarly, we can estimate t_{PT} for **2** to be ~2 s. The dimer formation is insignificant (<15%) over the entire frequency range, and efficient D incorporation was observed even at f as high as 300 Hz, suggesting that the typical t_{HAT} should be shorter than 0.003 s (estimated from the highest experimentally tested f of 300 Hz). Table 3 summarizes the estimated t_{PT} and t_{HAT} for **1** and **2** from the f-dependent product and D-incorporation analysis in Fig. 2 and the predicted relative t_{PT} and t_{HAT} from Table 1, revealing a quantitative agreement between the experimentally determined values and predicted ones.

From the analysis above, we found that (1) efficient HAT and low overoxidation were critical to achieving high D incorporation, and (2) slow deprotonation had some adverse but not critical effects on the final D incorporation. These findings suggest that the competition between the different reaction pathways of α-amino radicals (HAT, dimerization, and overoxidation) is the key factor in determining the D-incorporation efficiency (Fig. 2E). For **1**, under any AC conditions, the dimerization of α-amino radicals outcompetes HAT, resulting in low D incorporation. Under DC conditions, dimerization is insignificant because the concentration of α-amino radicals decreases due to concentration polarization (the amine substrate is depleted near the electrode surface under a constant potential, thus lowering the radical concentration). Therefore, the α-amino radicals have enough time to complete HAT, which is a slow process for **1**. However, for **2**, at $f <$ 0.5 Hz, the overoxidation of α-amino radicals dominates, leading to low D incorporation and yield. This is not surprising because α-amino radicals of pyrrolidine can easily undergo overoxidation to form iminium cations or further oxidized products in an oxidizing environment in the presence of a base.[23] At high f, the slow deprotonation does affect the availability of α-amino radicals due to the reverse reaction of amine cation radicals to the amine. However, due to the highly efficient HAT, α-amino radicals are effectively deuterated whenever available. Therefore, the final D incorporation remains relatively high at high f as long as the reaction time is sufficient.

Guided by our findings using **1** and **2** as the model substrates, we continued to determine the optimal conditions for the HIE of various tetrahydroisoquinolines, pyrrolidines, and a piperidine (**3–24**). We calculated their BDEs and predicted their relative k_{HAT}. Fig. 3A and B show that the k_{HAT} values for N-aryl tetrahydroisoquinoline substrates **3–13** and pyrrolidine/piperidine substrates **14–24** are comparable to those of **1** and **2**, respectively. The slow HAT kinetics of

Table 3 Estimated time scale for deprotonation (t_{PT}) and HAT (t_{HAT}) for **1** and **2**

	t_{PT}	t_{HAT}
1	~0.05 s	$\geq$10 s
2	~2 s	<0.003 s
Experimental $t(\mathbf{1})/t(\mathbf{2})$	~40 : 1	>3 × 10^3 : 1
Predicted $t(\mathbf{1})/t(\mathbf{2})$	20 : 1	3 × 10^4 : 1

tetrahydroisoquinoline substrates requires performing the HIE reaction under DC conditions. In contrast, HAT should not be the limiting step for the HIE reaction of pyrrolidine/piperidine substrates. The limiting factor should be the undesired overoxidation of α-amino radicals, which can be reduced by increasing f. Thus, we adopted $f = 0.5$ Hz considering the similarity between **2** and the other pyrrolidine/piperidine substrates in terms of their BDEs, unless the D incorporation was not satisfactory, in which case we further increased f.

Fig. 4 shows the substrate scope. Tetrahydroisoquinolines with N-aryl rings bearing electron-donating groups such as methoxy, methyl, tertiary butyl, dimethyl, phenyl and dioxane (**3–9**) and electron-withdrawing substituents such as fluoro and trifluoromethyl (**10–13**) worked well under DC conditions affording high D incorporation of 70–93%. During the substrate-scope development, we varied the reaction time for different N-aryl tetrahydroisoquinoline substrates between 15 and 48 h to improve the yield under the DC electrolysis conditions. However, due to the overoxidation of the benzylic position in N-aryl tetrahydroisoquinolines, the yield for the deuterated products is still limited (~10–60%). On the other hand, N-aryl pyrrolidines bearing electron-donating groups such as methyl, tertiary butyl, phenyl, dioxane, ethyl acetate, and ethanol (**14, 16–21**) afforded a moderate to good D incorporation of 52–89% at 0.5 Hz. Due to the steric hindrance, methyl groups at the *ortho* position of the N-aryl ring (**15**) led to low D incorporation of 36%. The N-aryl pyrrolidines substituted with electron-withdrawing groups such as Cl and Br (**22** and **23**) resulted in less D incorporation (23 and 36%) at 0.5 Hz, possibly due to a dehalogenation reaction. In the case of N-aryl piperidine with a methoxy group on the ring (**24**), we observed poor D incorporation (23%) and low yield (42%) at 0.5 Hz but significantly improved D incorporation of 50% and better yield of 51% at 50 Hz. The improvement at higher f is because the impaired overlap between the lone pair of the amine and the unpaired electron at the α-carbon significantly destabilizes the α-amino radical intermediate, leading to fast overoxidation kinetics that requires high f to suppress the overoxidation. How to quantitively determine the

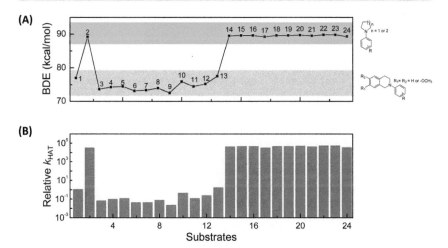

Fig. 3   (A) Calculated BDE values and (B) predicted relative HAT rates at α-amino C(sp^3)–H sites for different substituted tetrahydroisoquinolines (**1, 3–13**) and pyrrolidines (**2, 14–23**) and a piperidine (**24**).

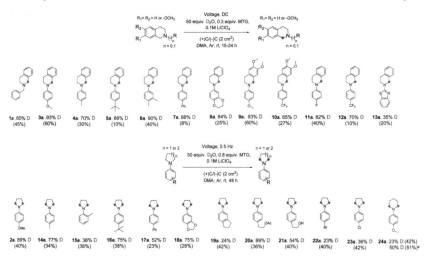

Fig. 4 Substrate scope of the electrochemical HIE reaction. [a]Reaction was conducted at 50 Hz. D incorporation was determined by ^{1}H NMR peak integration relative to an unlabeled compound. All isolated yields are provided in parentheses. The experimental voltage for each substrate is provided in the ESI.†

overoxidation kinetics for predicting the optimal f of HIE reactions for different types of amine substrates is still being studied.

Conclusions

In conclusion, we reported an electrochemical protocol for the HIE of α-amino C(sp^3)–H bonds for the first time. Tetrahydroisoquinolines and pyrrolidines were selected as the two categories of model substrates in this study. We achieved the best D incorporation at $f = 0$ Hz (i.e., DC electrolysis conditions) for tetrahydroisoquinolines and at 0.5 Hz for pyrrolidines. The two different optimal HIE frequencies arise from their different PT and HAT kinetics. Because the HAT rate of tetrahydroisoquinolines is four orders of magnitude slower than that of the pyrrolidines, the HIE of tetrahydroisoquinolines requires DC conditions, under which the dimerization pathway is suppressed, and HAT has sufficient time to be completed. For pyrrolidines, the high HAT rate allows the HIE to occur at high f, but the overoxidation of its α-amino radical intermediates outcompetes the HAT at low f, resulting in low D incorporation and low yield. The AC-frequency-dependent HIE for different amines can be utilized for selective isotope labeling of α-amino C(sp^3)–H sites in one drug molecule, which will be significant and advantageous to the pharmaceutical industry.

Methods

Chemicals and materials

Tertiary amines were synthesized following the protocol in the ESI.† LiClO$_4$ (>95%, MilliporeSigma), methyl thioglycolate (MTG, 95%, MilliporeSigma), D$_2$O (99 atom% D, MilliporeSigma) and N,N'-dimethylacetamide (DMA) (anhydrous, 99.8%, MilliporeSigma) were used as received.

Experimental setups for electrochemical HIE reactions

IKA setup. The IKA ElectraSyn 2.0 purchased from IKA has a reaction vial, vial cap, graphite electrodes, electrode holder, and base unit with a stirrer and vial holder (Fig. S1†). Both graphite electrodes (3 mm in thickness, ~1 cm in width, and ~5 cm in length) are attached to the vial cap with an electrode–electrode separation of ~5 mm and inserted in a vial containing the reaction mixture in 4 mL DMA. The vial is connected to a power supply through the vial holder. Then the output voltage and time can be applied as per the reaction conditions.

Our setup. In this setup, a waveform generator, amplifier, two glassy carbon plate electrodes (3 mm in thickness, 1 cm in width, and ~10 cm in length) and a Schlenk tube were used for the HIE reaction (Fig. S2†). The electrodes were inserted into the reaction flask with an electrode–electrode separation of ~1 mm in 4 mL of DMA and partially immersed (~2 cm) in the solution. Then the electrodes were connected to a waveform generator linked to the amplifier by applying the output voltage and frequency.

General procedures for DC and AC electrolysis

DC electrolysis. An oven-dried 5 mL ElectraSyn reaction vial containing a magnetic stir bar was charged with tertiary amine (0.25 mmol, 1.0 equiv.), LiClO$_4$ (0.5 mmol, 2.0 equiv.), MTG (0.15 mmol, 0.3 equiv.), D$_2$O (12.5 mmol, 50 equiv.) and anhydrous DMA (4.0 mL) under an argon atmosphere. Using two graphite electrodes, the reaction mixture was electrolyzed at a constant voltage of 3.2 to 3.9 V for 15–24 hours. After completion of the reaction, the electrodes were removed, and the reaction mixture was diluted with ethyl acetate (10 mL). The organic layer was washed with a saturated NaHCO$_3$ solution (15 mL) and brine (saturated NaCl, 15 mL). The separated organic layer was dried over Na$_2$SO$_4$ and concentrated under reduced pressure. The crude product was purified using flash silica-gel chromatography with ethyl acetate and hexane to afford the desired products.

AC electrolysis. An oven-dried 10 mL Schlenk tube containing a triangular magnetic stir bar was charged with tertiary amine (0.25 mmol, 1.0 equiv.), LiClO$_4$ (0.5 mmol, 2.0 equiv.), MTG (0.2 mmol, 0.8 equiv.), D$_2$O (12.5 mmol, 50.0 equiv.), and anhydrous DMA (4.0 mL) under an argon atmosphere. Two glassy carbon plate electrodes were then inserted into the reaction flask and connected to a waveform generator linked to the amplifier. The output voltage and frequency were set to 3–4 V and 0–300 Hz, and the reaction mixture was allowed to stir at room temperature for 48 h. Then, the electrodes were removed, and the reaction mixture was diluted with ethyl acetate (10 mL). The organic layer was washed with brine (15 mL), and the separated organic layer was dried over Na$_2$SO$_4$ and then concentrated under reduced pressure. The crude product was purified using flash silica-gel chromatography with ethyl acetate and hexane to afford the desired products.

Cyclic voltammetry. All cyclic voltammograms were collected using a Schlenk tube fitted with a 3 mm-diameter glassy carbon disk electrode as the working electrode, a glassy carbon plate as the counter electrode, and an Ag/Ag$^+$ electrode as the reference electrode under an argon atmosphere. The Ag/Ag$^+$ electrode was prepared by filling the glass tube with 10 mM AgNO$_3$ and LiClO$_4$ (0.1 M, supporting electrolyte) in an anhydrous DMA solution. The Ag/Ag$^+$

reference electrode potential was calibrated using ferrocene/ferrocenium ($E_{Fc/Fc^+} = 0.11$ vs. Ag/Ag$^+$) (Fig. S3†). The glassy carbon disk electrode was cleaned by polishing with a series of alumina powders (0.3 and 0.05 μm) and then sonicated and washed with a large amount of deionized water and methanol before use. All the electrodes were dried in air before the electrochemical measurements.

NMR analysis for H/D exchange quantification. D incorporation was quantified by the decrease in ^{1}H NMR integral intensities at the specified positions compared to the unlabeled starting material.

Computational methods. We computed the α-C(sp^3)–H BDEs using spin-polarized density functional theory as implemented in the Quantum Espresso Package.[32] The calculations employed the projector augmented wave method[33] and the Perdew–Burke–Ernzerhof functional.[34] The neutral molecules and their radicals generated from homolytic cleavage of the C–H bonds at the benzylic and α-amine positions were fully optimized in a sufficiently large box of $30 \times 30 \times 30$ Å using an energy cutoff of 50 Ry and Γ-point sampling until the Hellmann–Feynman forces were less than 0.04 eV Å^{-1}, after which their total energies were computed. The hydrogen BDEs were defined as:

$$BDE = E(R^{\cdot}) + E(H^{\cdot}) - E(R-H),$$

where R–H is the neutral molecule, R$^{\cdot}$ is the radical after the cleavage of the C–H bond, and H$^{\cdot}$ is the hydrogen radical.

Author contributions

The manuscript was written through the contributions of all authors. All authors have given approval to the final version of the manuscript. JF and ZL performed the theoretical calculations.

Conflicts of interest

There are no conflicts to declare.

Acknowledgements

NB, DG, JM, and LL gratefully acknowledge support from the NIH (1R35 GM142590-01), start-up funds, a Rumble Fellowship, and a Faculty Competition for Postdoctoral Fellows award from Wayne State University. The computational part of this work was supported by the U.S. Department of Energy (DOE), Office of Science, Basic Energy Sciences, under award no. DE-SC0023324, and used resources of the National Energy Research Scientific Computing Center (NERSC), a DOE Office of Science User Facility supported by the Office of Science of the U.S. DOE under contract no. DE-AC02-05CH11231 using NERSC award BES-ERCAP0023653. JF acknowledges a Rumble Fellowship and the A. Paul and Carole C. Schaap Endowed Distinguished Graduate Award in Chemistry from Wayne State University. The authors also acknowledge the valuable feedback from Dr Jingwei Li and Zheng Huang of Merck & Co.

References

1 J. Atzrodt and V. Derdau, *J. Labelled Compd. Radiopharm.*, 2010, **53**, 674–685.
2 D. H. Chace, T. Lim, C. R. Hansen, B. W. Adam and W. H. Hannon, *Clin. Chim. Acta*, 2009, **402**, 14–18.
3 A. Nakanishi, Y. Fukushima, N. Miyazawa, K. Yoshikawa, T. Maeda and Y. Kurobayashi, *J. Agric. Food Chem.*, 2017, **65**, 5026–5033.
4 N. Penner, L. Xu and C. Prakash, *Chem. Res. Toxicol.*, 2012, **25**, 513–531.
5 T. Pirali, M. Serafini, S. Cargnin and A. A. Genazzani, *J. Med. Chem.*, 2019, **62**, 5276–5297.
6 T. G. Gant, *J. Med. Chem.*, 2014, **57**, 3595–3611.
7 E. Levernier, K. Tatoueix, S. Garcia-Argote, V. Pfeifer, R. Kiesling, E. Gravel, S. Feuillastre and G. Pieters, *JACS Au*, 2022, **2**, 801–808.
8 W. J. Kerr, G. J. Knox and L. C. Paterson, *J. Labelled Compd. Radiopharm.*, 2020, **63**, 281–295.
9 M. Daniel-Bertrand, S. Garcia-Argote, A. Palazzolo, I. Mustieles Marin, P.-F. Fazzini, S. Tricard, B. Chaudret, V. Derdau, S. Feuillastre and G. Pieters, *Angew. Chem., Int. Ed.*, 2020, **59**, 21114–21120.
10 W. N. Palmer and P. J. Chirik, *ACS Catal.*, 2017, **7**, 5674–5678.
11 H. Yang, C. Zarate, W. N. Palmer, N. Rivera, D. Hesk and P. J. Chirik, *ACS Catal.*, 2018, **8**, 10210–10218.
12 R. Pony Yu, D. Hesk, N. Rivera, I. Pelczer and P. J. Chirik, *Nature*, 2016, **529**, 195–199.
13 C. Taglang, L. M. Martínez-Prieto, I. del Rosal, L. Maron, R. Poteau, K. Philippot, B. Chaudret, S. Perato, A. Sam Lone, C. Puente, C. Dugave, B. Rousseau and G. Pieters, *Angew. Chem., Int. Ed.*, 2015, **54**, 10474–10477.
14 H. Sajiki, N. Ito, H. Esaki, T. Maesawa, T. Maegawa and K. Hirota, *Tetrahedron Lett.*, 2005, **46**, 6995–6998.
15 Y. Y. Loh, K. Nagao, A. J. Hoover, D. Hesk, N. R. Rivera, S. L. Colletti, I. W. Davies and D. W. C. MacMillan, *Science*, 2017, **358**, 1182–1187.
16 Q. Shi, M. Xu, R. Chang, D. Ramanathan, B. Peñin, I. Funes-Ardoiz and J. Ye, *Nat. Commun.*, 2022, **13**, 4453.
17 M. Lepron, M. Daniel-Bertrand, G. Mencia, B. Chaudret, S. Feuillastre and G. Pieters, *Acc. Chem. Res.*, 2021, **54**, 1465–1480.
18 K. Hayashi, J. Griffin, K. C. Harper, Y. Kawamata and P. S. Baran, *J. Am. Chem. Soc.*, 2022, **144**, 5762–5768.
19 J. Fährmann and G. Hilt, *Angew. Chem., Int. Ed.*, 2021, **60**, 20313–20317.
20 S. Rodrigo, D. Gunasekera, J. P. Mahajan and L. Luo, *Curr. Opin. Electrochem.*, 2021, **28**, 100712.
21 N. E. Tay, D. Lehnherr and T. Rovis, *Chem. Rev.*, 2022, **122**, 2487–2649.
22 J. Zhong, C. Ding, H. Kim, T. McCallum and K. Ye, *Green Synthesis and Catalysis*, 2022, **3**, 4–10.
23 D. Gunasekera, J. P. Mahajan, Y. Wanzi, S. Rodrigo, W. Liu, T. Tan and L. Luo, *J. Am. Chem. Soc.*, 2022, **144**, 9874–9882.
24 E. O. Bortnikov and S. N. Semenov, *J. Org. Chem.*, 2021, **86**, 782–793.
25 Y. Kawamata, K. Hayashi, E. Carlson, S. Shaji, D. Waldmann, B. J. Simmons, J. T. Edwards, C. W. Zapf, M. Saito and P. S. Baran, *J. Am. Chem. Soc.*, 2021, **143**, 16580–16588.

26 C. Schotten, C. J. Taylor, R. A. Bourne, T. W. Chamberlain, B. N. Nguyen, N. Kapur and C. E. Willans, *React. Chem. Eng.*, 2021, **6**, 147–151.

27 L. Zeng, Y. Jiao, W. Yan, Y. Wu, S. Wang, P. Wang, D. Wang, Q. Yang, J. Wang and H. Zhang, *Nat., Synth.*, 2023, 1–10.

28 D. Wang, T. Jiang, H. Wan, Z. Chen, J. Qi, A. Yang, Z. Huang, Y. Yuan and A. Lei, *Angew. Chem., Int. Ed.*, 2022, **61**, e202201543.

29 Y. Yuan, J.-C. Qi, D.-X. Wang, Z. Chen, H. Wan, J.-Y. Zhu, H. Yi, A. D. Chowdhury and A. Lei, *CCS Chem.*, 2022, **4**, 2674–2685.

30 E. O. Bortnikov, B. S. Smith, D. M. Volochnyuk and S. N. Semenov, *Chem. – Eur. J.*, 2023, **29**, e2022038.

31 J. M. Mayer, *Acc. Chem. Res.*, 2011, **44**, 36–46.

32 P. Giannozzi, O. Andreussi, T. Brumme, O. Bunau, M. Buongiorno Nardelli, M. Calandra, R. Car, C. Cavazzoni, D. Ceresoli, M. Cococcioni, N. Colonna, I. Carnimeo, A. Dal Corso, S. de Gironcoli, P. Delugas, R. A. DiStasio, A. Ferretti, A. Floris, G. Fratesi, G. Fugallo, R. Gebauer, U. Gerstmann, F. Giustino, T. Gorni, J. Jia, M. Kawamura, H. Y. Ko, A. Kokalj, E. Küçükbenli, M. Lazzeri, M. Marsili, N. Marzari, F. Mauri, N. L. Nguyen, H. V. Nguyen, A. Otero-de-la-Roza, L. Paulatto, S. Poncé, D. Rocca, R. Sabatini, B. Santra, M. Schlipf, A. P. Seitsonen, A. Smogunov, I. Timrov, T. Thonhauser, P. Umari, N. Vast, X. Wu and S. Baroni, *J. Phys.: Condens. Matter*, 2017, **29**, 465901.

33 P. E. Blöchl, *Phys. Rev. B: Condens. Matter Mater. Phys.*, 1994, **50**, 17953.

34 J. P. Perdew, K. Burke and M. Ernzerhof, *Phys. Rev. Lett.*, 1996, **77**, 3865.

Faraday Discussions

ROYAL SOCIETY
OF CHEMISTRY

Electrochemical synthesis of the protected cyclic (1,3;1,6)-β-glucan dodecasaccharide†

Akito Shibuya,[a] Yui Ishisaka,[b] Asuka Saito,[b] Moeko Kato,[b] Sujit Manmode,[a] Hiroto Komatsu,[c] Md Azadur Rahman,[a] Norihiko Sasaki,[ad] Toshiyuki Itoh[ad] and Toshiki Nokami [iD] *[ad]

Received 20th February 2023, Accepted 15th March 2023

DOI: 10.1039/d3fd00045a

Automated electrochemical assembly is an electrochemical method to synthesise middle-sized molecules, including linear oligosaccharides, and some linear oligosaccharides can be electrochemically converted into the corresponding cyclic oligosaccharides effectively. In this study, the target cyclic oligosaccharide is a protected cyclic (1,3;1,6)-β-glucan dodecasaccharide, which consists of two types of glucose trisaccharides with β-(1,3)- and β-(1,6)-glycosidic linkages. The formation of the protected cyclic dodecasaccharide was confirmed by the electrochemical one-pot dimerisation–cyclisation of the semi-circular hexasaccharide. The yield of the protected cyclic dodecasaccharide was improved by using a stepwise synthesis via the linear dodecasaccharide.

Introduction

Electrochemical transformations of small molecules have been used as a powerful set of tools in organic synthesis for many decades.[1] Recent progress in this area has enabled the synthesis of complex molecules such as natural products,[2] peptides,[3] and oligosaccharides.[4] We have been interested in the automated synthesis of oligosaccharides using electrochemical methods and have developed a method named 'automated electrochemical assembly' (AEA), which is based on electrochemical generation of a glycosylation intermediate and its subsequent coupling with alcohols, including oligosaccharides.[5]

Cyclic oligosaccharides such as cyclodextrins (CDs), which contain 1,4-α-linked D-glucopyranose, have attracted the interest of researchers for more than

---

[a]Graduate School of Engineering, Tottori University, Japan. E-mail: tnokami@tottori-u.ac.jp

[b]Graduate School of Sustainable Science, Tottori University, Japan

[c]Department of Chemistry and Biotechnology, Faculty of Engineering, Tottori University, Japan

[d]Centre for Research on Green Sustainable Chemistry, Faculty of Engineering, Tottori University, Japan

† Electronic supplementary information (ESI) available. See DOI: https://doi.org/10.1039/d3fd00045a

a century because of their unique structures and properties.[6] To our knowledge, δ-CD (nonasaccharide) is the largest CD that has been chemically synthesised to date.[7] With regard to cyclic oligosaccharides containing other glycosidic linkages and monosaccharides, cyclic oligo-1,6-β-D-glucosamines up to the hepta-saccharide were synthesised by the Nifantiev group[8] and our group.[9] More recently, our group reported the synthesis of cyclic oligo-1,4-α-N-acetylglucos-amine 'cyclokasaodorin' through an electrochemical polyglycosylation–isomer-isation–cyclisation process.[10] In this case, however, only hexasaccharide and heptasaccharide were obtained. Therefore, the chemical synthesis of large cyclic oligosaccharides remains challenging.

We then focused on a natural oligosaccharide isolated from *Bradyrhizobium japonicum* MTCC120.[11] The oligosaccharide has a cyclic dodecasaccharide struc-ture that consists of two types of glucose trisaccharides with β-(1,3)- and β-(1,6)-glycosidic linkages. Here, we report the electrochemical synthesis of the protected cyclic (1,3;1,6)-β-glucan dodecasaccharide as a potential precursor of the natural cyclic dodecasaccharide.

Results and discussion

Retrosynthetic analysis

Protected cyclic dodecasaccharide **1** has a symmetric structure that consists of β-(1,3)- and β-(1,6)-glycosidic linkages (Fig. 1). Thus, we envisioned that an ideal approach to synthesise protected cyclic dodecasaccharide **1** would be through dimerisation of the semi-circular hexasaccharide building block **2** followed by cyclisation in the same pot. These semi-circular hexasaccharides **2a** and **2b** were considered suitable building blocks because they both have a protecting-group-free hydroxy group (**2a**: 3-OH, **2b**: 6-OH) and thioaryl (SAr, **2a**: Ar = 4-FC₆H₄, **2b**: Ar = 4-ClC₆H₄) leaving group at the anomeric position (C-1). To examine the hypothesis, we synthesised the semi-circular hexasaccharide building block **2a**, bearing two β-(1,3)-glycosidic linkages and three β-(1,6)-glycosidic linkages.[12] Although **2a** was prepared under the electrochemical conditions, its total yield was very low. Moreover, **2a** had a protecting-group-free 3-OH which must be less

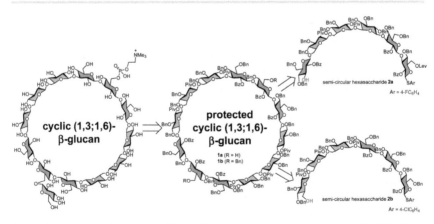

Fig. 1 Semi-circular hexasaccharides building blocks for the protected cyclic (1,3;1,6)-β-glucan.

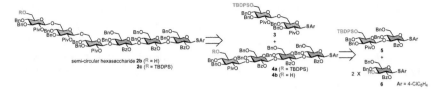

Fig. 2 Retrosynthesis of semi-circular hexasaccharide **2b** and its building blocks **3–6**.

reactive than the 6-OH group. Therefore, we designed semi-circular hex-asacharide **2b** as a building block equipped with a protecting-group-free 6-OH. Semi-circular hexasaccharide **2b** could be disconnected to disaccharide building block **3** and tetrasaccharide building block **4b** with β-(1,6)-glycosidic and β-(1,3)-glycosidic linkages, respectively (Fig. 2). Tetrasaccharide **4a**, as the precursor of **4b**, derived from disaccharide building block **5**, with a β-(1,3)-glycosidic linkage, and two equivalents of monosaccharide building block **6**, equipped with the protecting-group-free 3-OH.

Protocol of automated electrochemical assembly

Automated electrochemical assembly was performed using a middle-size divided electrolysis cell (15 mL for anode and cathode) equipped with carbon fiber anode and platinum plate cathode under argon atmosphere (Fig. 3a). The electrolysis cell was placed in the cooling bath of the 1st generation automated electro-chemical synthesizer (Fig. 3b). The synthesizer was composed of chiller with a cooling bath, stable direct current (DC) power supply, syringe pump, magnetic stirrer, and personal computer (PC), and these devices were controlled by Lab-VIEW installed on the notebook PC. The schedule of a single cycle is shown in Fig. 3c. The DC power supply applied a constant current (13 mA, 1.05 F mol^{-1}) during the anodic oxidation and the electrolysis time (4677 s = *ca.* 78 min) depended on both reaction scale (0.60 mmol) and current value (13 mA). The chiller kept two temperatures −50 °C and −30 °C during anodic oxidation and

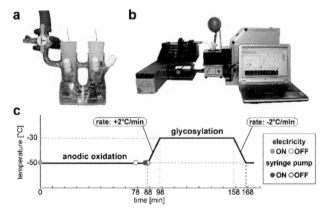

Fig. 3 Devices for automated electrochemical assembly. (a) Divided electrolysis cell equipped with platinum plate cathode and carbon fibre anode. (b) The 1st generation automated electrochemical synthesizer. (c) Schedule of synthesizer for a single cycle.

glycosylation, respectively. In some cases, we switched off the chiller before quenching of the reaction and raised the temperature up to 0 °C to complete the glycosylation. Two gastight syringes were filled with the solution of a building block and solvent for the anodic chamber and cathodic chamber, respectively. They were set to the syringe pump and solutions were added at rate of 1.0 mL min^{-1} after electrolysis.

Synthesis of the disaccharide building blocks

Monosaccharide building blocks **6**, **7a**, and **7b** were prepared from D-glucose pentaacetate according to the reported procedures (Fig. 4). Disaccharide building block **3**, with a β-(1,6)-glycosidic linkage, was prepared using AEA between **7a** and **7b** in the presence of tetrabutylammonium triflate (Bu$_4$NOTf) as an electrolyte. Glycosyl dioxalenium ion intermediate **8** was generated by anodic oxidation of **7a** under constant current conditions at −50 °C. Subsequent coupling of **8** and building block **7b** (1.2 equiv.) afforded disaccharide **3** in 81% yield. This is a standard AEA protocol, and the details of reaction conditions and structures of possible intermediates have been omitted from the following figures (see the ESI† for details of reaction conditions). Although disaccharide building block 5, with a β-(1,3)-glycosidic linkage, was also prepared using AEA of **7a** and **6** (1.2 equiv.) in 85% yield, the one-pot synthesis of tetrasaccharide **4a** from monosaccharide building block **7a** using AEA with three cycles was sluggish.

Optimisation of electrolyte for AEA

The electrolyte for AEA was optimised using the electrochemical formation of β-(1,3)-glycosidic linkages using monosaccharides **9** and **6** (1.2 equiv.) as model building blocks (Table 1). Whereas the use of tetraethylammonium triflate (Et$_4$-NOTf) afforded disaccharide **10** in moderate yield (entry 1), Bu$_4$NOTf, which has been used as a standard electrolyte for AEA, gave the product **10** in good yield (entry 2). We also examined the use of ionic liquids (entries 3–6). The initial voltage of anodic oxidation was significantly influenced by the electrolyte;

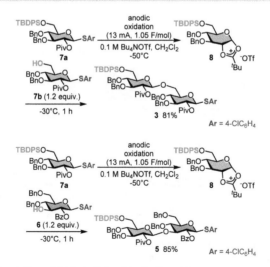

Fig. 4   Synthesis of disaccharide building blocks **3** and **5**.

Table 1 Optimisation of electrolyte of AEA

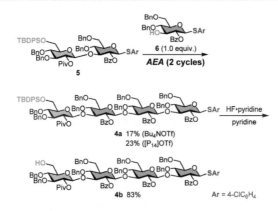

Entry	Electrolyte	Initial voltage[a]	Yield[b]
1	Et$_4$NOTf	26 V	61
2[a]	Bu$_4$NOTf	14 V	79
3[a]	[Bmim]OTf	59 V	70
4	[P$_{1MOM}$]OTf	93 V	81
5	[P$_{1MEM}$]OTf	26 V	89
6	[P$_{14}$]OTf	51 V	97

[a] Inter-electrode voltage. [b] Determined by NMR.

however, there was no clear relationship between the initial voltage and the product yield. Amongst these ionic liquids, 1-butyl-1-methylpyrrolidinium triflate ([P$_{14}$]OTf) afforded the desired disaccharide **10** in the highest yield (entry 6). Therefore, we used ionic liquid [P$_{14}$]OTf as an electrolyte in the following glycosylation reactions. It has not been clarified why [P$_{14}$]OTf gave the best yield; however, the oxidation potential (E_{ox}) of monosaccharide building block **9** measured with [P$_{14}$]OTf (E_{ox} = 1.67 V vs. SCE) was slightly lower than that measured with Bu$_4$NOTf (E_{ox} = 1.70 V vs. SCE). We assume that electrolytes may influence the structure of the electrical double layer and the process of single electron transfer.

Synthesis of the tetrasaccharide building block

The synthesis of tetrasaccharide **4a**, with three β-(1,3)-glycosidic linkages, from disaccharide **5** was carried out using AEA with two consecutive glycosylation

Fig. 5 Synthesis of tetrasaccharide building block **4b**.

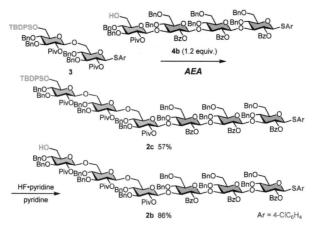

Fig. 6 Synthesis of semi-circular hexasaccharide **2b**.

cycles with monosaccharide building block **6** (1.0 equiv). The process was still challenging; however, performing the reaction sequence in the presence of [P$_{14}$] OTf gave a slightly better yield than with Bu$_4$NOTf (Fig. 5). Deprotection of the *tert*-butyldiphenylsilyl (TBDPS) group of **4a** was achieved successfully in the presence of hydrogen fluoride pyridine complex (HF·pyridine) to obtain tetrasaccharide building block **4b**, equipped with a protecting-group-free 6-OH, in 83% yield.

Synthesis of the semi-circular hexasaccharide and its dimerisation

Semi-circular hexasaccharide building block **2b** was prepared using AEA and subsequent TBDPS deprotection (Fig. 6). Disaccharide **3** and tetrasaccharide **4b** (1.2 equiv.) were assembled to prepare TBDPS-protected semi-circular hexasaccharide

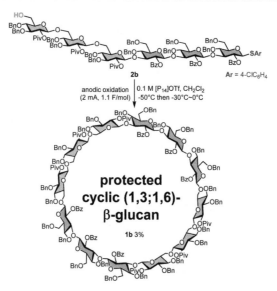

Scheme 1 One-pot dimerisation–cyclisation process for the preparation of protected cyclic dodecasaccharide **1b**.

2c in the presence of [P$_{14}$]OTf as an electrolyte. Deprotection of the TBDPS group at 6-OH was carried out under the standard reaction conditions with HF·pyridine, and the desired semi-circular hexasaccharide **2b** was obtained in 86% yield. Thus-obtained **2b**, equipped with a protecting-group-free 6-OH, was used as a building block in the one-pot dimerisation–cyclisation process to synthesise protected cyclic dodecasaccharide **1b** (Scheme 1). Although the yield of **1b** was very low (3%), protected cyclic (1,3;1,6)-β-glucan dodecasaccharide was obtained, together with by-products such as cyclic hexasaccharide and larger cyclic oligosaccharides, which were detected by MALDI-TOF-MS (see the ESI† for details).

Synthesis of the protected cyclic dodecasaccharide *via* AEA

The results of the one-pot dimerisation–cyclisation process encouraged us to synthesise protected cyclic dodecasaccharide **1b** using AEA (Scheme 2). Two semi-

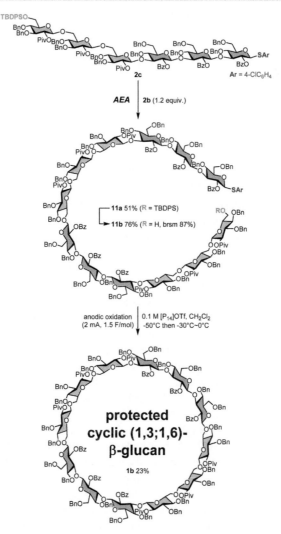

Scheme 2 Synthesis of protected cyclic dodecasaccharide **1b**.

circular hexasaccharide building blocks, **2b** and **2c** (1.2 equiv.), were assembled using AEA to prepare linear dodecasaccharide **11a** in 51% yield. The major by-product of the reaction was the hydroxy sugar of **2c**, which was detected by MALDI-TOF-MS. The TBDPS group on the 6-OH of **11a** was then deprotected to obtain **11b** as a precursor of protected cyclic dodecasaccharide **1b**. Finally, the intramolecular electrochemical glycosylation of **11b** was performed at a low concentration (5 mM) to synthesise **1b** in 23% yield. The three-step yield of **1b** was *ca.* 10%, which was three times higher than that of the one-pot process shown in Scheme 1.

Conclusions

We have synthesised the protected precursor of cyclic (1,3;1,6)-β-glucan dodeca-saccharide, which is the core structure of the natural oligosaccharide isolated from *Bradyrhizobium japonicum* MTCC120. We designed a semi-circular hex-asaccharide and its reactivity was confirmed by the electrochemical one-pot dimerisation–cyclisation process. Finally, the linear precursor of cyclic dodeca-saccharide was prepared using AEA of linear hexasaccharides and subsequent electrochemical intramolecular glycosylation afforded the protected cyclic dodecasaccharide in a higher yield. Further optimisation of the electrochemical intramolecular glycosylation and global deprotection to obtain the cyclic dodec-asaccharide are in progress in our laboratory.

Experimental

General procedure for automated electrochemical assembly and subsequent deprotection of TBDPS group

The automated synthesis of linear dodecasaccharide **11a** was carried out in an H-type divided cell equipped with a carbon felt anode and a platinum plate cathode (20 mm × 20 mm). In the anodic chamber were placed hexasaccharide building block **2c** (0.135 mmol, 405 mg), $[P_{14}]$OTf (0.76 mmol, 0.175 mL) and CH_2Cl_2 (3.9 mL). In the cathodic chamber were placed trifluoromethanesulfonic acid (0.20 mmol, 18 µL), $[P_{14}]$OTf (0.50 mmol, 0.12 mL) and CH_2Cl_2 (4.9 mL). The constant current electrolysis (2.0 mA) was carried out at −50 °C with stirring until 1.1 F mol^{-1} of electricity was consumed. After the electrolysis, hexasaccharide building block **2b** (0.162 mmol, 450 mg) dissolved in CH_2Cl_2 (0.9 mL) was subsequently added by the syringe pump under an argon atmosphere at −50 °C and then −30 °C kept for 60 min. After elevation of the reaction temperature to −5 °C, Et$_3$N (0.4 mL) was added, and the reaction mixture was filtered through a short column (4 × 3 cm) of silica gel to remove electrolyte Bu$_4$NOTf. After removal of the solvent under reduced pressure, the crude product was purified with silica gel chromatography (eluent: hexane/EtOAc 3 : 1) and preparative recycling GPC (eluent: CHCl$_3$). Target linear dodecasaccharide **11a** was obtained in 51% isolated yield (0.069 mmol, 389 mg). Thus-obtained **11a** was used as a starting material for the next step without detailed structural characterisation.

Linear dodecasaccharide **11a** (0.069 mmol, 389 mg) was dissolved in pyridine (0.53 mL) and the solution was cooled to 0 °C. 70% HF·pyridine (0.10 mL) was added to the solution and the reaction mixture was stirred at 0 °C to room temperature for 2 h. Conversion of **11a** was confirmed by TLC (hexane/EtOAc 7 : 3) and aqueous sodium bicarbonate solution was added to quench the reaction. The aqueous solution was extracted with chloroform and the combined organic layer was washed with aqueous sodium bicarbonate solution and 1 N aqueous hydrochloric acid. The reaction mixture was dried over Na_2SO_4 and concentrated under reduced pressure to obtain the crude product (430 mg). The thus-obtained crude product was purified by silica gel chromatography (eluent: hexane/EtOAc 7 : 3) and **11b** (0.053 mmol, 284 mg) was obtained in 76% yield (87% conversion). 4-Chlorophenyl 3,4-di-O-benzyl-2-O-pivaloyl-β-D-glucopyranosyl-(1→6)-3,4-di-O-benzyl-2-O-pivaloyl-β-D-glucopyranosyl-(1→6)-3,4-di-O-benzyl-2-O-pivaloyl-β-D-glucopyranosyl-(1→3)-2-O-benzoyl-4,6-di-O-benzyl-β-D-glucopyranosyl-(1→3)-2-O-benzoyl-4,6-di-O-benzyl-β-D-glucopyranosyl-(1→3)-2-O-benzoyl-4,6-di-O-benzyl-β-D-glucopyranosyl-(1→6)-3,4-di-O-benzyl-2-O-pivaloyl-β-D-glucopyranosyl-(1→6)-3,4-di-O-benzyl-2-O-pivaloyl-β-D-glucopyranosyl-(1→3)-2-O-benzoyl-4,6-di-O-benzyl-β-D-glucopyranosyl-(1→3)-2-O-benzoyl-4,6-di-O-benzyl-β-D-glucopyranosyl-(1→3)-2-O-benzoyl-4,6-di-O-benzyl-1-thio-β-D-glucopyranoside (**11b**); TLC (hexane/EtOAc 7 : 3) R_f = 0.20; ^{1}H NMR (600 MHz, CDCl$_3$) δ 7.80 (d, J = 7.2 Hz, 2H), 7.76–7.73 (m, 5H), 7.52–7.48 (m, 2H), 7.45 (d, J = 7.2 Hz, 1H), 7.41–7.39 (m, 2H), 7.36–6.99 (m, 141H), 6.89–6.86 (m, 1H), 5.08–5.05 (m, 2H), 4.99–4.80 (m, 16H), 4.74–4.14 (m, 55H), 4.10–4.00 (m, 6H), 3.90–3.82 (m, 5H), 3.71–3.62 (m, 6H), 3.60–3.16 (m, 41H), 3.04–3.00 (m, 1H), 2.17 (pseudo-t, J = 6.0 Hz, 1H), 1.10 (s, 18H), 1.08 (s, 9H), 1.06 (s, 9H), 1.05 (s, 9H), 1.04 (s, 9H); ^{13}C NMR (150 MHz, CDCl$_3$) δ 177.3, 177.2, 177.1, 176.7, 176.6, 176.5, 176.4, 164.52, 164.47, 163.9, 138.58, 138.54, 138.51, 138.45, 138.35, 138.31, 138.28, 138.21, 138.18, 138.15, 138.11, 138.07, 138.01, 137.96, 137.87, 137.8, 137.7, 133.6, 133.28, 133.25, 133.22, 133.14, 133.0, 132.93, 132.90, 132.1, 129.78, 129.72, 129.65, 129.55, 129.48, 129.45, 129.38, 129.32, 129.25, 129.18, 129.15, 128.8, 128.7, 128.6, 128.53, 128.46, 128.35, 128.29, 128.24, 128.21, 128.18, 128.14, 128.03, 127.97, 127.94, 127.92, 127.91, 127.76, 127.73, 127.69, 127.63, 127.60, 127.56, 127.43, 127.41, 127.37, 127.35, 127.31, 127.26, 127.22, 127.21, 127.1, 127.0, 126.9, 101.5, 100.8, 100.7, 100.6, 100.5, 100.2, 99.8, 99.6, 99.5, 86.0, 83.1, 82.99, 82.95, 82.93, 82.85, 82.5, 82.4, 80.4, 79.1, 79.0, 78.78, 78.60, 77.73, 77.65, 77.63, 77.43, 76.4, 76.3, 75.98, 75.93, 75.89, 75.87, 75.72, 75.59, 75.56, 75.44, 75.15, 75.11, 75.03, 74.85, 74.77, 74.65, 74.61, 74.51, 74.37, 74.34, 74.27, 74.15, 74.12, 74.02, 73.96, 73.87, 73.79, 73.30, 73.25, 73.17, 73.11, 73.09, 72.82, 72.80, 72.52, 72.44, 72.22, 69.9, 69.21, 69.17, 69.10, 67.6, 61.8, 38.68, 38.62, 38.58, 27.24, 27.19, 27.13, 27.09, 26.99, 26.80, 26.76; MS (MALDI) m/z calculated for $C_{318}H_{341}ClKO_{72}S$ [M + K]$^+$ 5417.21; found 5417.67.

Electrochemical intramolecular glycosylation

The intramolecular glycosylation of linear dodecasaccharide **11b** was carried out in an H-type divided cell equipped with a carbon felt anode and a platinum plate cathode (10 mm × 10 mm). In the anodic chamber were placed protected linear dodecasaccharide **11b** (0.028 mmol, 151 mg), [P$_{14}$]OTf (0.74 mmol, 0.17 mL) and CH$_2$Cl$_2$ (5.0 mL). In the cathodic chamber were placed trifluoromethanesulfonic acid (0.14 mmol, 12 μL), [P$_{14}$]OTf (0.50 mmol, 0.12 mL) and CH$_2$Cl$_2$ (5.1 mL). The constant current electrolysis (2.0 mA) was carried out at −50 °C with stirring until 1.5 F mol^{-1} of electricity was consumed and then −30 °C kept for 60 min. After elevation of the reaction temperature to −5 °C, Et$_3$N (0.1 mL) was added to both chambers, and the reaction mixture was dissolved in CHCl$_3$ and washed with water to remove electrolyte [P$_{14}$]OTf. The thus-obtained organic layer was dried over Na$_2$SO$_4$ and concentrated under reduced pressure to obtain the crude product (160 mg). Silica gel chromatography (eluent: hexane/EtOAc 3 : 1) and preparative recycling GPC (eluent: CHCl$_3$) afforded target protected cyclic dodecasaccharide **1b** in 23% yield (6.9 μmol, 36 mg). Cyclobis-(1→6)-(3,4-di-*O*-benzyl-2-*O*-pivaloyl-β-D-glucopyranosyl)-(1→6)-(3,4-di-*O*-benzyl-2-*O*-pivaloyl-β-D-glucopyranosyl)-(1→6)-(3,4-di-*O*-benzyl-2-*O*-pivaloyl-β-D-glucopyranosyl)-(1→3)-(2-*O*-benzoyl-4,6-di-*O*-benzyl-β-D-glucopyranosyl)-(1→3)-(2-*O*-benzoyl-4,6-di-*O*-benzyl-β-D-glucopyranosyl)-(1→3)-(2-*O*-benzoyl-4,6-di-*O*-benzyl-β-D-glucopyranosyl) (**1b**); TLC (hexane/EtOAc 3 : 1) R_f = 0.30; ^{1}H NMR (600 MHz, CDCl$_3$) δ 7.79 (pseudo-t, J = 6.0 Hz, 4H), 7.67–7.62 (m, 4H), 7.59 (d, J = 7.2 Hz, 4H), 7.46–7.43 (m, 6H), 7.36–7.14 (m, 126H), 7.05 (d, J = 6.0 Hz, 2H), 5.09–5.00 (m, 12H), 4.96–4.88 (m, 6H), 4.74 (d, J = 7.8 Hz, 2H), 4.70 (dd, J = 10.8, 3.0 Hz, 2H), 4.66–4.61 (m, 6H), 4.56–4.42 (m, 26H), 4.38–4.25 (m, 12H), 4.15 (d, J = 7.8 Hz, 2H), 4.09 (pseudo-t, J = 9.0 Hz, 2H), 4.02 (pseudo-t, J = 9.0 Hz, 2H), 4.00–3.96 (m, 4H), 3.79–3.74 (m, 4H), 3.67–3.27 (m, 48H), 3.21–3.15 (m, 4H), 1.11 (s, 36H), 1.08 (s, 18H); ^{13}C NMR (150 MHz, CDCl$_3$) δ 177.0, 176.5, 176.4, 164.7, 164.4, 163.8, 138.7, 138.6, 138.43, 138.38, 138.2, 138.1, 138.04, 137.95, 137.91, 133.30, 133.24, 133.20, 129.82, 129.75, 129.5, 129.4, 129.3, 128.8, 128.61, 128.56, 128.43, 128.35, 128.25, 128.20, 128.15, 128.10, 128.05, 128.01, 127.96, 127.7, 127.6, 127.45, 127.39, 127.35, 127.31, 127.26, 127.21, 127.13, 127.0, 126.9, 126.8, 100.7, 100.5, 100.43, 100.35, 100.14, 99.3, 83.0, 82.9, 82.8, 82.7, 79.75, 79.66, 78.3, 78.2, 78.1, 77.6, 77.5, 76.3, 76.1, 75.6, 75.3, 75.2, 75.1, 74.85, 74.80, 74.72, 74.59, 74.54, 74.48, 74.33, 74.21, 73.84, 73.76, 73.3, 73.2, 73.1, 72.9, 72.2, 69.7, 69.15, 69.07, 67.7, 66.9, 38.65, 38.63, 38.59, 27.3, 27.1, 26.9; MS (MALDI) *m*/z calculated for C$_{312}$H$_{336}$KO$_{72}$ [M + K]$^+$ 5273.23; found 5273.04.

Author contributions

A. Shibuya, N. S., T. I. and T. N. organized the research. A. Shibuya, Y. I., A. Saito, M. K., S. M., H. K. and M. A. R. synthesized and characterized the compounds. A. Shibuya and T. N. principally wrote the manuscript according to the suggestions and discussion with all authors.

Conflicts of interest

There are no conflicts to declare.

Acknowledgements

The authors deeply acknowledge the Grant-in-Aid for Scientific Research on Innovative Areas: 'Middle Molecular Strategy' (No. 2707) from the JSPS (No. JP15H05844), Grant-in-Aid for Scientific Research (C) from JSPS (No. JP19K05714), and JST CREST (No. JPMJCR18R4) for financial support.

Notes and references

1 (*a*) J. Yoshida, K. Kataoka, R. Horcajada and A. Nagaki, *Chem. Rev.*, 2008, **108**, 2265; (*b*) J. Yoshida, Y. Ashikari, K. Matsumoto and T. Nokami, *J. Synth. Org.*

Chem., Jpn., 2013, **71**, 1136; (c) J. Yoshida, A. Shimizu, Y. Ashikari, T. Morofuji, R. Hayashi, T. Nokami and A. Nagaki, *Bull. Chem. Soc. Jpn.*, 2015, **88**, 763; (d) M. Yan, Y. Kawamata and P. S. Baran, *Chem. Rev.*, 2017, **117**, 13230; (e) J. Yoshida, A. Shimizu and R. Hayashi, *Chem. Rev.*, 2018, **118**, 4702; (f) K. D. Moeller, *Chem. Rev.*, 2018, **118**, 4817; (g) C. Zhu, N. W. J. Ang, T. H. Meyer, Y. Qiu and L. Ackermann, *ACS Cent. Sci.*, 2021, **7**, 415.

2 (a) P. Hu, B. K. Peters, C. A. Malapit, J. C. Vantourout, P. Wang, J. Li, L. Mele, P.-G. Echeverria, S. D. Minteer and P. S. Baran, *J. Am. Chem. Soc.*, 2020, **142**, 20979; (b) Y. Gao, D. E. Hill, W. Hao, B. J. McNicholas, J. C. Vantourout, R. G. Hadt, S. E. Reisman, D. G. Blackmond and P. S. Baran, *J. Am. Chem. Soc.*, 2021, **143**, 9478; (c) K. Hayashi, J. Griffin, K. C. Harper, Y. Kawamata and P. S. Baran, *J. Am. Chem. Soc.*, 2022, **144**, 5762; (d) S. J. Harwood, M. D. Palkowitz, C. N. Gannett, P. Perez, Z. Yao, L. Sun, H. D. Abruña, S. L. Anderson and P. S. Baran, *Science*, 2022, **375**, 745.

3 S. Nagahara, Y. Okada, Y. Kitano and K. Chiba, *Chem. Sci.*, 2021, **12**, 12911.

4 (a) S. Manmode, K. Matsumoto, T. Itoh and T. Nokami, *Asian J. Org. Chem.*, 2018, **7**, 1719; (b) A. Shibuya and T. Nokami, *Chem. Rec.*, 2021, **21**, 2389; (c) K. Yano, N. Sasaki, T. Itoh and T. Nokami, *J. Synth. Org. Chem., Jpn.*, 2021, **79**, 839.

5 (a) T. Nokami, R. Hayashi, Y. Saigusa, A. Shimizu, C.-Y. Liu, K.-K. Mong and J. Yoshida, *Org. Lett.*, 2013, **15**, 4520; (b) T. Nokami, Y. Isoda, N. Sasaki, A. Takaiso, S. Hayase, T. Itoh, R. Hayashi, A. Shimizu and J. Yoshida, *Org. Lett.*, 2015, **17**, 1525.

6 (a) M. Davis and M. Brewster, *Nat. Rev. Drug Discovery*, 2004, **3**, 1023; (b) G. Crini, *Chem. Rev.*, 2014, **114**, 10940.

7 M. Wakao, K. Fukase and S. Kusumoto, *J. Org. Chem.*, 2002, **67**, 8182.

8 (a) M. L. Gening, D. V. Titov, A. A. Grachev, A. G. Gerbst, O. N. Yudina, A. S. Shashkov, A. O. Chizhov, Y. E. Tsvetkov and N. E. Nifantiev, *Eur. J. Org. Chem.*, 2010, 2465; (b) D. V. Titov, M. L. Gening, A. G. Gerbst, A. O. Chizhov, Y. E. Tsvetkov and N. E. Nifantiev, *Carbohydr. Res.*, 2013, **381**, 161; (c) M. L. Gening, Y. E. Tsvetkov, D. V. Titov, A. G. Gerbst, O. N. Yudina, A. A. Grachev, A. S. Shashkov, S. Vidal, A. Imberty, T. Saha, D. Kand, P. Talukdar, G. B. Pier and N. E. Nifantiev, *Pure Appl. Chem.*, 2013, **85**, 1879.

9 S. Manmode, S. Tanabe, T. Yamamoto, N. Sasaki, T. Nokami and T. Itoh, *ChemistryOpen*, 2019, **8**, 869.

10 H. Endo, M. Ochi, M. A. Rahman, T. Hamada, T. Kawano and T. Nokami, *Chem. Commun.*, 2022, **58**, 7948.

11 (a) M. W. Breedveld and K. J. Miller, *Microbiol. Rev.*, 1994, **58**, 145; (b) A. V. Nair, S. N. Gummadi and M. Doble, *Biotechnol. Lett.*, 2016, **38**, 1519; (c) E. Cho, D. Jeong, Y. Choi and S. Jung, *J. Inclusion Phenom. Macrocyclic Chem.*, 2016, **85**, 175.

12 (a) S. Manmode, M. Kato, T. Ichiyanagi, T. Nokami and T. Itoh, *Asian J. Org. Chem.*, 2018, **7**, 1802; (b) A. Shibuya, M. Kato, A. Saito, S. Manmode, N. Nishikori, T. Itoh, A. Nagaki and T. Nokami, *Eur. J. Org. Chem.*, 2022, **19**, e202200135.

Faraday Discussions

DISCUSSIONS

Selective organic electrosynthesis: general discussion

Mickaël Avanthay, [iD] Belen Batanero, [iD] Dylan G. Boucher,
Christoph Bondue, [iD] Pim Broersen, Richard C. D. Brown, [iD]
Robert Francke, [iD] Toshio Fuchigami, [iD] Alexander Kuhn, [iD]
Kevin Lam, [iD] Chia-Yu Lin, [iD] T. Leo Liu, Long Luo, [iD]
Shelley D. Minteer, [iD] Kevin Moeller, Toshiki Nokami, [iD]
Robert Price, [iD] Shahid Rasul [iD] and Eniola Sokalu [iD]

DOI: 10.1039/d3fd90036c

Shahid Rasul opened the discussion of the introductory Spiers Memorial Lecture by Toshio Fuchigmai: From your overview (https://doi.org/10.1039/d3fd00129f), which top three electrochemical reactions for electrosynthesis are most suitable for commercialization, considering the significant challenge of scaling in electrochemistry?

Toshio Fuchigami answered: In consideration of the significant challenge of scaling in electrochemistry, (1) anodic homo- and cross-coupling of arenes (Scheme 13 in my Lecture article (https://doi.org/10.1039/d3fd00129f)), (2) micro-flow cell, thin-layer flow cell, and a parallel laminar flow cell (Fig. 4 in my Lecture article), as well as (3) PTFE-fiber-coated electrodes, are most suitable for commercialization.

Mickaël Avanthay opened the discussion of the paper by Alexander Kuhn: Does the magnetic fields contribute to the chemistry in any way other than moving the swimmer (*i.e.* heating, *etc.*)?

Alexander Kuhn answered: There is no heating because local currents are too low to generate Joule heating. We have performed control experiments to check whether more exotic effects like chiral induced spin selectivity (CISS) effects might contribute to the enantioselectivity, but inverting the direction of the magnetic field leads to the same %ee, and we think that the enhanced efficiency is solely due to the magnetohydrodynamic (MHD) effect which provides a more efficient flux of prochiral starting compounds to the swimmer surface. The magnetic field doesn't change the electron transfer kinetics but impacts only the mass transfer.

Mickaël Avanthay asked: Have you ever tried to tether the microswimmers?

Alexander Kuhn answered: We have performed experiments, also described in our paper, where we have used a rather macroscopic version of such a core–shell wire (https://doi.org/10.1039/d3fd00041a). We have fixed it at the bottom of the beaker and have added tracer particles in order to visualize the local MHD effect. From the supplementary file videos one can clearly see clockwise and anti-clockwise motion of the tracer particles at the two extremities of the wire.

Pim Broersen asked: When you are not applying the magnetic field during the reaction, is there more hydrogen production due to substrate mass transfer being limited?

Alexander Kuhn responded: We did not quantify how much the ratio of electrons that are used for the enantioselective reduction *versus* the electrons that are used for hydrogen evolution, is changing when applying the magnetic field or not. On the one hand, a first guess might be that in the absence of a magnetic field, more electrons are used for hydrogen evolution because there is not enough prochiral educt present at the oligomer surface due to mass transport limitations. However, on the other hand, in the absence of the MHD effect the produced hydrogen bubbles also have a tendency to stick for a longer time at the swimmer surface, due to the absence of hydrodynamic motion, thus partially blocking the interface and therefore also slowing down the arrival of protons. Thus, it is difficult to predict the final impact of the magnetic field on the ratio between these two competing processes. However, it is obvious from our measurements, that the presence of the magnetic field greatly enhances the production rate of the chiral molecule.

Pim Broersen continued: When comparing the performance of the swimmers with and without the magnetic field, did you observe a different rate of the reaction, and does this lead to a different rate at which the zinc 'fuel' is consumed.

Alexander Kuhn answered: Indeed, when applying the magnetic field the global reaction rate increases significantly due to enhanced mass transport. This means intrinsically that zinc has to dissolve faster in order to provide the electrons necessary for the reduction of the prochiral starting compound.

Christoph Bondue asked: You mentioned that enantiomeric excess does not change when applying magnetic field – do you think this is because the magnets are weak?

Alexander Kuhn answered: Right now the magnets we are using have a strength of around 200 mT. We could try to use stronger magnets, but I think that this won't affect the %ee, because it will just change the speed of the swimmers as the Lorentz force will be stronger and thus increase the hydrodynamic motion helping to provide fresh prochiral starting molecules to the surface of the swimmers. Therefore the global yield for a given reaction time, or in other words the production rate, will increase but not the ratio between the two enantiomers. There could be one reason to see a change in %ee: if some kind of spin selectivity were involved in the reaction mechanism, but we have no indication for such an additional effect so far.

Kevin Moeller queried: Can you change the potential the swimmer reduces at by altering the oligomer?

Alexander Kuhn responded: The potential at which the enantioselective reduction reaction is carried out can be controlled by changing the chemical nature of the oligomer layer. Our Italian collaborators have a series of different monomers that can be used to deposit oligomer films having chiral features. Depending on their chemical structure, they will be redox active at different potentials.

Eniola Sokalu commented: You have shown that the magnet can generate an electric field in the solution and thus, influencing bubble removal, meaning you can expose active sites for further reactions. Have you had any thoughts on potentially optimising the shape of the magnet or finding a way to generate magnetic 'local hotspots'?

Alexander Kuhn responded: We don't generate an electric field in solution with the magnet, but the redox reactions at the swimmer generate an electron and ion flux, which under the influence of the magnetic field induces the Lorentz force. The latter triggers a boost of motion of the swimmer which in turn helps in removing the bubbles from the surface for more efficient delivery of the prochiral starting compound to the swimmer surface for the catalytic transformation. Replacing a homogeneous magnetic field by inhomogeneous magnetic fields can have some beneficial effects, especially if the magnetic pattern is changing as a function of time because the swimmers will always try to move towards positions with the highest magnetic field strength, a phenomenon that we already observed. This would allow to further control their trajectory and eventually further enhance their speed.

Shelley D. Minteer asked: Considering that in the modern day we are focusing on sustainability, can you replace the sacrificial anode with something that will be greener?

Alexander Kuhn answered: This is a very important comment. Indeed the current system is based on the gradual consumption of the zinc wire which acts as a fuel, simultaneously for the electroreduction of the prochiral starting compound, but also to trigger motion *via* the hydrogen bubble evolution. This type of energy source can be easily replaced by other reactions which are more sustainable/greener. An interesting option is to use for example an enzymatically driven oxidation reaction (*e.g.* of abundant fuel molecules such as glucose). In this case the swimmers don't get consumed like in the present case, but can in principle work for a long time if the enzyme is stable and substrate is available. We've already explored this kind of concept by combining an enzymatic reaction with the inherently chiral oligomers (see ref. 1) in order to design an autonomous swimmer. However in this case reactions went the other way round: a target molecule was oxidized on the chiral oligomer surface and the electrons were delivered to bilirubin oxidase for reducing oxygen. The ion flux that accompanies this electron transfer leads then to self-electrophoretic motion, which in addition shows a directionality that depends on the chirality of the target molecule. We are currently further exploring such combinations.

1 S. Arnaboldi, G. Salinas, A. Karajić, P. Garrigue, T. Benincori, G. Bonetti, R. Cirilli, S. Bichon, S. Gounel, N. Mano and A. Kuhn, *Nat. Chem.*, 2021, 13, 1241–1247, DOI: 10.1038/s41557-021-00798-9.

Dylan G. Boucher said: As the inspiration for this system is more efficient mass-transfer, it seems the long-term vision is a reactor with many micro-swimmers operating simultaneously. What happens when they hit each other? Do they short circuit? What are the design principles for preventing collisions in reactors with large concentrations of swimmers?

Alexander Kuhn answered: This is an interesting point. For the moment we perform our experiments only with a small amount of swimmers present simultaneously, so the probability of them running into each other is very small. Also they have more of a tendency to avoid each other, most likely due to a small meniscus present around every swimmer. Actually, we are actively investigating this type of interaction because we would like to understand the swarming behavior of these objects, that means how they self-organize when there are (too) many of them. The interesting point is that if you think about these systems in more detail, actually every swimmer is an individual tiny magnet because of the local electric current, which obviously generates, according to the corkscrew rule, a tiny magnetic field around every swimmer. So if there are no other perturbations (like the bubble generation in our present system) it should be possible to see the effect of these small local magnetic fields on the self-organization of the swimmers. One possibility to reach this situation, is to replace the present dissolution of zinc (or magnesium) with another type of oxidation reaction, for example based on enzymatic reactions which won't produce bubbles.

T. Leo Liu queried: Can mechanic stirring work instead of a magnetic field? What is the effect of magnetic field strength? Can the magnetic field be used for a flow cell?

Alexander Kuhn replied: Mechanic stirring can for sure also increase the production rate, but it is not as efficient as magnetic field-induced stirring (*i.e.*, MHD). We compared in previous work,[1] the efficiency of both and found that even when mechanically stirring the solution around a static wire, the production rate of chiral molecules is lower than with the swimmers of an equivalent length. The reason for this resides most likely in the fact that mechanical stirring is a top-down approach and the solution will not be stirred right at the interface because of the hydrodynamic boundary layer. In contrast, MHD induces liquid motion right at the interface as this is the place where the driving force (Lorentz force) is generated, so it is a bottom-up phenomenon. Obviously in this latter case, the stronger the magnetic field, the stronger the Lorentz force (as it is the cross product between the local current and the magnetic field), and thus the stronger the generated hydrodynamics. In the present work, we use moderate magnetic fields of around 200 mT, so there is still the possibility to enhance the effect using stronger magnets.

I don't see any reason why this phenomenon can't be used in a flow cell.

1 S. Arnaboldi, G. Salinas, G. Bonetti, P. Garrigue, R. Cirilli, T. Benincori and A. Kuhn, *Angew. Chem.*, 2022, **134**, e202209098, DOI: 10.1002/ange.202209098.

Chia-Yu Lin commented: The magnetic field could induce the induction heating to modify the kinetics of the electrochemical reactions. Did you examine this effect?

Alexander Kuhn replied: We didn't investigate more precisely whether local heating might contribute to the enhanced reaction kinetics. However we estimated the local currents typically flowing through the swimmers and these values are very small, so I don't think that a related Joule effect would lead to a significant increase in temperature of the swimmer or around the swimmer.

Shahid Rasul asked: Do these factors – process intensification, applied magnetic field, and improved mass transfer – have any observable impact on overpotential in electrochemical reactions?

Alexander Kuhn replied: In our specific case the magnetic field does not change the overpotential of the electron transfer. It helps to improve the mass transport of electroactive species towards the catalytically active interface of the swimmers.

Shahid Rasul added: Do different interfaces such as platinum and zinc significantly influence selectivity in electrochemical reactions?

Alexander Kuhn answered: The selectivity can be influenced by the overall driving force which is directly related to the composition of the swimmers. If a very reactive metal, such as magnesium, is used as the core material of the swimmers, the driving force is very high and this can have a negative impact on the enantioselectivity. Therefore in general a metal core with a modest reactivity leads to a better selectivity.

T. Leo Liu queried: Is the magnetic stirring approach scalable? Can it be used in a flow cell?

Alexander Kuhn answered: Yes I think it is scalable. For the moment we've only worked with a small number of swimmers in the context of these proof-of-principle experiments, but obviously the number of swimmers can be increased. However, above a certain threshold concentration they might interact with each other and efficiency won't scale anymore with the number of swimmers. I don't see how and why they should be integrated in a flow cell because first of all they might be washed out with the flow, and second, their effect might be limited as there is already hydrodynamics in the cell.

Belen Batanero further questioned: How possible is it to scale up the process?

Alexander Kuhn replied: The process might be scaled up using many swimmers simultaneously, ideally as a kind of powder of active particles that is added to the solution containing the prochiral starting compound and then exposed to the external magnetic field.

Belen Batanero asked: In these magnetic swimmer processes what time scale order is handled? Can they continue overnight?

How general is the oligomer with respect to other asymmetric reductions? Can enantioselectivity can be improved? Do you know the detailed induction mechanism?

Alexander Kuhn replied: The swimmers can stay active as long as there is "fuel" available, that means as long as the zinc wire is not completely dissolved they will swim. However enantioselectivity slowly decays as a function of time. The BT_2T_4 oligomer that we use right now can be employed not only for this stereospecific reaction transforming acetophenone into phenylethanol, but also for other enantioselective reactions as has been shown by our Italian collaborators in a long series of publications. We have tested our swimmer concept for example, for the reduction of phenylglyoxylic acid into mandelic acid.[1] In addition, other oligomers with similar stereoselective features are available, which further opens up the spectrum of possible asymmetric reactions. We don't know for the moment the detailed molecular mechanism that leads to this outstanding stereoselectivity (hydrogen bonding, π–π interactions *etc.*), but we are actively working on its elucidation.

1 S. Arnaboldi, G. Salinas, G. Bonetti, P. Garrigue, R. Cirilli, T. Benincori and A. Kuhn, *Angew. Chem.*, 2022, **134**, e202209098, DOI: 10.1002/ange.202209098.

Christoph Bondue opened the discussion of the paper by Long Luo: The AC method helps prevent overoxidation, keeping a radical, how does it help achieve site selectivity?

Long Luo answered: Yes, it does help. Check out our recent paper[1] and our review paper.[2] We have another paper discussing this topic coming out soon as well. Stay tuned!

1 D. Gunasekera, J. P. Mahajan, Y. Wanzi, S. Rodrigo, W. Liu, T. Tan and L. Luo, *J. Am. Chem. Soc.*, 2022, **144**, 9874–9882, DOI: 10.1021/jacs.2c02605.
2 S. Rodrigo, C. Um, J. C. Mixdorf, D. Gunasekera, H. M. Nguyen and L. Luo, *Org. Lett.*, 2020, **22**, 6719–6723, DOI: 10.1021/acs.orglett.0c01906.

Pim Broersen asked: In your paper, when selecting the potential range between which you oscillate the AC, you base the reductive potential on the potential required to reduce the thiol (https://doi.org/10.1039/d3fd00044c). However, the species you need to reduce for catalytic turnover is a thiyl radical. Do you not expect the reduction potential of the radical to be different than of the thiol itself?

Long Luo answered: Great question. I am sure the reduction potentials for thiyl radical and thiol are different. However, we still maintain at a voltage bias that can reduce thiol because there is no thiyl radical at the very beginning of the reaction. Without applying such voltage bias, the reaction cannot be initiated.

Robert Price said: As your work employs AC frequencies, have you performed any AC impedance spectroscopy in order to evaluate your process?

Long Luo replied: No, we haven't but it is something interesting to explore. In principle, the impedance value should reflect the electrochemical processes in our system.

Richard C. D. Brown commented: The work described in your paper involves a complex interplay between electrochemical and chemical steps as well as mass transport with species diffusing towards and away from, the electrode as its potential is switched. Have you considered, or are you using simulations, to help understand the experimental outcomes or even guide optimisation of the conditions?

Long Luo responded: Yes, computational simulations will definitely help.

Kevin Moeller said: My thoughts are to do with the rate of the processes and whether the radical can be reduced to an anion by the back electron transfer? This would allow the trapping of electrophiles.

Long Luo responded: It will be interesting to investigate this question. In our system, we fixed the applied voltage between the two electrodes but not the electrode potentials. We will need to find out the electrode potential to test if the potential is negative enough to reduce radical to anion. There are a few species that can be reduced, including the oxidized amine, thiol, and possibly the α-amino radical.

Mickaël Avanthay asked: How does the speed of synthesis of polysaccharide on the e-synthesiser compare to the conventional, commercial synthesiser?

Toshiki Nokami replied: The synthesizer developed by Professor Ye and co-workers[1] is based on a conventional chemical method in solution phase and the reaction is extremely fast. But the reaction rate depends on the current value and the reaction scale. So, the e-synthesizer is also fast when we compare with other methods based-on solid-phase synthesis.

1 X.-S. Ye, *et al.*, *Nat. Synth.*, 2022, **1**, 854–863, DOI: 10.1038/s44160-022-00171-9.

T. Leo Liu opened a general discussion of the paper by Toshiki Nokami: Does the coupling chemical step take place in the electrochemical cell (https://doi.org/10.1039/d3fd00045a)? What is the yield of each oxidation-coupling step?

Toshiki Nokami replied: Both oxidation (activation) and coupling occur in the anodic chamber of the electrochemical cell. Conversion of the starting material by anodic oxidation is almost quantitative in most cases; however, accumulation of the glycosylation intermediate is not 100%. Moreover, the yield of the coupling step between the accumulated intermediate and the building block, which is a thioglycoside with a protecting-group-free hydroxyl group, depends on the combination.

T. Leo Liu asked: What are side products in terms of 3% overall yield of your reaction?

Toshiki Nokami replied: The corresponding molecular ion peak of the cyclic hexasaccharide was observed as a major peak; however, the yield was less than 10%. I think that the major by-products may be hydroxy sugars which are reaction products between glycosyl triflate intermediates and water.

Belen Batanero said: You use an extremely low temperature, does it not function at room temperature, and what is the problem with working at 25 °C?

Toshiki Nokami responded: Some glycosylation intermediates can be treated at 0 °C; however, most glycosyl triflate intermediates are unstable at elevated temperature. Side reactions such as elimination of HOTf must occur at room temperature.

Belen Batanero asked: Do you need a previous activation of the electrode? How is the solution conductivity at low temperature? You have many protecting groups and they are so large that I wonder if the electrode surface was passivated.

Toshiki Nokami responded: We use the carbon fiber electrode after drying at 120 °C for several hours; however, pre-activation of the surface of the electrode is not needed. Conductivity of the solution at low temperature must be very low. So, we added more than 0.1 M of electrolyte Bu_4NOTf in dichloromethane (DCM). Of course, using DCM as a solvent is not ideal. But DCM is the only solvent which can dissolve the electrolyte at low temperature.

Kevin Lam suggested: Have your tried a weakly coordinating anion such as $B(C_6F_5)_4^-$ in your supporting electrolyte? Would that allow the generation of the cation pool at room temperature?

Toshiki Nokami replied: Cation pools can be generated with $B(C_6F_5)_4^-$. Professor Yoshida also reported generation of a glycosylation intermediate with $B(C_6F_5)_4^-$; however, there is a catch. TfOH is added to the cathodic chamber and the triflate anion can move to the anodic chamber during electrolysis. So, I think that the accumulated species must be a glycosyl triflate, not glycosyl cations. Of course, we can generate cations at room temperature, but accumulation is difficult at that temperature. It depends on the stability of cations whether we can accumulate generated species or not. Some glycosylation intermediates which have neighboring groups at 2-OH can be accumulated at 0 °C because they form stable glycosyl dioxalenium ions.

Kevin Lam added: What happens when you try to generate a cation pool in the presence of a weakly coordinating anion? Does it decompose?

Toshiki Nokami answered: In many cases of cation pools a weakly coordinating anion is necessary to accumulate cations. Glycosylation intermediates are quite different from carbocations. Although glycosyl triflates are also unstable, there are covalent species and detectable by NMR at low temperature. The major decomposition pathways of glycosyl cations might be hydrolysis, elimination, intramolecular reaction with a protected hydroxyl group, and intermolecular reaction with anions of electrolytes. All our attempts to generate and accumulate glycosyl cations using weakly coordinating anions failed. Although we did not check by-products carefully, glycosyl fluorides were produced exclusively in the presence of BF_4 anions. This means glycosyl cations are very reactive and I have never read any reports about detection of glycosyl cations. There are some modified ones, but they are stabilized.

Robert Francke said: Your comparison between different supporting electrolytes (triflate salts) showed a strong impact of the cation on cell voltage and product yield (Table 1 in your paper). What is the explanation for this?

Toshiki Nokami replied: It is still difficult to explain at this stage. We should investigate the effects of electrolytes on electrolysis (first step) and glycosylation (second step) separately. I think that electrolytes may influence the stability and reactivity of glycosyl triflate intermediates. We will investigate the stability of glycosyl triflate intermediates using several electrolytes.

Robert Francke added: The electrolyte effects shown in Table 1 are very interesting and may be worth investigating more closely. In addition to electrochemical double layer effects, differences in ion pair dissociation should be taken into account as a possible explanation. Conductivity measurements are suggested. It may also be worth analyzing electrolysis and chemical follow-up reactions separately to figure out in which of the two processes the salt effect comes into play.

Toshiki Nokami responded: These are very important comments. We will measure conductivity and viscosity as well. It is also important to investigate the influence of the electrolyte on electrolysis and the following chemical reaction separately.

Shelley D. Minteer addressed all: Since many electrosynthesis systems are actually a cascade of reactions, is it possible to use time dependent control of potential/current or magnetic fields to improve flux through a cascade of reactions?

Alexander Kuhn replied: In our specific case we have no control over the potential/current as a function of time. We could vary the magnetic field as a function of time, but I don't think that this alone will improve the flux through a cascade of reactions if the magnetic field is homogeneous. However, temporal variations of the spatial distribution of the magnetic field strength might have a positive effect on the efficiency of such reactions, as the swimmers always try to move along magnetic field gradients towards positions having the highest field.

Long Luo replied: It is possible. AC frequency can be used to control the mediator concentration. We will need to look at particular reactions to figure out how to do it exactly.

Conflicts of interest

There are no conflicts to declare.

Faraday Discussions

PAPER

Photoelectrocatalyzed undirected C–H trifluoromethylation of arenes: catalyst evaluation and scope†

Julia Struwe and Lutz Ackermann [iD] *

Received 10th April 2023, Accepted 5th June 2023

DOI: 10.1039/d3fd00076a

During the last few years, photoelectrocatalysis has evolved as an increasingly viable tool for molecular synthesis. Despite several recent reports on the undirected C–H functionalization of arenes, thus far, a detailed comparison of different catalysts is still missing. To address this, more than a dozen different mediators were employed in the trifluoromethylation of (hetero-)arenes to compare them in their efficacies.

Introduction

Electrochemical synthesis has emerged as a powerful approach in molecular syntheses.[1] Strategies to replace the stoichiometric oxidant through the hydrogen evolution reaction and to enable novel reactivities have gained significant attention.[2] Likewise, visible-light photoredox catalysis has been recognized as a valuable strategy to allow chemical transformations under mild conditions,[3] particularly with efficient single electron transfer (SET).

The merger of electrocatalysis and photochemistry with pioneering work in the 1970s and 1980s[4] has the power to allow chemistry at extreme potentials[5] and enabled a diverse set of desirable transformations, such as oxidative C–C,[6] C–N[7] and C–O[8] bond formations through radical mechanisms. In addition, reductive and isohypsic, thus redox-neutral, reactions have also been achieved under photoelectrocatalysis.[5a] While different photoelectrocatalysts have been employed (Fig. 1a), to the best of our knowledge, a systematic comparison of different photocatalysts has thus far proven elusive.

Thus, given the diversity of different photo- and redox-catalysts, from which some have also been employed in photoelectrocatalyzed reactions, we wished to elucidate the impact of the nature of the photoelectrocatalyst on the reaction efficacies. As a representative testing ground, we selected the photoelectrochemical C–H

Institut für Organische und Biomolekulare Chemie, Georg-August-Universität Göttingen, Tammannstrasse 2, 37077 Göttingen, Germany. E-mail: Lutz.Ackermann@chemie.uni-goettingen.de

† Electronic supplementary information (ESI) available. See DOI: https://doi.org/10.1039/d3fd00076a

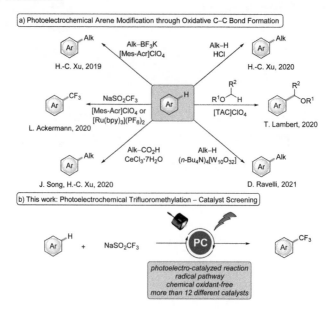

Fig. 1   (a) Overview of photoelectrochemical C–C bond formations and the employed photocatalysts; (b) performance study of different mediators in the photoelectrocatalyzed trifluoromethylation of arenes.

trifluoromethylation of arenes (Fig. 1b)[9] with the Langlois reagent (2, $NaSO_2CF_3$)[10] as the trifluoromethyl source.

Results and discussion

We commenced our study by testing various photoredox catalysts with 5.0 mol% catalyst loading using 1,3,5-trimethoxybenzene (1) as the model substrate, while transition metal-based catalysts were employed at 2.0 mol% (Table 1). The frequently used ruthenium(II) trisbipyridine photocatalyst provided the desired functionalized arene in an excellent conversion yield of 91% (Table 1, entry 1). Additionally, other transition metal catalysts have been explored, and to our delight, the earth-abundant iron- and nickel-derivatives of the trisbipyridine ruthenium(II) catalyst gave comparably good results under otherwise identical reaction conditions (entries 2 and 3). Also, the photocatalyst $(n\text{-}Bu_4N)_4[W_{10}O_{32}]$ enabled good reactivity, although slightly diminished yields of the decorated arenes 3 and 3′ were obtained (entry 4). Cerium trichloride was also able to catalyze the reaction with comparable conversions, with irradiation at 450 nm or 390 nm wavelength (entries 5 and 6). Furthermore, halide salts were suitable mediators, with strongly enhanced reactivities under light irradiation compared to solely electrochemical conditions,[11] and the products 3 and 3′ were obtained in yields up to 97% with significant amounts of difunctionalized product 3′ (entries 7–9). Also, other organic photoelectrocatalysts were identified as more sustainable alternatives to the metal-containing ones, furnishing the desired products 3 and 3′ in excellent yields of around 90% (entries 10–14). To our delight, the trisaminocyclopropenium-based catalyst [TAC]ClO_4 provided the highest mono-

Table 1 Functionalization of trimethoxybenzene **1** with different catalysts[a]

Entry	PEC	Deviation	Conversion (3 : 3′)
1	$[Ru(bpy)_3](PF_6)_2$	2 mol%	91% (3.3 : 1)
2	$[Fe(bpy)_3](PF_6)_2$	2 mol%	87% (3.8 : 1)
3	$[Ni(bpy)_3]Br_2$	2 mol%	89% (5.2 : 1)
4	$(n\text{-}Bu_4N)_4[W_{10}O_{32}]$	2 mol%	85% (5.5 : 1)
5	$CeCl_3 \cdot 7H_2O$	—	90% (3.5 : 1)
6	$CeCl_3 \cdot 7H_2O$	390 nm	93% (2.6 : 1)
7	TBAI	—	95% (2.2 : 1)
8	TBABr	—	97% (1.5 : 1)
9	TBACl	—	95% (1.4 : 1)
10	DCA	—	96% (5.0 : 1)
11	DCN	—	90% (4.3 : 1)
12	DDQ	—	93% (1.7 : 1)
13	[TAC]ClO_4	—	89% (6.4 : 1)
14	[Mes-Acr]ClO_4	—	95% (4.9 : 1)
15	[Mes-Acr]ClO_4	$Zn(SO_2CF_3)_2$ (**4**)	63% (20 : 1)
16	—	—	9% (—)
17	[Mes-Acr]ClO_4	Without light	7% (—)
18	[Mes-Acr]ClO_4	Without current	4% (—)

[a] Reaction conditions: undivided cell, GF anode (10 mm × 15 mm × 6 mm), Pt cathode (10 mm × 15 mm × 0.25 mm), constant-current electrolysis at 4.0 mA. **1** (0.25 mmol), $NaSO_2CF_3$ (**2**, 0.50 mmol) or $Zn(SO_2CF_3)_2$ (**4**, 0.25 mmol), PEC (5.0 mol%), $LiClO_4$ (0.1 M), MeCN (4.0 mL), 30–35 °C, 8 h, under N_2, blue LEDs (450 nm); conversions were determined by ^{1}H-NMR using dimethyl-terephthalate as an internal standard. The ratio is given as (mono : di) selectivity. GF = graphite felt, PEC = photoelectrocatalyst, LED = light-emitting diode, bpy = 2,2-bipyridine, TBA = tetra-n-butyl ammonium, DCA = 9,10-dicyanoanthracene, DCN = 1,4-dicyanonaphthalene, DDQ = 2,3-dichloro-5,6-dicyano-1,4-benzoquinone, TAC = trisaminocyclopropenium, Mes-Acr = 9-mesityl-10-methylacridinium.

selectivity for the functionalization, while DDQ resulted in an optimal difunctionalization of substrate **1** (entries 12 and 13). In addition, the sulfinate $Zn(SO_2CF_3)_2$ (**4**) could be also employed as the trifluoromethyl source under photoelectrocatalytic conditions, albeit with a significantly diminished conversion (entry 15). In sharp contrast, the reaction failed to proceed in the absence of a photocatalyst (entry 16), thereby reflecting the key influence of the catalytic mediator. Likewise, the essential role of visible light as well as an electric current was verified in control experiments (entries 17 and 18).

Moreover, the impact of the irradiation with visible light was analyzed by comparing the photoelectrocatalytic approach to electrooxidative trifluoromethylations[12] (Table 2). With the user-friendly undivided cell set-up for the electrooxidative transformation, the functionalized arene **6** was obtained at a low conversion of only 21% (Table 2, entry 1). In sharp contrast, the photoelectrocatalysis was found to be uniquely powerful in furnishing arene **6** (Table 2, entry 2).[11]

Table 2 Electrooxidation *vs.* photoelectrocatalysis[a]

Entry	Sulfinate	Conditions	Yield (6)
1	$Zn(SO_2CF_3)_2$	Electrooxidation	(26%)
2	$Zn(SO_2CF_3)_2$	Photoelectrocatalysis	(80%)
3	$NaSO_2CF_3$	Electrooxidation	(21%)
4	$NaSO_2CF_3$	Photoelectrocatalysis	70% (81%)

[a] Reaction conditions: electrooxidation: undivided cell, GF anode (10 mm × 15 mm × 6 mm), GF cathode (10 mm × 15 mm × 6 mm), constant-current electrolysis at 4.0 mA. 5 (0.25 mmol), 2 (0.5 mmol), n-Bu_4NClO_4 (0.15 M), DMSO (5.0 mL), 8 h. Photoelectrocatalysis: undivided cell, GF anode (10 mm × 15 mm × 6 mm), Pt cathode (10 mm × 15 mm × 0.25 mm), constant-current electrolysis at 4.0 mA. 5 (0.25 mmol), 2 (0.5 mmol), [Mes-Acr]ClO_4 (5.0 mol%), $LiClO_4$ (0.1 M), MeCN (4.0 mL), 30–35 °C, 8 h, under N_2, blue LEDs (450 nm). Yields refer to the isolated product, conversions were determined by ^{19}F-NMR using 1-fluorononane as internal standard.

Further studies on the efficiency of the different catalysts were performed through the analysis of the reaction mixture in a time-resolved manner. Caffeine (5) was chosen as the model substrate to avoid any difficulties in interpretation because of difunctionalization (Fig. 2). A comparison of the data showed only minor differences during the reaction process. However, slight differences in the catalytic performance of the catalysts were reflected in the product yields, which differ by around 15% after 8 h reaction time, with the best results for DCN and TBACl. The determined faradaic efficiency of this undirected C–H functionalization ranged from 40% to 47% with a current density of 1/400 A cm^{-2}.

The undirected C–H functionalization with the Langlois reagent was proposed as shown in Scheme 1.[9] In the case of the organic dye [Mes-Acr]ClO_4, the SET was enabled by the excited-state photocatalyst species, which was generated after absorption of blue light, resulting in the formation of a trifluoromethyl radical after fragmentation. Likewise, the reduced photocatalyst was formed and subsequently reoxidized at the anode. Subsequent modification of the arenes follows a radical pathway with a Wheland complex as the key intermediate.

With a set of viable photoredox catalysts for the undirected C–H trifluoromethylation of arenes in hand, we became interested in further substantiating the catalytic performance of representative photoelectrocatalysts with a series of arenes 7 (Scheme 2 top). Besides triethylbenzene, several heterocyclic systems were tested, among those quinolines and pyrimidine derivatives. These examples of nitrogen containing heterocycles furnished the corresponding products 10–12 in moderate to good yields being similar for both catalysts. Furthermore, furane as well as thiophene scaffolds were fully tolerated in both catalytic systems to access 13–15, although the decorated heterocyclic amides 14 and 15 were obtained as a mixture of two regioisomers. Moreover, several heterocyclic compounds associated with natural products were employed as well.

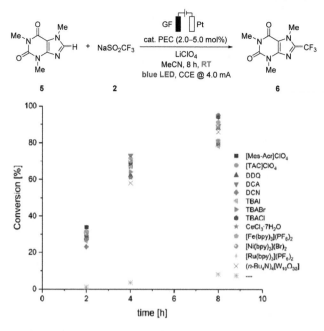

Fig. 2 Comparison of different catalysts for the trifluoromethylation of caffeine (5). Reaction conditions: undivided cell, GF anode (10 mm × 15 mm × 6 mm), Pt cathode (10 mm × 15 mm × 0.25 mm), constant-current electrolysis at 4.0 mA. 5 (0.30 mmol), 2 (0.60 mmol), catalyst (2.0 mol% or 5.0 mol%), $LiClO_4$ (0.1 M), MeCN (4.0 mL), 30–35 °C, 8 h, under N_2, blue LEDs (450 nm). Conversions were determined after 2, 4 and 8 hours by ^{19}F-NMR using 1-fluorononane as an internal standard.

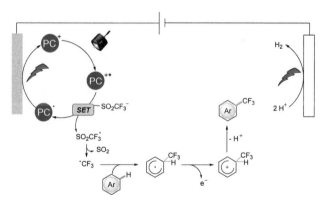

Scheme 1 Proposed catalytic cycle for the undirected functionalization with [Mes-Acr] ClO_4 as the photocatalyst.

Among those, a gallic acid derivative, pentoxifylline as well as a purine base and modified tryptophan were smoothly converted into the corresponding products 16–19 with moderate to good yields. For all the depicted examples, the isolated yield was comparable for both catalysts, with differences of up to 7%.

Interestingly, in the case of arene 20, which is a building block in the synthesis of the anti-cancer kinase inhibitor KAN0438757,[13] the reaction outcome was

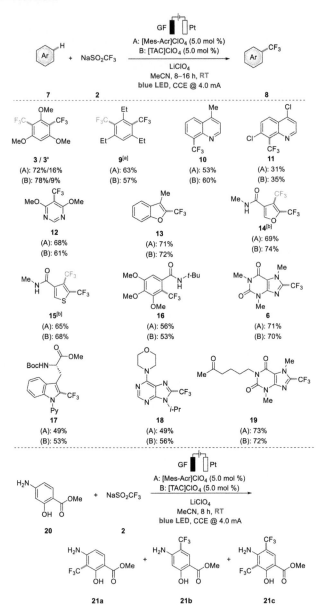

Scheme 2 Comparison of [Mes-Acr]ClO₄ and [TAC]ClO₄ for different arenes. Reaction conditions: undivided cell, GF anode (10 mm × 15 mm × 6 mm), Pt cathode (10 mm × 15 mm × 0.25 mm). **7** (0.25 mmol), **2** (0.50 mmol), [Mes-Acr]ClO₄ or [TAC]ClO₄ (5.0 mol %), LiClO₄ (0.1 M), MeCN (4.0 mL), 30–35 °C, 8–16 h, under N₂, blue LEDs (450 nm); isolated yields. [a]Product was obtained together with the difunctionalized product. [b]Product was obtained as a mixture of regioisomers. [c]Reaction conditions: undivided cell, GF anode (10 mm × 15 mm × 6 mm), Pt cathode (10 mm × 15 mm × 0.25 mm), constant-current electrolysis at 4.0 mA, **20** (0.25 mmol), **2** (0.50 mmol), [Mes-Acr]ClO₄ or [TAC]ClO₄ (5.0 mol%), LiClO₄ (0.1 M), MeCN (4.0 mL), 30–35 °C, 8 h, under N₂, blue LEDs (450 nm); isolated yields.

significantly influenced by the choice of the catalyst in terms of the yield and the ratio of the obtained products **21a–c** (Scheme 2, bottom). For this substrate, possessing sensitive free hydroxyl- and amine functionalities, [TAC]ClO$_4$ outperformed [Mes-Acr]ClO$_4$ with an isolated total product yield of 76%. Thus, the result indicates the potential of [TAC]ClO$_4$ for the functionalization of sensitive substrates.

Conclusions

In conclusion, we reported on an evaluation of various photoelectrocatalysts for the undirected C–H trifluoromethylation of arenes. Thereby, we identified two powerful metal-free photoelectrocatalysts, [Mes-Acr]ClO$_4$ and [TAC]ClO$_4$, for direct C–H trifluoromethylation without directing groups, enabling efficient photoelectrocatalysis under mild conditions with ample scope. However, the exact mode of action of the catalyst might differ between different catalysts, dependent on the properties.

Conflicts of interest

There are no conflicts to declare.

Acknowledgements

Generous support by DZHK and the DFG (Gottfried Wilhelm Leibniz prize to L. A.) is gratefully acknowledged.

References

1 (*a*) M. C. Leech and K. Lam, *Nat. Rev. Chem.*, 2022, **6**, 275–286; (*b*) N. Chen and H.-C. Xu, *Chem. Rec.*, 2021, **21**, 2306–2319; (*c*) L. F. T. Novaes, J. Liu, Y. Shen, L. Lu, J. M. Meinhardt and S. Lin, *Chem. Soc. Rev.*, 2021, **50**, 7941–8002; (*d*) C. Zhu, N. W. J. Ang, T. H. Meyer, Y. Qiu and L. Ackermann, *ACS Cent. Sci.*, 2021, **7**, 415–431; (*e*) F. Wang and S. S. Stahl, *Acc. Chem. Res.*, 2020, **53**, 561–574; (*f*) W. R. Browne, *Electrochemistry*, Oxford Chemistry Press, Oxford, 2019; (*g*) M. Elsherbini and T. Wirth, *Acc. Chem. Res.*, 2019, **52**, 3287–3296; (*h*) A. Scheremetjew, T. H. Meyer, Z. Lin, L. Massignan and L. Ackermann, in *Science of Synthesis: Electrochemistry in Organic Synthesis*, ed. L. Ackermann, Thieme, Stuttgart, 2019, pp. 3–32; (*i*) A. Wiebe, T. Gieshoff, S. Möhle, E. Rodrigo, M. Zirbes and S. R. Waldvogel, *Angew. Chem., Int. Ed.*, 2018, **57**, 5594–5619; (*j*) M. Yan, Y. Kawamata and P. S. Baran, *Chem. Rev.*, 2017, **117**, 13230–13319; (*k*) R. Francke and R. D. Little, *Chem. Soc. Rev.*, 2014, **43**, 2492–2521.

2 (*a*) C.-Y. Cai, Z.-J. Wu, J.-Y. Liu, M. Chen, J. Song and H.-C. Xu, *Nat. Commun.*, 2021, **12**, 3745; (*b*) M. Hielscher, E. K. Oehl, B. Gleede, J. Buchholz and S. R. Waldvogel, *ChemElectroChem*, 2021, **8**, 3904–3910; (*c*) M. Stangier, A. M. Messinis, J. C. A. Oliveira, H. Yu and L. Ackermann, *Nat. Commun.*, 2021, **12**, 4736; (*d*) H.-J. Zhang, L. Chen, M. S. Oderinde, J. T. Edwards, Y. Kawamata and P. S. Baran, *Angew. Chem., Int. Ed.*, 2021, **60**, 20700–20705; (*e*) F. Xu, H. Long, J. Song and H.-C. Xu, *Angew. Chem., Int. Ed.*, 2019, **58**,

9017–9021; (f) N. Sauermann, T. H. Meyer, C. Tian and L. Ackermann, *J. Am. Chem. Soc.*, 2017, **139**, 18452–18455; (g) E. J. Horn, B. R. Rosen, Y. Chen, J. Tang, K. Chen, M. D. Eastgate and P. S. Baran, *Nature*, 2016, **533**, 77–81.

3 (a) A. Y. Chan, I. B. Perry, N. B. Bissonnette, B. F. Buksh, G. A. Edwards, L. I. Frye, O. L. Garry, M. N. Lavagnino, B. X. Li, Y. Liang, E. Mao, A. Millet, J. V. Oakley, N. L. Reed, H. A. Sakai, C. P. Seath and D. W. C. MacMillan, *Chem. Rev.*, 2022, **122**, 1485–1542; (b) N. A. Romero and D. A. Nicewicz, *Chem. Rev.*, 2016, **116**, 10075–10166; (c) J. M. R. Narayanam and C. R. J. Stephenson, *Chem. Soc. Rev.*, 2011, **40**, 102–113.

4 (a) R. Scheffold and R. Orlinski, *J. Am. Chem. Soc.*, 1983, **105**, 7200–7202; (b) J.-C. Moutet and G. Reverdy, *J. Chem. Soc., Chem. Commun.*, 1982, 654–655; (c) J.-C. Moutet and G. Reverdy, *Tetrahedron Lett.*, 1979, **20**, 2389–2392.

5 (a) H. Huang, K. A. Steiniger and T. H. Lambert, *J. Am. Chem. Soc.*, 2022, **144**, 12567–12583; (b) J. P. Barham and B. König, *Angew. Chem., Int. Ed.*, 2020, **59**, 11732–11747; (c) J. Galczynski, H. Huang and T. H. Lambert, in *Science of Synthesis: Electrochemistry in Organic Synthesis*, ed. L. Ackermann, Thieme, Stuttgart, 2019, pp. 325–362.

6 (a) C.-Y. Cai, X.-L. Lai, Y. Wang, H.-H. Hu, J. Song, Y. Yang, C. Wang and H.-C. Xu, *Nat. Catal.*, 2022, **5**, 943–951; (b) L. Capaldo, L. L. Quadri, D. Merli and D. Ravelli, *Chem. Commun.*, 2021, **57**, 4424–4427; (c) H. Huang, Z. M. Strater and T. H. Lambert, *J. Am. Chem. Soc.*, 2020, **142**, 1698–1703; (d) X.-L. Lai, X.-M. Shu, J. Song and H.-C. Xu, *Angew. Chem., Int. Ed.*, 2020, **59**, 10626–10632; (e) P. Xu, P.-Y. Chen and H.-C. Xu, *Angew. Chem., Int. Ed.*, 2020, **59**, 14275–14280; (f) H. Yan, Z.-W. Hou and H.-C. Xu, *Angew. Chem., Int. Ed.*, 2019, **58**, 4592–4595.

7 (a) T. Shen and T. H. Lambert, *Science*, 2021, **371**, 620–626; (b) T. Shen and T. H. Lambert, *J. Am. Chem. Soc.*, 2021, **143**, 8597–8602; (c) L. Niu, C. Jiang, Y. Liang, D. Liu, F. Bu, R. Shi, H. Chen, A. D. Chowdhury and A. Lei, *J. Am. Chem. Soc.*, 2020, **142**, 17693–17702; (d) H. Huang, Z. M. Strater, M. Rauch, J. Shee, T. J. Sisto, C. Nuckolls and T. H. Lambert, *Angew. Chem., Int. Ed.*, 2019, **58**, 13318–13322; (e) F. Wang and S. S. Stahl, *Angew. Chem., Int. Ed.*, 2019, **58**, 6385–6390; (f) L. Zhang, L. Liardet, J. Luo, D. Ren, M. Grätzel and X. Hu, *Nat. Catal.*, 2019, **2**, 366–373.

8 (a) H. Huang and T. H. Lambert, *Angew. Chem., Int. Ed.*, 2021, **60**, 11163–11167; (b) H. Huang and T. H. Lambert, *J. Am. Chem. Soc.*, 2021, **143**, 7247–7252.

9 Y. Qiu, A. Scheremetjew, L. H. Finger and L. Ackermann, *Chem.–Eur. J.*, 2020, **26**, 3241–3246.

10 (a) J. Shen, J. Xu, L. He, C. Liang and W. Li, *Chin. Chem. Lett.*, 2022, **33**, 1227–1235; (b) Y. Ji, T. Brueckl, R. D. Baxter, Y. Fujiwara, I. B. Seiple, S. Su, D. G. Blackmond and P. S. Baran, *Proc. Natl. Acad. Sci. U. S. A.*, 2011, **108**, 14411–14415; (c) B. R. Langlois, E. Laurent and N. Roidot, *Tetrahedron Lett.*, 1991, **32**, 7525–7528.

11 For detailed information, see the ESI†

12 A. G. O'Brien, A. Maruyama, Y. Inokuma, M. Fujita, P. S. Baran and D. G. Blackmond, *Angew. Chem., Int. Ed.*, 2014, **53**, 11868–11871.

13 N. M. S. Gustafsson, K. Färnegårdh, N. Bonagas, A. H. Ninou, P. Groth, E. Wiita, M. Jönsson, K. Hallberg, J. Lehto, R. Pennisi, J. Martinsson, C. Norström, J. Hollers, J. Schultz, M. Andersson, N. Markova, P. Marttila, B. Kim, M. Norin, T. Olin and T. Helleday, *Nat. Commun.*, 2018, **9**, 3872.

PAPER

Impact of sodium pyruvate on the electrochemical reduction of NAD⁺ biomimetics†

Chase Bruggeman, Karissa Gregurash and David P. Hickey ⓘ *

Received 20th February 2023, Accepted 14th March 2023

DOI: 10.1039/d3fd00047h

Biomimetics of nicotinamide adenine dinucleotide (mNADH) are promising cost-effective alternatives to their natural counterpart for biosynthetic applications; however, attempts to recycle mNADH often rely on coenzymes or precious metal catalysts. Direct electrolysis is an attractive approach for recycling mNADH, but electrochemical reduction of the oxidized mimetic (mNAD⁺) primarily results in the formation of an enzymatically inactive dimer. Herein, we find that aqueous electrochemical reduction of an NAD⁺ mimetic, 1-n-butyl-3-carbamoylpyridinium bromide (1^+), to its enzymatically active form, 1,4-dihydro-1-n-butyl nicotinamide (1H), is favored in the presence of sodium pyruvate as a supporting electrolyte. Maximum formation of 1H is achieved in the presence of a large excess of pyruvate in combination with a large excess of a co-supporting electrolyte. Formation of 1H is found to be favored at pH 7, with an optimized product ratio of ~50/50 dimer/1H observed by cyclic voltammetry. Furthermore, sodium pyruvate is shown to promote electroreductive generation of the 1,4-dihydro form of several additional mNADH as well as NADH itself. This method provides a general strategy for regenerating 1,4-dihydro-nicotinamide mimetics of NADH from their oxidized forms.

Introduction

Recently there has been a significant expansion in the use of bioelectrocatalysis for environmentally friendly organic synthesis.[1] In bioelectrocatalysis, redox enzymes drive highly selective redox reactions under mild aqueous conditions, and the corresponding enzyme cofactor is regenerated with an electrode as the terminal oxidant/reductant. One cofactor, nicotinamide adenine dinucleotide (phosphate) (NAD(P)⁺/NAD(P)H), holds special appeal for bioelectrocatalysis, because it is used by half of documented oxidoreductases;[2] however, its commercial utilization is limited by its high cost. A promising step towards low-

Department of Chemical Engineering and Materials Science, Michigan State University, East Lansing, MI 48824-1226, USA. E-mail: Hickeyd6@msu.edu

† Electronic supplementary information (ESI) available: Derivation of convolution voltammetry equations, NMR spectra. See DOI: https://doi.org/10.1039/d3fd00047h

cost redox cofactors was demonstrated with simple NADH mimetics (mNADH), such as 1-*n*-butyl-1,4-dihydronicotinamide (**1H**), that were shown to exhibit activity with enoate reductase, hydroxybiphenyl monooxygenase, and BM3 cyto-chrome P450.[3-7] Both the natural cofactor, NADH, and its mimetics commonly function as hydride donors, undergoing a concerted two-electron/one-proton transfer to yield the oxidized form, NAD$^+$ (Scheme 1); however, direct electro-chemical reduction of NAD$^+$ and its mimetics proceeds *via* sequential single electron/proton transfers. These reductions can be intercepted after the first electron transfer, with rapid formation of enzymatically inactive dimers (rate constant for dimerization exceeds 10^7 M^{-1} s^{-1}).[8] The first one-electron reduction of NAD$^+$ occurs between -0.9 V and -1.2 V, and the resulting neutral pyridine radical can be further reduced near -1.6 V to form the enzymatically active 1,4-dihydropyridine (all potentials reported *vs.* SCE).[8,9]

In practice, however, the electrochemical yields of enzymatically-active NADH from constant potential electrolysis are near 65–76%, and no NADH is observed at moderate reduction potentials (*ca.* -1.2 V).[10] While methods do exist to optimize the yield of electrochemically generated NADH above even 90%, for example Rh hydride-transfer catalysts[6,11] and Pt-patterned glassy carbon electrodes,[12] these methods are limited by slow turnover rates and a sensitivity to the precise voltage/ metal surface coverage. While the operational simplicity of direct electrochem-istry and its high theoretical (m)NADH regeneration rate (limited only by mass transport) hold promise for bioelectrocatalysis, its use is still limited by the undesired dimerization reaction.

While it is known that the supporting electrolyte can play a critical role in governing the product distribution of electrochemically-coupled chemical

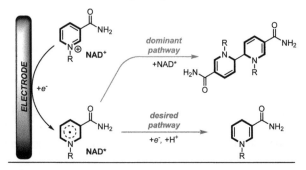

Scheme 1 (A) Example biological reaction involving NADH, where alcohol dehydrogenase catalyzes the reduction of an aldehyde to an alcohol. The reaction proceeds *via* hydride transfer. (B) Mechanism for direct electrochemical reduction of NAD$^+$. The reaction proceeds *via* sequential electron–proton transfer and is intercepted after the first electron to form dimer rather than the desired biologically active NADH.

reactions,[13-18] little attention has been given to the role of supporting electrolyte in the electrochemical reduction of NAD^+ or its mimetics.[19] In light of this, we considered the possibility that dimerization could be minimized in electrochemical $(m)NAD^+$ reduction through judicious choice of supporting electrolyte. We chose **1H** as a model cofactor, hypothesizing that supporting electrolyte could lower the relative rate of dimerization and increase the relative rate of **1H** formation during electrochemical reduction of the mimetic's oxidized form, 1-*n*-butyl nicotinamide bromide (**1⁺**). To test this hypothesis, we screened the electrochemistry of the bromide salt of **1⁺** in 28 different supporting electrolytes at a boron-doped-diamond electrode, and we compared this against the electrochemistry of **1H** formed by dithionite reduction. We used cyclic voltammetry for its operational simplicity, and for the ease with which information about concentration profiles and product distributions could be extracted.

Results and discussion

The cyclic voltammogram (CV) for 2 mM **1⁺** in aqueous 200 mM sodium bicarbonate (Fig. 1) reveals an irreversible reduction peak at −1.3 V and an irreversible oxidation peak at −0.05 V (ox1) (all potentials reported *vs.* SCE). The potential of ox1 is consistent with that of a dimer formed by electrochemically generated neutral radical species of **1⁺** (*i.e.*, **1***).[20] By contrast, CVs of the chemically prepared, pure form of **1H** result in no peak at 0 V during the initial scan, but instead a large irreversible oxidation peak is observed near 0.44 V (ox2) followed by a reduction peak at −1.3 V matching that of **1⁺**. After the first cycle, an oxidation peak consistent with ox1 becomes visible. A summary of this general reaction mechanism is given by:

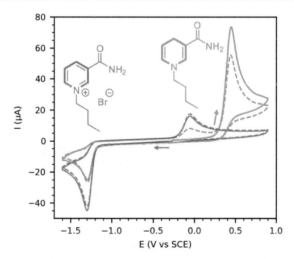

Fig. 1 Representative cyclic voltammograms of 2 mM 1-*n*-butyl nicotinamide bromide, **1⁺**, (−) and 2 mM 1,4-dihydro-1-*n*-butyl nicotinamide, **1H**, (−). Solid curves represent the first CV scans, while dashed lines represent the second CV scans. Experiments were performed using a 2 mm boron doped diamond working electrode, saturated calomel reference electrode, and platinum wire counter electrode at 800 mV s⁻¹, with 200 mM aqueous NaHCO₃ purged with N₂ at 25 °C.

$$1^+ \xrightarrow{\ -e^-\ } 1^* \tag{1}$$

$$1^* + 1^* \xrightarrow{\ -e^-\ } 1_{dim} \tag{2}$$

$$1^* \xrightarrow{\ H^+/-e^-\ } 1H \tag{3}$$

$$1_{dim} \xrightarrow{\ -e^-\ } 1^* + 1^+ \ (ox1) \tag{4}$$

$$1H \xrightarrow{\ -e^-\ } 1^* + H^+ \ (ox2) \tag{5}$$

where 1_{dim} is the electrochemically dimerized mimetic. The presence of two distinct peaks corresponding to electrochemical oxidation of the dimer (ox1) and oxidation of the 1,4-dihydropyridine $1H$ (ox2) provides a convenient means to measure the relative amounts of dimer and $1H$ generated upon electrochemical reduction under variable conditions.

To determine the impact of supporting electrolyte on the distribution of peaks corresponding to dimer and dihydro species (ox1 and ox2, respectively), CVs of 2 mM 1^+ were performed in 28 different supporting electrolytes, including Na^+, K^+, Li^+, Ca^{2+}, Mg^{2+}, NH_4^+, or organic ammonium cations with halide, sulfate, phosphate, nitrate, carbonate, azide, or borate anions. Additionally, sodium and potassium salts of several organic anions were tested, including acetate, citrate, oxalate, gluconate, propionate, and pyruvate. The resulting voltammograms are summarized in Table 1. For most of the electrolytes studied, 1_{dim} was the only observable reduction product. The CVs were qualitatively nearly identical with only small variations in the reduction event ($E_{pc} = -1.3$ V, $i_{pc} = 35$–45 mA) and moderate variations in ox1 ($E_{pa1} = -0.07$ to 0 V, $i_{pa1} = 11$–19 mA). A small additional oxidation event ($E_{pa2} = 0.35$ V, $i_{pa2} < 0.1$ mA) was observed in the most acidic solutions (NaH_2PO_4 and $NH_4H_2PO_4$, where pH < 5), indicating a general pH dependence for the product distribution of electrochemically reduced 1^+.

Among the electrolytes tested, sodium pyruvate stood out for its ability to promote $1H$ formation, even at neutral pH ($E_{pa2} = 0.49$ V, $i_{pa2} = 4$ mA at pH 7.0). To better understand the unique effect of sodium pyruvate on the electrochemistry of 1^+, we used convolution voltammetry to compare the product distributions in the presence and absence of pyruvate (see ESI† for a brief description of convolution voltammetry).[21-23]

Convolution voltammetry shows the effective concentration change of a redox-active substrate at the electrode surface, assuming that the substrate participates only in diffusion and electron transfer. The height of the convolution integral is proportional to the number of electrons transferred per molecule. Fig. 2 shows CVs (top, solid curves) and convolution integrals (bottom, dashed curves) of 1^+ at pH 7, in 100 mM sodium phosphate (blue) and in 100 mM sodium phosphate plus 100 mM sodium pyruvate (red). Both graphs show the appearance of ox2 when sodium pyruvate is added. The convolution integral, when normalized for a one-electron reduction (Fig. 2B), shows that the concentration of 1^+ drops by 2 mM (*i.e.*, the bulk concentration) when reduced without sodium pyruvate, and that it returns to the bulk concentration after ox1. By contrast, the effective concentration change with sodium pyruvate is over 4 mM after the reduction event. Part of this change is attributable to the change in mechanism, from a one-

Table 1 CV output data from inorganic (top) and organic (bottom) electrolytes tested with 1^+, including the peak currents and peak potentials for 1^+ reduction, 1_{dim} oxidation (ox1), and 1H oxidation (ox2)[a]

Electrolyte	Peak current/µA (peak potential/V vs. SCE[b])				pH
	Dimer oxidation (ox1)		1H oxidation (ox2)		
Na_2CO_3	17.4 ± 0.1	(-0.049)	—	—	11.2
NaN_3	18.22 ± 0.01	(-0.056)	—	—	10.1
$Na_2B_4O_7$[c]	15.62 ± 0.06	(-0.049)	—	—	9.6
K_2HPO_4	16.7 ± 0.1	(-0.036)	—	—	9.2
Na_2HPO_4	14.81 ± 0.05	(-0.001)	—	—	9.1
$NaHCO_3$	17.41 ± 0.05	(-0.052)	—	—	8.9
$CaCl_2$	13.35 ± 0.08	(0.017)	—	—	8.8
$(NH_4)_2HPO_4$	15.14 ± 0.02	(-0.019)	—	—	8.3
Na_2SO_4	16.89 ± 0.09	(-0.012)	—	—	7.7
KNO_3	17.2 ± 0.04	(-0.029)	—	—	7.3
LiBr	16.73 ± 0.09	(-0.051)	—	—	7.3
KCl	16.95 ± 0.07	(-0.020)	—	—	7.2
$MgSO_4$	16.4 ± 0.02	(-0.041)	—	—	7.2
KBr	16.59 ± 0.01	(-0.032)	—	—	7.1
NaCl	18.06 ± 0.09	(-0.056)	—	—	6.9
NaBr	17.78 ± 0.01	(-0.039)	—	—	6.6
$NaNO_3$	17.0 ± 0.2	(-0.038)	—	—	6.5
NH_4Cl	16.31 ± 0.02	(0.000)	—	—	6.2
$NH_4H_2PO_4$	11.92 ± 0.02	(0.012)	0.04	(0.350)	4.7
NaH_2PO_4	10.88 ± 0.03	(-0.025)	0.07	(0.350)	4.6
Citrate[d]	16.4 ± 0.2	(-0.036)	—	—	8.9
Oxalate[e]	18.1 ± 0.1	(-0.033)	—	—	7.8
Bu_4NBr	14.42 ± 0.04	(-0.001)	—	—	7.2
Pr_4NBr	15.21 ± 0.03	(-0.025)	—	—	7.2
Pyruvate[d]	12.66 ± 0.01	(-0.008)	4.4 ± 0.1	(0.489)	7.0
Acetate[d]	15.39 ± 0.09	(-0.039)	—	—	6.9
Propionate[d]	14.85 ± 0.02	(-0.025)	—	—	6.7
Gluconate[e]	15.38 ± 0.05	(-0.037)	—	—	6.7

[a] Values are reported as the average and one standard deviation with $n = 3$. Experiments were performed using 2 mM 1^+ with 200 mM supporting electrolyte purged with N_2 and at 25 °C and 800 mV s^{-1}. [b] Unless otherwise noted, the standard deviation for a given peak potential was less than 1 mV. [c] 100 mM sodium tetraborate decahydrate was used. [d] Sodium counterion was used. [e] Potassium counterion was used.

electron reduction (1^+ to 1_{dim}) to a two-electron reduction (1^+ to 1H). However, part of the change is due to the irreversible reduction of a species other than 1^+ in the presence of sodium pyruvate, evidenced by an offset in the convolution integral of ~0.5 mM from the baseline after both oxidation events. With sodium pyruvate, the concentration changes after ox1 and ox2 are roughly equal, showing that the concentrations of 1_{dim} and 1H are approximately the same after the reduction event (corresponding to ~30% selectivity for 1H formation).

To better understand the mechanism of action of sodium pyruvate, we studied the dependence of 1^+ reduction on the concentration of pyruvate, the reduction potential, and the pH of the solution. Voltammetry of 1^+ with variable pyruvate concentration (using sodium bicarbonate as a co-supporting electrolyte to

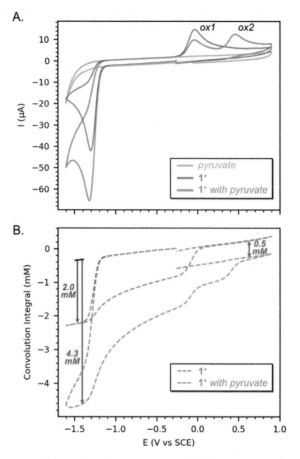

Fig. 2 (A) Representative cyclic voltammograms and (B) the corresponding convolution integrals of 2 mM 1^+ in the absence (—) or presence (—) of 100 mM sodium pyruvate, or 100 mM sodium pyruvate alone (—). Experiments were performed using a 2 mm boron doped diamond working electrode with 100 mM sodium phosphate supporting electrolyte at 800 mV s^{-1}, purged with N_2 and at pH 7.0, and 25 °C.

maintain a total supporting electrolyte concentration of 200 mM) revealed that trace amounts of **1H** are formed with as little as 2 mM sodium pyruvate (1 eq. with respect to 1^+) (Fig. 3). Although raising the concentration of sodium pyruvate increased the yield of **1H** at low pyruvate concentrations, too much sodium pyruvate was detrimental to the yield of **1H**. For example, a 50 : 50 mixture of sodium bicarbonate/sodium pyruvate yielded more **1H** than sodium pyruvate alone, which might suggest a synergistic interaction between pyruvate and the co-supporting electrolyte. A similar result was also observed with a 50/50 mixture of sodium phosphate/sodium pyruvate, which yielded more **1H** than sodium pyruvate alone.

We next sought to determine the impact of switching potential on the peak currents corresponding to ox1 and ox2 (i_{pa1} and i_{pa2}, respectively) exhibited in CVs of 1^+. In sodium pyruvate (200 mM, pH 7.0), both 1_{dim} and **1H** are formed in the same reduction event, with **1H** favored at lower potentials and 1_{dim} favored at

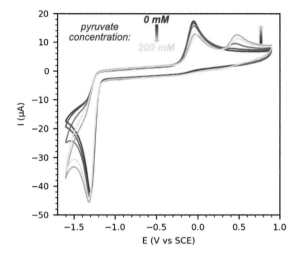

Fig. 3 Representative cyclic voltammograms of 2 mM 1^+ with varying ratios of sodium bicarbonate/sodium pyruvate: 100/0 (darkest), 99/1, 90/10, 50/50, and 0/100 (lightest). Experiments were performed using a 2 mm boron doped diamond working electrode at 25 °C and 800 mV s^{-1}. Solutions were purged with nitrogen before scanning.

higher potentials (Fig. 4a). At a turnaround potential of −1.25 V (green curve), only a trace amount of 1^+ is generated during ox2, but this amount grows significantly as the potential is lowered to −1.6 V (blue curve). By contrast, the amount of 1^+ generated during ox1 grows more slowly at lower turnaround potentials. Without sodium pyruvate (200 mM sodium phosphate, pH 9.1), no observable **1H** is formed by −1.6 V. Additionally, there is minimal change in product distribution between −1.6 V and −2.1 V (Fig. 4). Below −2.1 V, the product distribution shifts away from dimer, yet **1H** is not formed exclusively: at least three new oxidation peaks are observed, and dimer is still the major product. The comparison of these two electrolytes shows a general pH dependence for the electrochemical reduction of 1^+ to **1H**, a reaction that happens both more selectively and more quickly with sodium pyruvate than without it.

We also explored the pH dependence of 1^+ reduction in 100 mM sodium phosphate, with and without 100 mM sodium pyruvate. In the absence of pyruvate, the peak current for ox1 changed weakly with pH, and only trace evidence of ox2 was observed at more acidic pH (see Table 1 above). Under neutral or slightly basic conditions, there was no impact on the reduction peak and only ox1 was observed. In slightly acidic solutions, a small oxidation peak near 0.38 V could be seen; however, this peak represents a negligible fraction of the total products formed during the electrochemical reduction of 1^+. In the presence of 100 mM sodium pyruvate, by contrast, the peak reduction current had a pronounced pH dependence (Fig. 5A, i_{pc} grows from 43 μA at pH 9 to 112 μA at pH 5). The two-electron product **1H** was clearly visible at every pH, although the product ratio **1H** : $\mathbf{1_{dim}}$ exhibited a maximum value at pH 7 (Fig. 5B). At more basic pH, the formation of **1H** increased with the peak reduction current, but at more acidic pH, even though the peak reduction current continued to grow, the amount of **1H** fell and dimer was again the major product. It should be noted that the sum of i_{pa1}

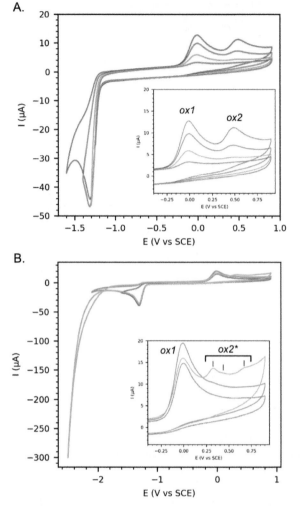

Fig. 4 (A) Representative cyclic voltammograms of 2 mM **1⁺** with 200 mM sodium pyruvate with variable turnaround potentials: −1.6 V (−), −1.4 V (−), −1.3 V (−), and −1.25 V (−). (B) Representative cyclic voltammograms of 2 mM **1⁺** with 200 mM Na$_2$HPO$_4$ with different turnaround potentials: −1.6 V (−), −2.1 V (−), and −2.5 V (−). Experiments were performed using a 2 mm boron doped diamond working electrode at 25 °C and 800 mV s^{-1}. Solutions were purged with nitrogen before scanning.

and i_{pa2} at each pH was nearly unchanged; assuming the diffusivities of **1$_{dim}$** and **1H** are independent of pH, this suggests that **1*** is reacting *via* only two competitive reaction pathways, one pathway that forms **1$_{dim}$** and one that forms **1H**. The decrease in ox1 at pH 5 may be due to acid-promoted decomposition of the dimer; for example, the dimer of 1-methyl nicotinamide has a first-order decomposition rate constant of 6400 s^{-1} at pH 4.4.[24]

While the total amount of **1H** and **1$_{dim}$** formed is approximately constant as the pH is lowered from 9 to 5, the peak reduction current continues to increase, showing that the reduction of the species other than **1$^+$** (described as causing the

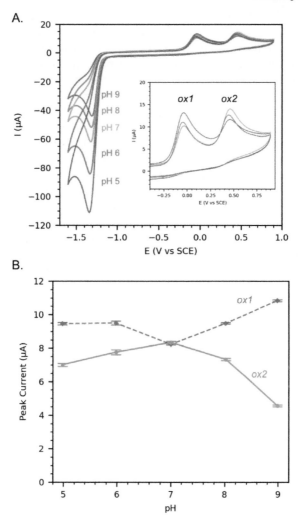

Fig. 5 (A) Representative cyclic voltammograms of 2 mM 1^+ in 100 mM sodium phosphate with 100 mM sodium pyruvate, at pH 9 (−), 8 (−), 7 (−), 6 (−), and 5 (−). (B) Normalized peak currents for ox1 and ox2 at the same pH values, showing a maximum of ox2 (**1H**) and a minimum of ox1 (**1$_{dim}$**) at pH 7. Error bars represent one standard deviation from the mean where $n = 3$. Experiments were performed using a 2 mm boron doped diamond working electrode at 25 °C and 800 mV s^{-1}. Solutions were purged with nitrogen before scanning.

offset of 0.5 M in the convolution integral in Fig. 2B above) is acid-promoted. Additionally, the pH maximum for **1H** formation indicates that the competing reduction event has a higher-order dependence on proton concentration than the reduction of 1^+ to **1H**, because its rate grows more rapidly with proton concentration than the rate of formation of **1H**. Further inspection of the CVs with and without sodium pyruvate indicates that this competing reaction may be pyruvate reduction. Control experiments indicate that sodium pyruvate is electrochemically active, with a peak reduction potential of −2.2 V at 400 mV s^{-1} in pH 7 phosphate buffer. Although this potential is well beyond the −1.6 V cutoff for the CVs of **1⁺** shown here, the onset of the pyruvate reduction wave is still visible in background scans (this can be seen in Fig. 2 above). An alternative explanation is that a synergistic interaction between **1⁺** and pyruvate may enable electrocatalytic proton reduction. However, we could not observe an interaction between sodium pyruvate and **1⁺** in bulk solution by ¹H-NMR (see ESI†). Ongoing research is aimed at understanding this secondary reaction/interaction.

Finally, we explored the ability for other pyridinium electrolytes to be regenerated by sodium pyruvate. To that end, we varied the 1-substituent to include ethyl (**2**), isopropyl (**3**), and allyl (**4**) derivatives, as well as NAD⁺ itself (**5**) (Fig. 6). All of these pyridinium compounds showed both ox1 and ox2 in a mixture of 100 mM sodium phosphate and 100 mM sodium pyruvate (pH 7). The effect of sodium pyruvate was most pronounced for the substrate with the smallest 1-substituent, 1-ethyl nicotinamide bromide (**2**), and it was least pronounced for the substrate with the largest 1-substituent, NAD⁺ (**5**). Although the 1-substituent can have a significant effect on the extent of interaction between pyruvate and the pyridinium electrolyte, the interaction appears to be general for nicotinamide-based

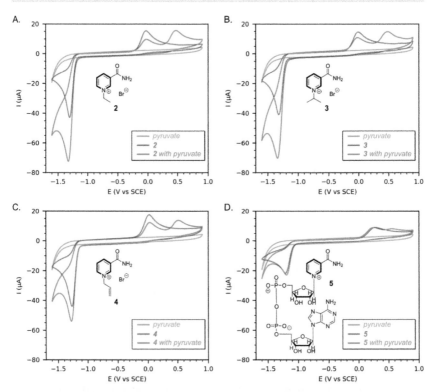

Fig. 6 Representative cyclic voltammograms of 2 mM pyridinium electrolyte in 100 mM sodium phosphate in the presence (−) or absence (−) of 100 mM sodium pyruvate, or 100 mM sodium pyruvate alone (−): (A) 1-ethyl nicotinamide bromide, **2**; (B) 1-isopropyl nicotinamide bromide, **3**; (C) 1-allyl nicotinamide bromide, **4**; (D) nicotinamide adenine dinucleotide, **5**. Experiments were performed using a 2 mm boron doped diamond working electrode at 25 °C and 800 mV s⁻¹. Solutions were purged with nitrogen before scanning.

electrolytes, suggesting that the ability of pyruvate to promote the electrochemical generation of **1H** from **1⁺** presents a general strategy for direct electrochemical regeneration of NADH analogues and NADH itself.

Conclusions

In conclusion, the electrochemical reduction of 1-*n*-butyl nicotinamide bromide (**1⁺**) in 28 different supporting electrolytes and compared against the electrochemical oxidation of 1,4-dihydro-1-*n*-butyl nicotinamide (**1H**). While **1⁺** reduction yielded only dimer in most of the electrolytes tested, electrochemical reduction in sodium pyruvate resulted in formation of **1H** alongside the dimer. Preference for **1H** was optimized when a large excess of sodium pyruvate was used in combination with an inert co-supporting electrolyte, such as bicarbonate or phosphate. Formation of **1H** was favored at low potentials and at neutral pH. Similar preferences for 1,4-dihydro species formation with several different nicotinamide derivatives, including NAD⁺, suggest that addition of sodium pyruvate may provide a general strategy for electrochemically regenerating both mNADH derivatives and NADH itself from their oxidized forms.

Experimental

Materials

All chemicals and solvents were purchased from Sigma Aldrich or Thermo Fisher Scientific, and were used as received without purification. All electrochemistry experiments were purged with ultra-high purity nitrogen. Boron doped diamond working electrodes were purchased from Fraunhofer USA.

Synthetic procedures

1-*n*-Butyl-3-carbamoylpyridinium bromide (1⁺). Nicotinamide (1.235 g, 10.1 mmol), acetonitrile (10 mL), 1-bromobutane (2.15 mL, 20.0 mmol), and a stir bar were added to a 20 mL scintillation vial with a pressure release cap. The mixture was stirred at 50 °C for 2 h and then at 80–100 °C for 10 h. The contents were quantitatively transferred with ethanol (~20 mL) to a 100 mL round bottom flask, and the solvent was removed *in vacuo*. The soapy residue was crystallized from ethanol by adding hexanes as a cosolvent (final ratio ~5 : 1 ethanol/hexanes) to afford the product as white crystals (1.457 g, 56%). ^{1}H NMR (500 MHz, DMSO-d$_6$) δ 9.48 (t, $J = 1.5$ Hz, 1H), 9.20 (d, $J = 6.0$ Hz, 1H), 8.91 (d, $J = 8.3$ Hz, 1H), 8.55 (s, 1H), 8.25 (dd, $J = 8.1, 6.1$ Hz, 1H), 8.16 (s, 1H), 4.64 (t, $J = 7.5$ Hz, 2H), 1.91 (p, $J = 7.5$ Hz, 2H), 1.29 (h, $J = 7.4$ Hz, 2H), 0.90 (t, $J = 7.4$ Hz, 3H).

Synthesis of 1-*n*-butyl-1,4-dihydronicotinamide (1H). The synthesis of **1H** was adapted from a previous procedure.[5] Water (10 mL), dichloromethane (5 mL), sodium bicarbonate (390 mg, 4.6 mmol), and **1⁺** (200 mg, 0.8 mmol) were added to a 25 mL round bottom flask with a stir bar. Dry sodium dithionite (540 mg, 3.1 mmol) was added to an addition funnel, which was clipped to the top of the reaction flask and sealed with a rubber stopper. The flask was lowered into an ice bath, and the contents were mixed while a Schlenk line was used to flush the system with nitrogen gas and remove oxygen. The addition funnel was opened and the dithionite added slowly over 10–15 min. After mixing for 2 h in the dark, the organic layer was washed three times with cold water, dried over $MgSO_4$, and concentrated *in vacuo* to yield the title product as a yellow-orange, semicrystalline residue (25 mg, 18%). ^{1}H NMR (500 MHz, DMSO-d_6) δ 6.84 (d, J = 1.5 Hz, 1H), 6.49 (s, 1H), 5.84 (dd, J = 8.0, 1.6 Hz, 1H), 4.55 (dt, J = 8.1, 3.4 Hz, 1H), 3.05 (t, J = 7.0 Hz, 2H), 2.93 (d, J = 2.0 Hz, 2H), 1.41 (p, J = 7.5 Hz, 2H), 1.24 (h, J = 7.4 Hz, 2H), 0.87 (t, J = 7.4 Hz, 3H).

Synthesis of 1-ethyl nicotinamide bromide (2). Nicotinamide (1.24 g, 10.1 mmol), bromoethane (1.50 mL, 20.1 mmol), and acetonitrile (10 mL) were combined in a Parr pressure reactor and heated at 90 °C for 41 h. The reaction mixture was cooled to room temperature and quantitatively transferred to a round-bottom flask with methanol (~20 mL), and the solvent was removed *in vacuo*. The solid was recrystallized from ethanol by adding hexanes as a cosolvent (final ratio ~5 : 1 ethanol/hexanes) to afford the product as white crystals (1.91 g, 82%). ^{1}H NMR (500 MHz, CD_3CN) δ 9.56 (s, 1H), 8.91 (d, J = 8.1 Hz, 1H), 8.81 (d, J = 6.1 Hz, 1H), 8.12 (t, J = 7.2 Hz, 1H), 7.88 (s, 1H), 6.61 (s, 1H), 4.68 (q, J = 7.4 Hz, 2H), 1.66 (t, J = 7.4 Hz, 3H).

Synthesis of 1-isopropyl-3-carbamoylpyridinium bromide (3). Nicotinamide (122.5 mg, 1.00 mmol), 2-bromopropane (0.1878 mL, 2.00 mmol), and acetonitrile (2 mL) were added to a Parr bomb reactor, and the contents were heated at 100 °C for 24 h. After cooling, the contents were quantitatively transferred with methanol to a round-bottom flask, the solvent was boiled away, and the residue was recrystallized from ethanol by adding hexanes as a cosolvent (final ratio ~5 : 1 ethanol/hexanes) to afford the product as a white powder (84.3 mg, 34%). ^{1}H NMR (500 MHz, DMSO-d_6) δ 9.46 (t, J = 1.6 Hz, 1H), 9.29 (d, J = 5.8 Hz, 1H), 8.90 (dt, J = 8.1, 1.4 Hz, 1H), 8.56 (s, 1H), 8.26 (dd, J = 8.0, 6.1 Hz, 1H), 8.17 (s, 1H), 5.08 (hept, J = 6.7 Hz, 1H), 1.62 (d, J = 6.7 Hz, 6H).

Synthesis of 1-allyl-3-carbamoylpyridinium bromide (4). Nicotinamide (1.241 g, 10.2 mmol), acetonitrile (10 mL), allyl bromide (0.840 mL, 9.9 mmol), and a stir bar were added to a 20 mL scintillation vial with a pressure-release cap. The mixture was stirred at 50 °C for 12 h. The contents were quantitatively transferred with ethanol to a 100 mL round bottom flask, and the solvent was removed *in vacuo*. The solid was recrystallized from ethanol by adding hexanes as a cosolvent (final ratio ~5 : 1 ethanol/hexanes) to afford the product as a light brown crystalline solid (1.509 g, 61%). ^{1}H NMR (500 MHz, DMSO-d_6) δ 9.45 (t, J = 1.6 Hz, 1H), 9.15 (d, J = 5.6 Hz, 1H), 8.97 (dt, J = 8.0, 1.3 Hz, 1H), 8.60 (s, 1H), 8.29 (dd, J = 8.1, 6.1 Hz, 1H), 8.18 (s, 1H), 6.22–6.12 (m, 1H), 5.47 (s, 1H), 5.45 (dt, J = 6.4, 1.0 Hz, 1H), 5.33 (d, J = 6.3 Hz, 2H).

Electrochemistry procedures

Electrochemical experiments were performed on a Biologic VSP potentiostat using a three-electrode cell with a 2 mm boron-doped-diamond working electrode, saturated calomel reference electrode, and platinum wire counter electrode. Unless otherwise noted, supporting electrolyte solutions were prepared by dissolving 2.00 mmol of electrolyte in a small amount of water and diluting to 10.0 mL. (Heat was needed to dissolve sodium tetraborate decahydrate.) Solutions of sodium acetate and sodium propionate were prepared by combining equal amounts of the acid and sodium bicarbonate.

For electrochemical experiments, 6 μmol (1.55 mg for 1^+, 1.08 mg for **1H**) of substrate was dissolved in 3 mL of supporting electrolyte solution. The solution was purged with nitrogen for several minutes before scanning and for ~10 seconds between scans. Unless otherwise noted, scans were run at 800 mV s^{-1} between −1.6 V and 0.9 V, starting from the open circuit potential.

Author contributions

The manuscript was written through contributions of all authors. CB and DPH contributed to conceptualization, formal analysis, methodology, visualization and writing of the original draft. CB and KG contributed to investigation. CB, KG, and DPH contributed to review and editing. All authors have given approval to the final version of the manuscript.

Conflicts of interest

There are no conflicts to declare.

Acknowledgements

The authors wish to acknowledge the ACS Petroleum Research Fund (PRF #65477-DNI4) for their support of this research.

Notes and references

1 H. Chen, F. Dong and S. D. Minteer, *Nat. Catal.*, 2020, **3**, 225–244.
2 L. Sellés Vidal, C. L. Kelly, P. M. Mordaka and J. T. Heap, *Biochim. Biophys. Acta, Proteins Proteomics*, 2018, **1866**, 327–347.
3 S. A. Löw, I. M. Löw, M. J. Weissenborn and B. Hauer, *ChemCatChem*, 2016, **8**, 911–915.
4 C. E. Paul, I. W. C. E. Arends and F. Hollmann, *ACS Catal.*, 2014, **4**, 788–797.
5 C. E. Paul, S. Gargiulo, D. J. Opperman, I. Lavandera, V. Gotor-Fernández, V. Gotor, A. Taglieber, I. W. C. E. Arends and F. Hollmann, *Org. Lett.*, 2013, **15**, 180–183.
6 J. Lutz, F. Hollmann, T. V. Ho, A. Schnyder, R. H. Fish and A. Schmid, *J. Organomet. Chem.*, 2004, **689**, 4783–4790.
7 J. D. Ryan, R. H. Fish and D. S. Clark, *ChemBioChem*, 2008, **9**, 2579–2582.
8 M. A. Jensen and P. J. Elving, *Biochim. Biophys. Acta, Bioenerg.*, 1984, **764**, 310–315.
9 P. J. Elving, W. T. Bresnahan, J. Moiroux and Z. Samec, *J. Electroanal. Chem. Interfacial Electrochem.*, 1982, **141**, 365–378.
10 J. N. Burnett and A. L. Underwood, *J. Org. Chem.*, 1965, **30**, 1154–1158.
11 F. Hollmann, A. Schmid and E. Steckhan, *Angew. Chem. Int. Ed.*, 2001, **40**(1), 169–171.
12 I. Ali, A. Gill and S. Omanovic, *Chem. Eng. J.*, 2012, **188**, 173–180.
13 S. Kaneco, K. Iiba, M. Yabuuchi, N. Nishio, H. Ohnishi, H. Katsumata, T. Suzuki and K. Ohta, *Ind. Eng. Chem. Res.*, 2002, **41**, 5165–5170.
14 N. J. Stone, D. A. Swelgart and A. M. Bond, *Organometallics*, 1986, **5**, 2553–2555.
15 L. J. Duić, Z. Mandić and F. Kovačiček, *J. Polym. Sci., Part A: Polym. Chem.*, 1994, **32**, 105–111.
16 I. Katsounaros and G. Kyriacou, *Electrochim. Acta*, 2007, **52**, 6412–6420.

17 B. K. Peters, K. X. Rodriguez, S. H. Reisberg, S. B. Beil, D. P. Hickey, Y. Kawamata, M. Collins, J. Starr, L. Chen, S. Udyavara, K. Klunder, T. J. Gorey, S. L. Anderson, M. Neurock, S. D. Minteer and P. S. Baran, *Science*, 2019, **363**, 838–845.

18 R. W. Murray and L. K. Hiller, *Anal. Chem.*, 1967, **39**, 1221–1229.

19 P. J. Elving, G. Milazzo, G. Dryhurst, J. Moiroux, A. J. Jensen, Y. Ohnishi, Y. Kikuchi and M. Kitami, *J. Am. Chem. Soc.*, 1981, **103**, 2379–2386.

20 C. O. Schmakel, K. S. V. Santhanam and P. J. Elving, *J. Electrochem. Soc.*, 1974, **121**, 345.

21 K. B. Oldham and J. Spanier, *J. Electroanal. Chem. Interfacial Electrochem.*, 1970, **26**, 331–341.

22 K. B. Oldham, *Anal. Chem.*, 1972, **44**, 196–198.

23 J. C. Imbeaux and J. M. Savéant, *J. Electroanal. Chem. Interfacial Electrochem.*, 1973, **44**, 169–187.

24 M. Miller, B. Czochralska and D. Shugar, *J. Electroanal. Chem. Interfacial Electrochem.*, 1982, **141**, 287–298.

PAPER

Effect of [Na$^+$]/[Li$^+$] concentration ratios in brines on lithium carbonate production through membrane electrolysis†

Walter R. Torres, [ID] [a] Nadia C. Zeballos[ab] and Victoria Flexer [ID] *[a]

Received 23rd February 2023, Accepted 5th April 2023

DOI: 10.1039/d3fd00051f

Lithium is a fundamental raw material for the production of rechargeable batteries. The technology currently in use for lithium salts recovery from continental brines entails the evaporation of huge water volumes in desert environments. It also requires that the native brines reside for not less than a year in open air ponds, and is only applicable to selected compositions, not allowing its application to more diluted brines such as those geothermally sourced or waters produced from the oil industry. We have proposed an alternative technology based on membrane electrolysis. In three consecutive water electrolyzers, fitted alternately with anion and cation permselective membranes, we have shown, at proof-of-concept level, that it is possible to sequentially recover lithium carbonate and several by-products, including magnesium and calcium hydroxide, sodium bicarbonate, H$_2$ and HCl. The big challenge is to bring this technology closer to practical implementation. Thus, the issue is how to apply relatively well-known electrochemical technology principles to large volumes and to a highly complex and saline broth. We have studied the application of this new methodology to ternary mixtures (NaCl, LiCl and KCl) with constant LiCl and KCl composition and increasing NaCl content. Results showed very similar behaviour for systems containing [Na$^+$]/[Li$^+$] concentration ratios ranging from 1.24 to 4.80. The voltage developed between the anode and cathode is almost the same in all systems at roughly 3.5 V when a constant current density of 50 A m^{-2} is applied. The three monovalent cations migrate with different rates across the cation exchange membrane, with Li$^+$ being the most sluggish and thus crystallization of Li$_2$CO$_3$ only occurs close to completion of the electrolysis. The dimensionless concentration profiles are almost indistinguishable despite the changes in total salinity. The solids crystallized from different feeds showed higher Na$^+$ and K$^+$ contents as the initial Na$^+$ concentration was increased. However, solids with over 99.9% purity in Li$_2$CO$_3$ could be obtained after a simple re-suspension treatment in hot water. The electrochemical energy consumption greatly

aCentro de Investigación y Desarrollo en Materiales Avanzados y Almacenamiento de Energía de Jujuy-CIDMEJu (CONICET-Universidad Nacional de Jujuy), Av. Martijena S/N, Palpalá, 4612, Argentina. E-mail: vflexer@unju.edu.ar

bInstituto Nacional de Tecnología Industrial (INTI) Sede Jujuy, Av. Martijena S/N, Palpalá, 4612, Argentina

† Electronic supplementary information (ESI) available. See DOI: https://doi.org/10.1039/d3fd00051f

increases with higher Na^+ concentrations, and the amount of fresh water that can be recovered is diminished.

1. Introduction

Nowadays, the concept of electrosynthesis is most often associated with the synthesis of organic molecules, particularly amongst academics.[1,2] However, at large industrial scales, electrochemical processes have been in use for several decades for the preparation of several inorganic compounds, such as bromine, fluorine, chlorate, perchlorate and persulfate, amongst others, with the chlor-alkali process leading the way.[3,4] The need to develop more sustainable hydro-metallurgical processes, together with astounding price increases for certain raw materials[5] pave the road for revisiting electrochemical technologies as an alternative for the production of inorganic compounds. Lithium and magnesium, classified by both the European Union and the USA as critical,[6] are examples of raw materials for which electrosynthetic strategies are worth exploring.

The evaporitic technology currently in place for the production of lithium carbonate from aqueous sources has been largely criticized both because of its poor sustainability and its techno-economic limitations.[5,7,8] The evaporitic technology fully relies on open air water evaporation to concentrate lithium rich brines. Over 90% of the original water content of said brines is lost through evaporation, a practice that is very controversial since it takes place in highly arid locations.[9,10] In addition, open air water evaporation is sluggish, taking between 10–24 months to reach a suitable concentration for processing, slowing down the start up of mining projects, and limiting the possibilities of adjusting production to market demands.[5]

On a second note, the evaporitic technology requires large amounts of chemicals for the processing of the concentrated brines. While these are not rare or particularly costly, the continuous shipping of large volumes of chemicals to very remote locations through winding roads brings along an extra logistical problem. Finally, open air evaporation is only efficient in desert locations, with very low annual rainfall, inexpensive flat land, and assuming a minimum Li^+ concentration in the aqueous source.[11,12] These requirements severely limit the application of the technology to relatively few deposits worldwide. Lithium rich geothermal and oilfield brines, which are available in many more countries worldwide, cannot be exploited through current technology.[5]

There is an open quest for the development of disruptive brine processing technologies that fully avoid evaporation, jointly named Direct Lithium Extraction technologies, DLE.[5,13–17] Several clever proposals based on electrochemical techniques have been put forward. These can be broadly divided into two large categories. A first group employs lithium insertion electrodes, with similar chemistries to those employed in lithium ion-batteries.[18,19] In a first step, Li^+ cations present in a highly complex brine are electrochemically inserted in materials such as lithium manganese oxide,[20,21] titanium oxide[22] and lithium iron phosphate,[23] amongst other examples. Subsequently, Li^+ is liberated in a pure LiCl solution when, alternately, the potential or the current is inverted. A second large group of technologies employs permselective membranes to move cations and anions in opposite directions.[17,24,25] Excellent results have been reported for the efficient separation of Li^+ and divalent cations, most importantly Mg^{2+}.[26,27] Unfortunately, the development of Li^+ selective membranes is sluggish, with reported examples only producing a limited selectively and no results reported for large Na^+/Li^+ concentration ratios. Thus, most electrodialysis setups today are only able to concentrate a mixture of the monovalent species, K^+, Na^+ and Li^+,[28,29] which is not suitable for the direct crystallization of pure lithium products.

In parallel, CO_2 capture in amine scrubbing systems is amongst the most efficient CO_2 capture systems. However, amine scrubbing requires the subsequent release and compression of CO_2.[30,31] Another proposal for CO_2 capture is its conversion to carbonate minerals, an approach that has the important added advantage of simultaneously storing CO_2.[32] High salinity aqueous sources, and especially those rich in multivalent cations, are abundant. However, CO_2 fixation rate and storage capacity can be largely increased if, in addition to a large source of cations, the aqueous source can also provide a large supply of hydroxyl anions to rapidly convert CO_2 to carbonate or bicarbonate anions. Hydroxyl anions can be easily produced *via* water reduction, and, with the acceleration in the quest for developing CO_2 capture and storage alternatives, several proposals have been put forward.[31,33–40] Sorimachi proposed the use of a NaOH and $CaCl_2$ solution, and the subsequent electrochemical regeneration of NaOH.[41] Llanos *et al.* achieved a 93% current efficiency for CO_2 fixation with simultaneous hypochlorite production.[42] Mustafa *et al.* recently reviewed the nascent field of concentrated brine management and CO_2 capture.[43]

We recently proposed to take advantage of the hydroxyl anions produced at the cathode of a water electrolyser to adsorb CO_2 as either bicarbonate or carbonate anions and crystallize $NaHCO_3$ and Li_2CO_3, see Fig. 1.[44,45] This novel approach simultaneously allows for chemical fixation and storage of CO_2 and for the production of Li_2CO_3 without the constant freighting of Na_2CO_3 as in the current evaporitic approach. Li^+ cations in natural sources are found in concentrations

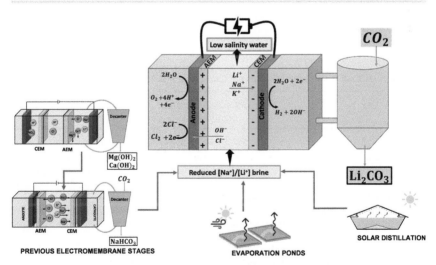

Fig. 1 (Top): Schematic representation of the electrolysis process studied in this work. (Bottom) three different pre-processing strategies that could provide the top process with a processed brine of reduced [Na^+]/[Li^+] ratio.

ranging from 10–150 ppm,[46] in geothermal and oilfield sources, through 500–900 ppm, in continental brines, with a few examples, such as the Atacama salt flat, reaching 1500 ppm.[12] Li_2CO_3 cannot be directly crystallized from natural sources, neither *via* addition of Na_2CO_3, nor *via* CO_2 absorption. Because of the relatively high water solubility of Li_2CO_3, our membrane electrolysis method for Li_2CO_3 crystallization cannot be directly used on native brines, which require some sort of pre-treatment.

In the work reported here, we studied how the ratio of concentrations of Na^+/Li^+ affects the crystallization of Li_2CO_3 in a membrane electrolysis system coupled to a crystallizer. Our study shows that at higher Na^+/Li^+ ratios the purity of Li_2CO_3 decreases. However, a high purity product can be produced following re-suspension of the crystals in hot water. The system is largely affected by electroosmosis, and thus, while water of very low salinity can always be recovered from the middle compartment of the electrolyser, the proportion of freshwater recovered decreases when the Na^+/Li^+ increases. K^+, Na^+ and Li^+ show differential mobility values across the cation exchange membrane, with Li^+ always being more sluggish than the other monovalent cations. Therefore, an increased Na^+/Li^+ concentration ratio implies that, in a batch system, the largest share of Li^+ cations are transferred to the cathodic compartment later during the electrolysis.

2. Experimental

2.1 Water electrolyzer

A three-compartment water electrolyzer was designed and made in-house. The design has previously been reported in ref. 44 and pictures are shown in Fig. S1.†️ The acrylic cell consisted of three identical compartments separated by an anion and a cation exchange membrane, with internal dimensions $5 \times 20 \times 2$ cm, and internal volume equal to 200 cm^3. The anode was a titanium mesh electrode coated with an iridium-based mixed metal oxide (IrO_2/TiO_2; 65/35%), with a perpendicular current collector (4.8×19.8 cm; 1 mm thickness, Magneto Special Anodes). The cathode was a stainless-steel wire mesh (4.8×19.8 cm) with

Table 1 Membrane characteristics, as specified by the manufacturer

	Cation exchange membrane	Anion exchange membrane
Supplier	Membranes international	
Manufacturer's coding	CMI-7000	AMI-7001
Polymer structure	Gel polystyrene cross-linked with divinylbenzene	
Functional group	Sulphonic acid	Quaternary ammonium
Electrical resistance (Ω cm^2) 0.5 M NaCl	<30	<40
Maximum current density (A m^{-2})	<500	<500
Permselectivity (%) 0.1 m KCl and 0.5 m KCl	94	90
Total exchange capacity (meq g^{-1})	1.6 ± 0.1	1.3 ± 0.1
Water permeability (mL h^{-1} ft^{-2}) at 5 psi	<3	<3
Mullen burst test strength (psi)	>80	>80
Thermal stability (°C)	90	90
Chemical stability range (pH)	1–10	1–10

a stainless-steel sheet metal current collector (distance between electrodes = 23 mm). Two plastic meshes were placed between the electrodes and the membranes to avoid direct contact, and to serve as turbulence promoters. The characteristics of the permselective membranes used to separate the three compartments are listed in Table 1.

All experiments were run in galvanostatic mode, in two electrode configurations (no reference electrodes were used), using a constant current/constant voltage power supply (simplest possible DC system, with potential for scaling up). The current density values are reported with respect to the geometrical area of the membranes (0.01 m^2 each). Each experiment lasted 8–16 days. For the first ~24 h, the applied current density was ~10 A m^{-2}, to stabilize the membranes. Next, for most of the duration of the experiment, the applied current density was (52 ± 6) A m^{-2}. The applied current did not fluctuate more than 0.1 A m^{-2}, but the applied current value varied slightly in the different experiments. When the DC power supply produced fluctuations, these were registered for the precise calculation of the total charge. When the ionic concentration in the middle compartment is severely decreased, the cell potential increases and eventually the voltage limit of the power supply is reached (30 V). As soon as this point was observed, the electrolysis was manually stopped (some experiments continued for about 2–4 hours in constant voltage mode at $E_{\text{source}} = 30$ V). The cell voltage was monitored directly from the digital screen in the power supply.

Experiments were run in batch recirculation mode at room temperature (20 ± 3) °C. The anodic compartment was connected to a 2000 cm^3 plastic drum, the middle compartment was connected to a 5000 cm^3 plastic drum, while the cathodic compartment was connected to a 1000 cm^3 Schott-bottle, serving as a decanter. The total initial volumes for anolyte, middle and catholyte were 2000 cm^3, 5000 cm^3, and 500 cm^3, respectively. These volumes were constantly recirculated through the cell compartments using a peristaltic pump (flow rate = 6 L h^{-1}, PC28 APEMA). Recirculation also assured stirring of the solutions.

CO_2 gas (technical quality) was bubbled in the Schott-bottle connected to the cathodic compartment. An electro valve and a pH-meter were connected to an Arduino UNO R3 card. The Arduino controlled the CO_2 bubbling, *i.e.* opening or closing the electro valve to allow for CO_2 bubbling only if pH > 11.

New membranes were immersed in NaCl 5% for 24 hours and then washed with de-ionized water, in order to hydrate and stabilize them. After each batch run, the reactor and the membranes were rinsed with de-ionized water. When not in use, the membranes were kept in artificial brine.

2.2 Test solutions

All solutions were prepared with de-ionized water. In these experiments, and for safety reasons, a buffer at pH 12 (1 M carbonate/bicarbonate, industrial grade) was used in the anodic compartment to disproportionate Cl_2 into Cl^- and ClO^-. In the cathodic compartment, a 0.0135 M Li_2CO_3 (0.5%) solution was used (battery grade, Sales de Jujuy, Argentina). Four different solutions mimicking pre-treated brines were alternately added to the middle compartment. The compositions of these solutions are detailed in Table 3 (LiCl: Biopack; KCl and NaCl Cicarelli, all reagents of analytical grade). Different experiments were run by changing the Na^+ concentration of the artificial brine, while keeping Li^+ and K^+

constant (with slight variations produced when weighing slightly different amounts).

2.3 Analytical measurements

0.5 cm^3 samples were taken from both the middle and cathodic compartments at regular intervals to measure conductivity, total dissolved solids (TDS), and concentrations evolution with time. TDS were measured by evaporating the water content of 1.000 cm^3 of the corresponding solution and weighing the remaining solids at a 0.1 mg precision. Conductivity measurements were carried out with a conductivity probe (HI763063 probe and HI99301 controller) on 1 : 100 dilutions of the original solutions. Li, K and Na compositions were determined by atomic absorption spectroscopy (Shimadzu AA7000).

At the end of the batch runs the recovered crystals were vacuum filtered and washed with 5 mL de-ionized water at 45 °C (the solid remained in the filter during washing), dried at 100 °C over 24 h, and weighed. Cation compositions were determined after re-dissolution in 5% HNO$_3$. Acid–base potentiometric titration was used to determine carbonate and bicarbonate concentrations in the solids re-dissolved in de-ionized water.

Next, the larger fraction of the original solids was re-suspended in the minimum amount of de-ionized water at 90 °C required to dissolve not more than 1% of Li$_2$CO$_3$, considering the purity values reported in Table 6. The re-suspended solids were sonicated for 15 minutes, and vacuum filtered. This purification procedure was repeated a total of 3 times. Cation compositions, and carbonate and bicarbonate concentrations were measured again after the purification procedure.

X-ray diffractograms were measured on a Rigaku Miniflex powder diffractometer. Cu Kα radiation, with $\lambda = 1.54056$ Å was used. The spectra were recorded over a 2θ range from 10–90°, using a step of 0.02°. Thermogravimetric analysis was carried out with a Shimadzu TGA-50, at 5°C min^{-1} scan rate from room temperature to 1000 °C in N$_2$ atmosphere with a flux of 50 mL min^{-1}.

2.4 Calculations

All symbols used in this work are summarized in Table 2.

In order to estimate the summation of all resistance elements within the electrolyzer, cell potential *vs.* current graphs for the first 50% of the electrolysis were plotted (Fig. S3†). Only the first half of the electrolysis was considered for this analysis, because we did not want to include data points where either solids could have formed on the membrane, and/or the ionic composition have become too low and hence R_{mC} become too high. The data was rather noisy, which was attributed both to the absence of a regular register (no automated data acquisition), and constant CO$_2$ sparging. Data points were fitted to a linear trend and the corresponding intercepts were interpreted as the summation of the thermodynamic equilibrium potential and the two electrodes' overpotentials. Thus, the cell potential, minus the intercept, divided by the applied current was interpreted as the summation of all resistance elements, following eqn (1):[4]

$$-E = -\Delta E_e + |\eta_C| + |\eta_A| + i \times (R_{CEM} + R_{AEM} + R_{aC}$$
$$+ R_{mC} + R_{cC} + R_{b,H_2} + R_{b,O_2} + R_{b,CO_2} + R_{Li_2CO_3,ads} + R_{cat} + R_{an})(1)$$

Table 2 List of variables used throughout the text

Variable	
$C_{k,0}$	Molar concentration of cation k in the middle compartment at time 0
V_0	Initial solution volume in the middle compartment
$C_{k,n}$	Molar concentration of cation k in the middle compartment at time n
V_n	Solution volume at time n in the middle compartment
F	Faraday constant
t	Electrolysis time
i	Applied current
$F_{M_1-M_2}$	Separation coefficient
c_{M_i}	Molar concentration of cation i
Q_n	Degree of advancement of the electrolysis at time n
ΔE_e	Thermodynamic cell potential
$|\eta_C|$	Cathodic overpotential
$|\eta_A|$	Anodic overpotential
R_{CEM}	Electrical resistance of the cation exchange membrane
R_{AEM}	Electrical resistance of the anion exchange membrane
R_{aC}	Electrical resistance of the anolyte
R_{mC}	Electrical resistance of the solution in the middle compartment
R_{cC}	Electrical resistance of the catholyte
R_{b,H_2}	Electrical resistance of the H_2 bubbles
R_{b,O_2}	Electrical resistance of the O_2 bubbles
R_{b,CO_2}	Electrical resistance of the CO_2 bubbles
$R_{Li_2CO_3,ads}$	Electrical resistance of Li_2CO_3 adsorbed on membrane
R_{an}	Electrical resistance of the anode
R_{cat}	Electrical resistance of the cathode

The coulombic efficiency for the passage of cations through the CEM (%) was calculated as:

$$\eta = \frac{\left(\sum_{k=1}^{j} C_{k,0} V_0 - \sum_{k=1}^{j} C_{k,n} V_n \right) \times F}{\int_0^n i \, dt} \times 100 \qquad (2)$$

where the summation is over all cations (Li^+, K^+, and Na^+), $C_{k,n}$ and $C_{k,0}$ (M) are the concentrations of the respective cation k in the middle compartment at time n and 0 respectively; V_0 and V_n (dm^3) are the initial solution volume and the volume at time n in the middle compartment, respectively; F is the Faraday constant; i (A) is the applied current; and t (s) is the electrolysis time.

The separation coefficient ($F_{M_1-M_2}$) is calculated as the ratio of concentrations in the middle compartment of coexisting cations (K^+/Li^+, Na^+/Li^+ and K^+/Na^+) at time t divided by the initial ratio of concentrations.

$$F_{M_1-M_2} = \frac{(c_{M_1}/c_{M_2})_t}{(c_{M_1}/c_{M_2})_0} \qquad (3)$$

where c_{M_i} is the concentration of cation i. $F_{M-Li} > 1$ indicates a preferential passage of Li^+ through the CEM. Conversely, $F_{M-Li} < 1$, indicates a preferential passage of M through the CEM.

Volume changes in each compartment were directly measured with volumetric glassware. The changes in water mass in all three compartments were calculated by first converting solution volumes to mass (the respective density values used in calculations are listed in Table S1†). Next, the total dissolved solids value measured in each compartment was subtracted from the mass of solution.

Data plotted in Fig. 6 was calculated as

$$\text{rate of concentration change} = \frac{\dfrac{C_{k,n}}{C_{k,0}} - \dfrac{C_{k,n-1}}{C_{k,0}}}{Q_n - Q_{n-1}} \quad (4)$$

where Q_n is the degree of advancement of the electrolysis at time n (in % of total circulated charge).

3.   Results and discussion

Lithium carbonate, Li_2CO_3, is one of the few common inorganic lithium products with a relatively low solubility in aqueous solution: 13.3 g kg^{-1} solution (0.18 mol kg^{-1}) in water at 20 °C.[47] Both Na_2CO_3 and K_2CO_3 display much higher water solubility values, 179 and 523 g kg^{-1} solution (1.69 and 3.80 mol kg^{-1}) at the same temperature, respectively.[47] Interestingly, Li_2CO_3 displays an inverse solubility with temperature. These differences allow for the crystallization of Li_2CO_3 even in the presence of larger amounts of Na^+ and K^+ in the feed solution. In order to crystallize high purity Li_2CO_3 the feed solution needs to have been fully depleted of multivalent species (notably Mg^{2+} and Ca^{2+}) which will readily co-crystallize as either hydroxides or carbonates together with Li_2CO_3. However, the presence of both Na^+ and K^+ is not only tolerated, but even purposely increased. Indeed in the classic evaporitic technology, Li_2CO_3 is crystallized from a mixture of LiCl, NaCl and KCl by addition of Na_2CO_3.[12]

We have recently proposed an overall membrane electrolysis concept for the joint recovery of Li_2CO_3 and several by-products, fully schematized to the left of Fig. 1. In a first stage, production of hydroxyl anions in the cathodic compartment of a water electrolyser secures the full depletion of Mg^{2+} and Ca^{2+}.[48,49] The next step is to produce the partial abatement of Na^+ cations via absorption of CO_2 in the cathodic compartment of another electrolyser.[48] In the final stage, Li_2CO_3 is crystallized.[44] Key to this approach is to carefully control the pH in stages 2 and 3, by an interplay between CO_2 flow and the applied current density.[24] Our laboratory scale experiments have proved the potential of this overall approach. The undisputed advantages of this concept are the production of several by-products, including fresh water, the zero chemical approach, the fixation of CO_2, its potential application to a large range of brine compositions, and the fact that it is weather independent and much faster than the evaporitic technology. These last two advantages are shared with most DLE technologies. On the down side, extensive process development is still needed, notably in dealing with the large production of solid products within electrochemical reactors, and the high energy cost, particularly in the depletion of Na^+ cations.

While we like to refer to the overall electrolysis concept, it is important to acknowledge that particularly the first and last stages could be implemented individually. Key to the application of the last stage, is that the process needs to be fed with a solution where the Na^+/Li^+ ratio of concentrations has been depleted from the original ratio found in all types of native brines (continental, geothermal, produced waters in the oil industry, and even sea water).[24] Alternatives to our full electrochemical approach include membranes that show preferential transport for Li[+] over Na[+] and K[+],[50–52] or thermal systems where brines are concentrated with concomitant water recovery, such as solar stills (bottom right of Fig. 1 (ref. 53)), or membrane distillation systems.[54–56] Eventually, even brines concentrated in solar ponds could be fed into the system (bottom-centre of Fig. 1), allowing for a certain amount of freshwater recovery, and a zero chemical approach. Thus, in the work reported here our aim was to analyse how the different Na[+]/Li[+] ratio of concentrations would affect the proposed membrane electrolysis concept.

In the water electrolyzer implemented here, the applied constant current forces cations to move towards the cathodic compartment and anions to move to the anodic compartment (Fig. 1, top). Fig. 2 depicts a monotonous decrease of the conductivity of the solution in the middle compartment (after a 1 : 100 dilution), Λ, while Fig. 4 shows the decrease in concentrations for K[+], Na[+] and Li[+] cations in the same compartment. The conductivity values of the diluted samples cannot be used to extrapolate the conductivity values of the original solutions. Fig. S2† shows, as expected, that the electrolytic conductance is not proportional to the total concentration of ions. At higher total ionic concentration, the conductivity is higher than at lower total ionic concentrations, but this increment is not linearly proportional. This non-linearity suggests ion pairing or multiple ion associations.[4] This is particularly noticeable for the experiments performed with [Na[+]]/[Li[+]] ratio of 4.80, and to a lesser extent, those of 3.74. Thus, while a higher conductivity in the middle compartment is best to decrease the cell voltage (Fig. S4†), the impact of this variable does not seem to be the most significant (see below).

In all four experiments the cell potentials remain fairly constant at roughly 3–4 V (Fig. S4†) when the applied current density was 50 A m^{-2}. The cell potential increases sharply during the last 90% of the batch experiment. No unexpected

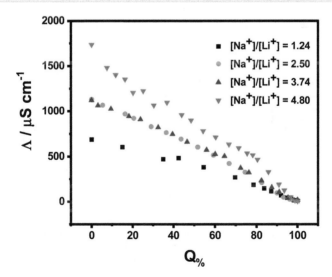

Fig. 2 Conductivity of the different test solutions in the middle compartment *vs.* the advancement of the electrolysis. All conductivity values correspond to 1 : 100 dilutions of the test solutions.

phenomena seem to occur, and the electric potential responds linearly to the applied current density (Fig. S3†). The intercept values in Fig. S3† are very similar for all experiments with varying $[Na^+]/[Li^+]$ ratios, which is to be expected considering that the electrochemical reactions (eqn (5) and (6)) are the same, and that the same electrodes were used.

$$\text{At the anode: } 4OH^- \rightarrow O_2(g) + 2H_2O + 4e^- \tag{5}$$

$$\text{At the cathode: } 2H_2O + 2e^- \rightarrow 2OH^-(aq) + H_2(g) \tag{6}$$

The thermodynamic E^0 values for eqn (5) and (6) above are $+0.64$ V (pH $= 10$) and -0.66 V (pH $= 11$), respectively. In turn, the competitive reaction at the anode is eqn (7), which has a higher $E^0 = 1.36$ V

$$2Cl^-(aq) \rightarrow Cl_2(g) + 2e^- \tag{7}$$

Thus, by comparing the resulting ΔE_e values for the two possible anodic reactions, data in Fig. S3 and S4† suggests that at pH $= 10$, the major reaction at the anode will be eqn (5). Otherwise, $-(\Delta E_e - |\eta_C| - |\eta_A|)$ should be larger. As previously discussed,[44] the use of a buffer to avoid Cl_2 formation is only viable at laboratory scale. Otherwise, the definition of zero chemical process no longer holds. An easy way out is to introduce a fourth compartment that avoids the arrival of chloride anions at the anode, albeit at the expense of an added resistance due to a third membrane.

The summation of all resistance elements was plotted *vs.* the advancement of the electrolysis, see Fig. 3. Although there is some fluctuation in the experimental data, the four curves are almost indistinguishable, suggesting that the prevailing resistance elements are most likely the same in all 4 systems. Overall cell

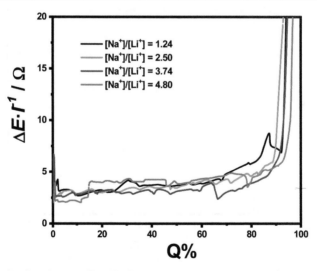

Fig. 3 Total cell resistance. The Y axis was calculated using the cell potential directly measured in the DC power supply, minus the estimated $\Delta E_e - |\eta_C| - |\eta_A|$ values, and the applied current.

resistance values fluctuate around (3.2 ± 1.1) Ω for the first 80% of the electrolysis. R_{AEM} and R_{CEM} are the same in all 4 experiments. The manufacturer lists values of 0.4 Ω and 0.3 Ω for R_{AEM} and R_{CEM}, respectively (Table 1). The membrane resistance values will also fluctuate according to the solutions' compositions that are in contact with each membrane. Bubbles are very likely to introduce non-negligible resistance values, and these are also present in all 4 systems. Kiuchi *et al.* estimated the resistance values of gas bubbles to be 4 Ω or higher.[57] While being non-conductive, and temporarily increasing the energy consumption of the electrolysis, the gas bubbles also promote a turbulent flow which favours mass transport, and renovate the diffusion layer in the vicinity of the membrane. There are two differences in the 4 experiments. Both the resistance of the solution in the middle compartment (R_{mC}) and the resistance of the solution in the cathodic compartment (R_{cC}) are different, at equal percentages of total circulated charge. However, for 80% of the electrolysis, values for R_{mC} are the lowest of the three compartments, since it still is the most concentrated compartment. Following Fig. 2, and considering the cell dimensions, R_{mC} can be estimated to be as low as 0.13 Ω (beginning of the experiment with highest TDS) and as high as about 1 Ω (80% completion of lowest TDS experiment). It is important to recall that the conductivity values were performed on 1 : 100 dilutions from the solution of interest and following Kohlrausch's laws the conductivity is not linearly dependent on concentration.[4] While we acknowledge that our home-made reactor setup is far from being an optimized design, understanding the different resistance components in a prospective electrochemical system is paramount for prospective technological implementation. Decreasing the width of the middle compartment, and using membranes with lower resistance values will be key in reducing the overall cell resistance, and hence the energy consumption of the process. Novel reactor designs should simultaneously address the issue of solid formation and intense sparging in the vicinity of the electrode surfaces. Towards the end of the batch runs, the total impedance values increase due to the formation of insulate carbonates which can be adsorbed on the membrane. Furthermore, this increase is caused by the depletion of ions in the middle compartment. At this point, water splitting on the membranes occurs, and is evidenced by a decrease of the pH in the brines, see below.

Table 3 shows initial and final composition for the different solutions fed to the middle compartment of the electrolysis cell. Na$^+$ initial concentrations were gradually increased while K$^+$ and Li$^+$ were kept constant. Thus, the initial value of total dissolved solids is increased. After completion of the electrolysis, it is observed that the TDS values in the middle compartment were reduced by three orders of magnitude indicating, once again, that most of the ions successfully migrated to the side compartments. In the middle compartment, neither the final concentrations of the cations, nor the TDS final values follow a clear trend with the initial solutions concentrations. This is attributed to the fact that some electrolysis experiments were not immediately stopped when the cell voltage reached 30 V, but a few hours later. TDS and ionic composition greatly increased in the catholyte; the magnitudes of these increments are correlated with the changes in solution volume.

A very clear trend can be observed for the volume changes in the middle and cathodic compartments, see Table 4. Unfortunately, the proportion of fresh water recovered in the middle compartment is lower when Na$^+$ concentration is highest,

Table 3 Cations' concentrations in the middle and cathodic compartments at the beginning and the end of the batch runs. Initial values are denoted by the sub-index i, while final values are those denoted with the sub-index f

Middle compartment

$[Na^+]/[Li^+]$ ratio	$[Li^+]_i$ g dm^{-3}	$[Li^+]_f$ g dm^{-3}	$[Na^+]_i$ g dm^{-3}	$[Na^+]_f$ g dm^{-3}	$[K^+]_i$ g dm^{-3}	$[K^+]_f$ g dm^{-3}	TDS_i g dm^{-3}	TDS_f g dm^{-3}	Difference in total cations concentration/ moles
1.24	1.349	$(30 \pm 1) \times 10^{-3}$	5.540	$(150 \pm 4) \times 10^{-3}$	14.030	$(10 \pm 1) \times 10^{-3}$	49.14 ± 0.04	0.584 ± 0.008	3.932
2.50	1.290	$(3.7 \pm 0.1) \times 10^{-3}$	10.613	$(36.1 \pm 0.4) \times 10^{-3}$	14.139	$(40.5 \pm 0.1) \times 10^{-3}$	61.82 ± 0.08	0.1916 ± 0.0005	5.551
3.74	1.257	$(0.2 \pm 0.1) \times 10^{-3}$	15.609	$(3.0 \pm 0.2) \times 10^{-3}$	13.992	$(0.5 \pm 0.1) \times 10^{-3}$	74.04 ± 0.03	0.010 ± 0.007	6.090
4.80	1.339	$(1.66 \pm 0.01) \times 10^{-3}$	21.305	$(19.51 \pm 0.01) \times 10^{-3}$	14.191	$(5.2 \pm 0.1) \times 10^{-3}$	89.40 ± 0.08	0.065 ± 0.006	7.409

Catholyte

$[Na^+]/[Li^+]$ ratio	$[Li^+]_i$ g dm^{-3}	$[Li^+]_f$ g dm^{-3}	$[Na^+]_i$ g dm^{-3}	$[Na^+]_f$ g dm^{-3}	$[K^+]_i$ g dm^{-3}	$[K^+]_f$ g dm^{-3}	TDS_i g dm^{-3}	TDS_f g dm^{-3}	Circulated charge C	Coulombic efficiency %
1.24	0.186	3.38 ± 0.05	—	36.48 ± 0.02	—	64.99 ± 0.02	0.990 ± 0.001	450 ± 40	385 351	99.7%
2.50	0.180	2.95 ± 0.01	—	49.02 ± 0.04	—	57.89 ± 0.01	0.9562 ± 0.0007	728 ± 7	509 927	>99.9%
3.74	0.186	2.05 ± 0.01	—	65.89 ± 0.01	—	55.35 ± 0.01	0.9902 ± 0.0007	266 ± 2	601 209	>99.9%
4.80	0.186	2.72 ± 0.01	—	88.81 ± 0.01	—	41.51 ± 0.01	0.9918 ± 0.0007	266 ± 2	725 875	>99.9%

Table 4 Changes in solution volumes for each of the three compartments measured from the beginning of the experiments to electrolysis completion[a]

$\frac{[Na^+]}{[Li^+]}$ ratio	$\Delta V_{cathodic}$ mL	ΔV_{middle} mL	ΔV_{anodic} mL	Final $mH_2O_{cathodic}$ g	Initial mH_2O_{middle} g	Final mH_2O_{middle} g	Final mH_2O_{anodic} g	Water contained in brine recovered as freshwater %	Circulated charge C
1.24	470 ± 10	−(1150 ± 10)	670 ± 10	940	4894	3850	2654	78.7	385 351
2.50	620 ± 10	−(1470 ± 10)	750 ± 10	1074	4871	3530	2734	72.5	509 927
3.74	670 ± 10	−(1750 ± 10)	815 ± 10	1131	4850	3250	2798	67.0	601 209
4.80	930 ± 10	−(2260 ± 10)	818 ± 10	1332	4853	2740	2801	56.5	725 875

[a] The changes in the mass of water for each of the solutions were calculated using the total salinity values and the corresponding solution density values. The initial mass of water for anolyte and catholyte for all 4 experiments was identical and equal to 500 g and 1988 respectively.

these values go from 78.7% to 56.5% of the original water content in the feed. A large electro-osmotic effect can explain this phenomenon, in which water molecules solvating the ions pass through membranes accompanying the migrating ions. The larger the absolute number of ions being transported across the membranes, the larger the extent of the electro-osmotic effect. There is a correlation between water loss in the middle compartment and water gain in the cathodic and the anodic compartments, and total number of charged particles initially present in each brine. It is still important to recall that even in the worst case scenario, our strategy allows for the recovery of over half of the original water content in the feed solutions, whereas the current evaporitic technology produces the loss of 100% of the water content.

Table 5 shows physico-chemical properties of the solutions in the middle compartment after electrolysis completion. Chloride concentrations at the end of the experiments also indicate electrolysis completion, e.g. the remaining Cl^- represents 0.03% of the original content in the compartment. The pH values are slightly alkaline; this could be explained by water splitting on membranes as shown in Fig. 7b, see below. In all cases, TDS at the end of the batch runs are lower than 2% (except for the first experiment) which is similar to the fresh water of wetlands. These TDS values are in good agreement with the conductivity measurements. In addition to specific upper limits for selected ions, fresh water is considered suitable for irrigation when its conductivity is in the range 150–500 μS cm^{-1}.[58] It can be observed that this is the case for three of the four experimental runs, the experiment with lowest total ionic composition does not fall far from the upper limit either. In order to more precisely compare the final results, it would have been necessary to precisely control the point when the experiments were considered as finished. It has to be acknowledged, however, that a techno-economic decision as to when it is more suitable to end an electrolysis, will not only be based on solution composition, but also on the overall operational cost. It is evident from Fig. 3 that the process becomes much more energy intensive for the last 20% of the electrolysis.

Coming back to Fig. 4, we observe that in all 4 cases studied, K^+ cations move across the membrane faster than both Na^+ and Li^+, while Na^+ moves faster than Li^+. For example, when 40% of the total charge has circulated, in the middle compartment the Li^+ concentration remains at about 90% of its initial concentration, while the K^+ concentration has decreased below 50%, for all the experiments with different initial Li^+/Na^+ concentration ratios. This behaviour is consistent with our previous work,[44,48] as well as reports by other authors.[27–29] Fig. S5† superposes data for the same cations and results are surprisingly similar

Table 5 Selected physico-chemical properties for solutions in the middle compartment after electrolysis completion. Λ values correspond to the undiluted solution

$[Na^+]/[Li^+]$ ratio	Density/g mL^{-1}	TDS_{final}/g L^{-1}	pH	$\Lambda/\mu S$ cm^{-1}	Cl^-/g L^{-1}
1.24	0.9940 ± 0.0001	0.584 ± 0.008	8.62 ± 0.02	660 ± 10	0.019 ± 0.003
2.50	0.997 ± 0.005	0.1916 ± 0.0005	7.42 ± 0.02	110 ± 10	$(2.5 \pm 0.5) \times 10^{-4}$
3.74	0.986 ± 0.009	0.010 ± 0.007	8.9 ± 0.1	16.2 ± 0.1	$(8 \pm 2) \times 10^{-3}$
4.80	0.995 ± 0.005	0.065 ± 0.006	9.85 ± 0.01	132.0 ± 0.1	0.030 ± 0.008

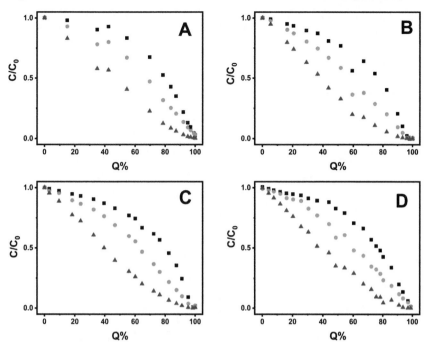

Fig. 4 Normalized concentrations (concentration at time t divided by initial concentration) in the middle compartment as a function of the extent of electrolysis. Black dots: Li^+, Red circles: Na^+. Blue triangles: K^+. Panels (A–D) correspond to increasing $[Na^+]/[Li^+]$ ratio.

despite the increasing Na^+ concentration. The separation coefficients shown in Fig. 5 compare the migration rates across the membrane. Fig. 6 shows the first derivative of data shown in Fig. 4, indicating how fast the concentrations for each cation are reduced in the middle compartment. The derivative of the summation of the dimensionless cation concentration is also shown (total cation concentration divided by total initial cation concentration, in molar units, green curves). In the first half of the experiments, K^+ disappears faster than Na^+ and Li^+. From roughly 55 to 75% of the total circulated charge, Na^+ moves faster than K^+ and Li^+, respectively. It is only in the last quarter of the electrolysis that Li^+ migration begins to increase and rapidly disappears from the middle compartment. This behaviour is interesting to consider when an electrolysis is performed because Li^+ passes through the CEM at the end of the experiment and can be recovered at this time.

The final catholyte composition (Table 3) shows that Li^+ and K^+ concentrations decrease when the $[Na^+]/[Li^+]$ ratio of concentrations increases. This is easily explained by the increase in the catholyte volume, *i.e.* the total amount of Li^+ and K^+ cations in solution is fairly constant. A precise mass balance was not calculated since all three cations are present in the catholyte, the first solid (prior to re-suspension in hot water), and the solutions used for washing and re-suspension, see below.

Fig. 7a shows the evolution of pH in the middle compartment for each batch run. The pH values of above 9 throughout most of the electrolysis are first

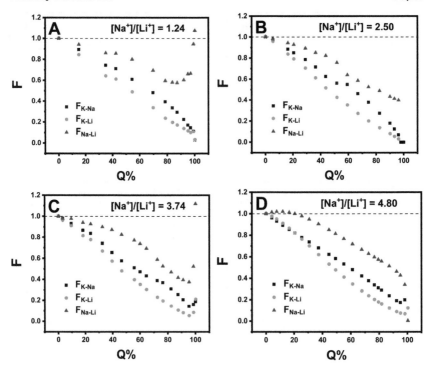

Fig. 5 Separation coefficients *vs.* the extent of electrolysis. Panels (A–D) correspond to increasing $[Na^+]/[Li^+]$ ratio.

attributed to the Lewis acid behaviour of Li^+ ions in aqueous solutions. The pH remains roughly constant for most of the electrolysis. It is also observed that the pH is slightly higher for experiments with higher total salinity. Only after about 90% of the total charge has circulated does the pH decrease. For the experiment with $[Na^+]/[Li^+] = 1.24$, the pH decreases to pH = 7.3 and this is observed when 95.0% of the total charge has circulated. For the other experiments, the pH goes down for circulated charge values of 93.9%, 96.4% and 96.9% for increasing $[Na^+]/[Li^+]$ ratios, reaching pH values close to 2. In all 4 experiments, towards the very end of the run, the pH value returns to about 8. Both the initial decrease and final increase in pH values in the middle compartment are attributed to water splitting on membranes.[59,60] Fig. 7b schematizes these processes.

In an aqueous solution with a much depleted ionic concentration, in order to allow for charge transfer due to an applied electric field, water can split into H^+ and OH^-.[61] The chemical nature of the ionic exchange membranes used in this work can allow for this splitting, and thus promote migration of, alternately, H^+ and OH^-, to the middle compartment. In the case of anion exchange membranes with quaternary ammonium residues, hydroxyl anions can migrate from the membrane to the anolyte.[59,62] In turn, protons are repelled from the membrane and migrate in the same direction of the electric field. The reduced pH in the middle compartment towards the end of the electrolysis is attributed to this phenomenon.

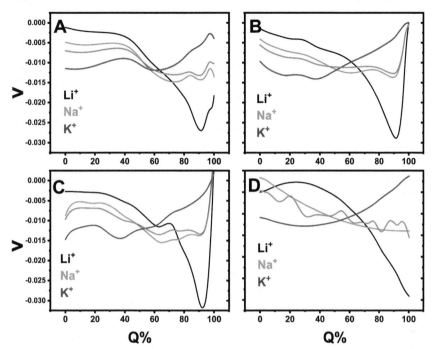

Fig. 6 Rate of concentration changes in the middle compartment with the extent of the electrolysis (derivatives of the corresponding data in Fig. 4). The green curves correspond to the derivatives of the dimensionless summation of all cations. Panels (A–D) correspond to increasing $[Na^+]/[Li^+]$ ratio.

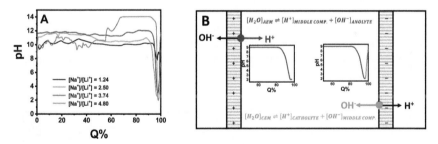

Fig. 7 (A) pH evolution in the middle compartment with the extent of electrolysis. (B) Schematic representation of the water splitting processes in both AEM and CEM.

Water splitting produces a reduced coulombic efficiency, since a share of the electric current will not result in cation transfer towards the cathodic compartment. Table 3 shows that the coulombic efficiency of the overall process (*i.e.* calculated over the whole duration of the electrolysis) is above 99.7% for all experiments. Therefore, we can conclude that while water splitting produces a highly noticeable effect in the middle compartment, which at this point has a very low total ionic composition, it is only a very minor process over the whole electrolysis. Moreover, water splitting did not produce any measurable pH changes in the cathodic or anodic compartments. This was attributed to the strong buffer concentration in the latter, and the constant production of hydroxyl anions at the cathode, which is certainly predominant over the few protons that might be generated at the membrane. Overall, no effects on the product purity could be attributed to the water splitting. However, potential detrimental consequences could be observed upon working under different conditions, such as higher current densities, or permselective membranes of different chemical nature, both of which could increase the extent of the water splitting phenomenon. The product yield was not quantified, and we do acknowledge that water splitting could potentially have an effect on this parameter as well.

According to the pH changes observed in our experiments, it seems that water splitting only occurs at the cation exchange membrane at a later stage of the batch run. Data in Fig. 6 and 7 correlate quite well. Fig. 6 shows that the rate of cation transfer across the cation exchange membrane is roughly constant (all 3 cations analysed together, within experimental error), except for approximately the last 5% of the electrolysis. Thus, the final pH increase corresponds to a marked diminution of the rate of consumption of all cations in the middle compartment. The advancement of the electrolysis is quantified in terms of total circulated charge, thus, a diminution in the rate of $K^+/Na^+/Li^+$ transfer must be associated with other species compensating for the charge transfer, *i.e.* water splitting at the CEM. This is also in line with the fact that the higher the total amount of ions in solution, the later the decrease in pH is observed, since at a given % of total circulated charge the absolute number of ions is different for the 4 tests. The last panel (ratio of concentrations = 4.80) does not seem to follow this trend, which is in principle attributed to experimental errors in the Li^+ concentrations; smalls errors become more evident in the derivatives. Water splitting at membranes is a complex phenomenon that should be studied more in depth in our current system.

The solids crystallized in the side vessel attached to the cathodic compartment were first analysed following vacuum filtration and a very simple washing step. The X-ray diffractograms shown in Fig. 8a suggest 4 pure solids were obtained.

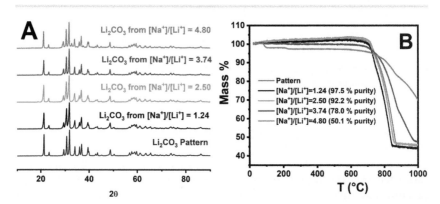

Fig. 8 X-ray diffractograms (A) and thermogravimetric analysis (B) for the 4 solids recovered in the cathodic compartments, and a battery grade commercial sample. Analysis performed before the solids were re-suspended in 90 °C de-ionized water and sonicated.

Table 6 Composition of solids crystallized in the cathodic compartment in the absence of solid washing (first 6 columns)[a]

[Na$^+$]/[Li$^+$] ratio	Experimental results before washing the solid						Expected values		Li$_2$CO$_3$ purity after thorough washing
	Moles g$_{solid}^{-1}$			CO$_3^{2-}$/moles g$_{solid}^{-1}$	HCO$_3^-$ moles g$_{solid}^{-1}$	Li$_2$CO$_3$ Purity	Li moles g$_{solid}^{-1}$	CO$_3^{2-}$/moles g$_{solid}^{-1}$	
	Potassium	Sodium	Lithium						
1.24	$(3.9 \pm 0.4) \times 10^{-5}$	$(1.0 \pm 0.2) \times 10^{-5}$	$(2.64 \pm 0.26) \times 10^{-2}$	$(1.34 \pm 0.01) \times 10^{-2}$	$(1.12 \pm 0.06) \times 10^{-4}$	97.5%	2.707×10^{-2}	1.353×10^{-2}	>99.9%
2.50	$(1.29 \pm 0.03) \times 10^{-5}$	$(1.7 \pm 0.1) \times 10^{-5}$	$(2.50 \pm 0.01) \times 10^{-2}$	$(1.36 \pm 0.02) \times 10^{-2}$	$(1.09 \pm 0.01) \times 10^{-4}$	92.2%	2.707×10^{-2}	1.353×10^{-2}	>99.9%
3.74	$(5.7 \pm 0.2) \times 10^{-5}$	$(82 \pm 3) \times 10^{-5}$	$(1.07 \pm 0.06) \times 10^{-2}$	$(1.32 \pm 0.04) \times 10^{-2}$	$(0.46 \pm 0.02) \times 10^{-4}$	78.0%	2.707×10^{-2}	1.353×10^{-2}	>99.9%
4.80	$(8.6 \pm 2) \times 10^{-4}$	$(2.5 \pm 1) \times 10^{-4}$	$(1.32 \pm 0.06) \times 10^{-2}$	$(1.14 \pm 0.02) \times 10^{-2}$	$(6.59 \pm 0.01) \times 10^{-4}$	50.1%	2.707×10^{-2}	1.353×10^{-2}	>99.9%

[a] The expected values are those calculated assuming the solid is pure Li$_2$CO$_3$. All concentrations correspond to moles per gram of solid. Chloride was below detection limit in all solids. Last column to the left: purity after 3 times re-suspension in 90 °C de-ionized water.

However, both the thermogravimetric analysis (Fig. 8b), and the concentrations determined by atomic spectroscopy and volumetric titration (Table 6) show otherwise, with purity rapidly decreasing as the $[Na^+]/[Li^+]$ ratio was increased in the feed solutions. Because no impurities are observed by X-ray diffraction, even when the solid is only 50% pure, it seems that the impurity salts were adsorbed on Li_2CO_3 rather than co-crystallization phenomena having occurred. Moreover, K_2CO_3, $NaHCO_3$, and Na_2CO_3 are all known to be much more water soluble than Li_2CO_3 particularly in hot water.[47] Thus, the solids were re-suspended in 90 °C de-ionized water and sonicated. After this procedure, Na^+, K^+ and HCO_3^- were all below detection limits, and quantification of Li^+ in the solid indicated Li_2CO_3 purity above 99.9%. This purification procedure was straightforward and could very easily be implemented at industrial scale, thus we can consider that the electrochemical methodology studied here is capable of producing battery grade Li_2CO_3.

4. Conclusions

We have proposed an overall electrochemical technology for the joint production of Li_2CO_3 and other raw materials from complex brines. Our proposed approach entails the electrochemical production of hydroxyl anions that will accelerate the absorption of CO_2. Li_2CO_3 crystallization will occur provided the pH is constantly monitored so that the applied current density can be adjusted to secure a range of pH where carbonate is the major carbon species.

Our electrochemical proposal aims at the production of Li_2CO_3 from a complex broth, *i.e.* not departing from pure precursors. Thus, we have studied its performance as the ratio of $[Na^+]/[Li^+]$ is increased. The as-crystallized product showed an important decrease in purity as the ratio of $[Na^+]/[Li^+]$ was increased. However, Li_2CO_3 with purity above 99.9% could be produced upon re-suspension in 90 °C de-ionized water, a readily scalable procedure. Upon increasing the $[Na^+]/[Li^+]$ 4 times, no major changes were observed in the electrochemical performance. The voltage difference between the anode and cathode suggested that the over-potentials remained fairly constant, while the summation of all resistance elements was mostly governed by the ion exchange membranes and the gas bubbles. Despite the increase in total salinity in the middle compartment, the voltage difference was very similar for all systems.

While the increase in total salinity did not impact the purity of the product upon washing, it will have two other consequences. At higher $[Na^+]/[Li^+]$ ratios, the electrical cost for the process will be largely increased. The potential difference will be the same, but a larger number of ions will need to be transferred across the membranes, requiring larger total electric charge. A major difference in this approach as compared to other hydrometallurgical processes is the production of fresh water as a by-product of a mining process. However, the proportion of water contained in the original feed that is successfully recovered decreases with higher total salinity in the feed. Not only is water lost as a consequence of water electrolysis, a process that could be reversed in a fuel cell, but water molecules are transferred to the concentrated compartments *via* electro-osmosis.

In this work, the coulombic efficiency has been calculated in terms of the amount of current successfully used to transfer cations across the membrane to the cathodic compartment. If we were to recalculate this parameter following the more classical definition of the percentage of electrons successfully used to synthesize the product of interest, the results would be much lower. This is, firstly, because a larger proportion of the electrons was used to transfer the co-existing cations, Na^+, K^+, that will remain in the catholyte and not yield a product of interest, and secondly, because Li_2CO_3 still shows a non-negligible water solubility and a part of the product will remain in solution. The low coulombic efficiency, in its classical definition, is the weakest point in our proposed methodology. With the high prices for battery grade Li_2CO_3, currently at about 70 000 US$ per tonne, almost any proposed technology for its production is profitable. While the results shown here have proved that as much as 5 times the molar ratio of Na^+/Li^+ still produces a high purity Li_2CO_3 product, any pre-processing reducing both Na^+ and K^+ content will reduce the production costs.

While this approach presents many potential benefits as compared to both traditional brine mining and other novel DLE approaches, lots of work remains to be undertaken if these ideas are to be taken to a higher technological readiness level. Key for the successful scaling up will be to further develop pre-processing strategies, and to re-configure the reactor setup so that crystals will not be formed in the vicinity of ion exchange membranes.

Author contributions

WRT: conceptualization, data curation, formal analysis, investigation, methodology, writing-original draft, funding acquisition. NCZ: investigation. VF: conceptualization, methodology, formal analysis, writing-original draft, writing-review & editing, funding acquisition.

Conflicts of interest

There are no conflicts of interest to declare.

Acknowledgements

WRT and VF are CONICET permanent research fellows. This work was supported by ANPCyT, AR, grant PICT 2019–1939, and PICT 2019–2363.

References

1 M. C. Leech and K. Lam, *Nat. Rev. Chem.*, 2022, **6**, 275–286.
2 E. C. R. McKenzie, S. Hosseini, A. G. C. Petro, K. K. Rudman, B. H. R. Gerroll, M. S. Mubarak, L. A. Baker and R. D. Little, *Chem. Rev.*, 2022, **122**, 3292–3335.
3 F. C. Walsh and D. Pletcher, in *Developments in Electrochemistry: Science Inspired by Martin Fleischmann*, 2014, pp. 95–111.
4 D. Pletcher and F. C. Walsh, *Industrial Electrochemistry*, Springer Netherlands, 1990.
5 M. L. Vera, W. R. Torres, C. I. Galli, A. Chagnes and V. Flexer, *Nat. Rev. Earth Environ.*, 2023, **4**, 149–165.
6 A. Koyamparambath, J. Santillán-Saldivar, B. McLellan and G. Sonnemann, *Resour. Policy*, 2022, **75**, 102465.

7 V. Flexer, C. F. Baspineiro and C. I. Galli, *Sci. Total Environ.*, 2018, **639**, 1188–1204.

8 W. Liu and D. B. Agusdinata, *Extr. Ind. Soc.*, 2021, **8**, 100927.

9 W. Liu and D. B. Agusdinata, *J. Cleaner Prod.*, 2020, **260**, 120838.

10 W. Liu, D. B. Agusdinata and S. W. Myint, *Int. J. Appl. Earth Obs. Geoinf.*, 2019, **80**, 145–156.

11 X. He, S. Kaur and R. Kostecki, *Joule*, 2020, **4**, 1357–1358.

12 D. E. Garrett, *Handbook of Lithium and Natural Calcium Chloride*, 2004, pp. 1–476.

13 A. Battistel, M. S. Palagonia, D. Brogioli, F. La Mantia and R. Trócoli, *Adv. Mater.*, 2020, **32**, 1905440.

14 J. Wang, X. Yue, P. Wang, T. Yu, X. Du, X. Hao, A. Abudula and G. Guan, *Renewable Sustainable Energy Rev.*, 2022, **154**, 111813.

15 F. Meng, J. McNeice, S. S. Zadeh and A. Ghahreman, *Miner. Process. Extr. Metall. Rev.*, 2021, **42**, 123–141.

16 S. Zavahir, T. Elmakki, M. Gulied, Z. Ahmad, L. Al-Sulaiti, H. K. Shon, Y. Chen, H. Park, B. Batchelor and D. S. Han, *Desalination*, 2021, **500**, 114883.

17 X. Li, Y. Mo, W. Qing, S. Shao, C. Y. Tang and J. Li, *J. Membr. Sci.*, 2019, **591**, 117317.

18 M. Pasta, A. Battistel and F. La Mantia, *Energy Environ. Sci.*, 2012, **5**, 9487–9491.

19 E. J. Calvo, *ACS Omega*, 2021, **6**, 35213–35220.

20 V. C. E. Romero, K. Llano and E. J. Calvo, *Electrochem. Commun.*, 2021, **125**, 106980.

21 F. Qian, B. Zhao, M. Guo, Z. Qian, Z. Wu and Z. Liu, *Mater. Des.*, 2020, **194**, 108867.

22 R. Chitrakar, Y. Makita, K. Ooi and A. Sonoda, *Dalt. Trans.*, 2014, **43**, 8933–8939.

23 C. Liu, Y. Li, D. Lin, P.-C. Hsu, B. Liu, G. Yan, T. Wu, Y. Cui and S. Chu, *Joule*, 2020, **4**, 1459–1469.

24 C. H. Díaz Nieto and V. Flexer, *Curr. Opin. Electrochem.*, 2022, **35**, 101087.

25 S. Gmar and A. Chagnes, *Hydrometallurgy*, 2019, **189**, 105124.

26 X.-Y. Nie, S.-Y. Sun, Z. Sun, X. Song and J.-G. Yu, *Desalination*, 2017, **403**, 128–135.

27 Z.-Y. Guo, Z.-Y. Ji, Q.-B. Chen, J. Liu, Y.-Y. Zhao, F. Li, Z.-Y. Liu and J.-S. Yuan, *J. Cleaner Prod.*, 2018, **193**, 338–350.

28 L.-M. Zhao, Q.-B. Chen, Z.-Y. Ji, J. Liu, Y.-Y. Zhao, X.-F. Guo and J.-S. Yuan, *Chem. Eng. Res. Des.*, 2018, **140**, 116–127.

29 Q.-B. Chen, Z.-Y. Ji, J. Liu, Y.-Y. Zhao, S.-Z. Wang and J.-S. Yuan, *J. Membr. Sci.*, 2018, **548**, 408–420.

30 B. Dutcher, M. Fan and A. G. Russell, *ACS Appl. Mater. Interfaces*, 2015, **7**, 2137–2148.

31 M. Rahimi, A. Khurram, T. A. Hatton and B. Gallant, *Chem. Soc. Rev.*, 2022, **51**, 8676–8695.

32 S. Ó. Snæbjörnsdóttir, B. Sigfússon, C. Marieni, D. Goldberg, S. R. Gislason and E. H. Oelkers, *Nat. Rev. Earth Environ.*, 2020, **1**, 90–102.

33 A. Mehmood, M. I. Iqbal, J.-Y. Lee, J. Hwang, K.-D. Jung and H. Y. Ha, *Electrochim. Acta*, 2016, **219**, 655–663.

34 K. J. Lamb, M. R. Dowsett, K. Chatzipanagis, Z. W. Scullion, R. Kröger, J. D. Lee, P. M. Aguiar, M. North and A. Parkin, *ChemSusChem*, 2018, **11**, 137–148.

35 R. J. Gilliam, B. K. Boggs, V. Decker, M. A. Kostowskyj, S. Gorer, T. A. Albrecht, J. D. Way, D. W. Kirk and A. J. Bard, *J. Electrochem. Soc.*, 2012, **159**, B627.

36 I. Sanjuán, V. García-García, E. Expósito and V. Montiel, *Electrochem. Commun.*, 2019, **101**, 88–92.

37 I. Sanjuán, D. Benavente, E. Expósito and V. Montiel, *Sep. Purif. Technol.*, 2019, **211**, 857–865.

38 T. Chen, J. Bi, Y. Zhao, Z. Du, X. Guo, J. Yuan, Z. Ji, J. Liu, S. Wang, F. Li and J. Wang, *Sci. Total Environ.*, 2022, **820**, 153272.

39 T. Chen, Y. Zhao, Y. Zhao, Y. Xie, Z. Ji, X. Guo, Y. Zhao and J. Yuan, *ACS Sustainable Chem. Eng.*, 2021, **9**, 8372–8382.

40 H. Xie, W. Jiang, Y. Wang, T. Liu, R. Wang, B. Liang, Y. He, J. Wang, L. Tang and J. Chen, *Environ. Earth Sci.*, 2015, **74**, 6481–6488.

41 K. Sorimachi, *Sci. Rep.*, 2022, **12**, 1694.

42 G. Acosta-Santoyo, L. F. León-Fernández, E. Bustos, P. Cañizares, M. A. Rodrigo and J. Llanos, *Chemosphere*, 2021, **285**, 131359.

43 J. Mustafa, A. A.-H. I. Mourad, A. H. Al-Marzouqi and M. H. El-Naas, *Desalination*, 2020, **483**, 114386.

44 W. R. Torres, C. H. Díaz Nieto, A. Prévoteau, K. Rabaey and V. Flexer, *J. Membr. Sci.*, 2020, **615**, 118416.

45 C. H. Díaz Nieto, N. A. Palacios, K. Verbeeck, A. Prévoteau, K. Rabaey and V. Flexer, *Water Res.*, 2019, **154**, 117–124.

46 B. Sanjuan, B. Gourcerol, R. Millot, D. Rettenmaier, E. Jeandel and A. Rombaut, *Geothermics*, 2022, **101**, 102385.

47 D. R. Lide, *CRC Handbook of Chemistry and Physics*, 2005.

48 C. H. Diaz Nieto, K. Rabaey and V. Flexer, *Sep. Purif. Technol.*, 2020, **252**, 117410.

49 C. H. Díaz Nieto, J. A. Kortsarz, M. L. Vera and V. Flexer, *Desalination*, 2022, **529**, 115652.

50 Y. Zhao, Y. Liu, C. Wang, E. Ortega, X. Wang, Y. F. Xie, J. Shen, C. Gao and B. Van der Bruggen, *J. Mater. Chem. A*, 2020, **8**, 4244–4251.

51 P. P. Sharma, V. Yadav, A. Rajput, H. Gupta, H. Saravaia and V. Kulshrestha, *Desalination*, 2020, **496**, 114755.

52 Z. Qian, H. Miedema, S. Sahin, L. C. P. M. de Smet and E. J. R. Sudhölter, *Desalination*, 2020, **495**, 114631.

53 C. F. Baspineiro, J. Franco and V. Flexer, *Sci. Total Environ.*, 2021, **791**, 148192.

54 S. H. Park, J. H. Kim, S. J. Moon, J. T. Jung, H. H. Wang, A. Ali, C. A. Quist-Jensen, F. Macedonio, E. Drioli and Y. M. Lee, *J. Membr. Sci.*, 2020, **598**, 117683.

55 A. Cerda, M. Quilaqueo, L. Barros, G. Seriche, M. Gim-Krumm, S. Santoro, A. H. H. Avci, J. Romero, E. Curcio and H. Estay, *J. Water Process. Eng.*, 2021, **41**, 102063.

56 M. Quilaqueo, G. Seriche, L. Barros, C. González, J. Romero, R. Ruby-Figueroa, S. Santoro, E. Curcio and H. Estay, *Desalination*, 2022, **537**, 115887.

57 D. Kiuchi, H. Matsushima, Y. Fukunaka and K. Kuribayashi, *J. Electrochem. Soc.*, 2006, **153**, E138.

58 M. Zaman, S. A. Shahid, and L. Heng, ed. M. Zaman, S. A. Shahid, and L. Heng, *Guideline for Salinity Assessment, Mitigation and Adaptation Using Nuclear and Related Techniques*, Springer International Publishing, Cham, 2018, pp. 113–131.

59 O. A. Rybalkina, K. A. Tsygurina, E. D. Melnikova, G. Pourcelly, V. V. Nikonenko and N. D. Pismenskaya, *Electrochim. Acta*, 2019, **299**, 946–962.

60 M. S. Kang, Y. J. Choi, H. J. Lee and S. H. Moon, *J. Colloid Interface Sci.*, 2004, **273**, 523–532.

61 R. Simons, *Electrochim. Acta*, 1985, **30**, 275–282.

62 D. Y. Butylskii, V. A. Troitskiy, M. V. Sharafan, N. D. Pismenskaya and V. V. Nikonenko, *Desalination*, 2022, **537**, 115821.

DISCUSSIONS

Interdisciplinary electrosynthesis: general discussion

Lutz Ackermann, Mickaël Avanthay, Belen Batanero,
Dylan G. Boucher, Pim Broersen, Emily Carroll, Victoria Flexer,
Robert Francke, Toshio Fuchigami, Rokas Gerulskis,
David P. Hickey, Bee Hockin, Alexander Kuhn,
Matthew J. Milner, Shelley D. Minteer, Kevin Moeller,
Zachary A. Nguyen, Toshiki Nokami, Shahid Rasul, Naoki Shida,
Eniola Sokalu, Kohei Taniguchi and Niklas von Wolff

DOI: 10.1039/d3fd90037a

Mickaël Avanthay opened discussion of the paper by Lutz Ackermann: Compared to direct, non-photochemical methods, do you see a higher conversion or more selectivity?

Lutz Ackermann replied: We primarily observed better conversions.

Belen Batanero enquired: Even when this is a mediated process and considering that radical coupling is facilitated at a platinum anode, why have you not tried with these anodes? Graphite felt (GF) morphology?

Lutz Ackermann answered: Thank you for the important question. We have explored different anode types, including carbon-based materials, in our previous related contribution.[1]

1 Y. Qiu, A. Scheremetjew, L. H. Finger and L. Ackermann, *Chem.-Eur. J.*, 2020, **26**, 3241–3246, DOI: 10.1002/chem.201905774.

Belen Batanero asked: What is the effect of the electrolyte? At the GF anode radicals can be further oxidized to cations, do you find cross contamination?

Lutz Ackermann responded: We have neither observed a specific effect nor cross contamination thus far.

Robert Francke said: Various catalysts were screened for the tri-fluoromethylation of trimethoxybenzene (Table 1 in the paper; https://doi.org/10.1039/d3fd00076a). These catalysts belong to different families (organometallics, metal salts, halides, organic dyes) and have different key-properties (light absorption, stability, ground and excited state oxidation potential). While the conversion is in all cases very good, strong differences in the catalyst's activity become apparent when comparing the degree of functionalization (mono- *vs.* difunctionalization). Which of the catalyst's properties are responsible for this?

Lutz Ackermann answered: That is a very good question, which we are currently trying to address in greater detail.

Kohei Taniguchi asked: Does the reaction proceed using sunlight instead of blue light?

Lutz Ackermann answered: Very good comment. Thus far, we have focused on the use of LED light sources.

Toshio Fuchigami enquired: CF_3COONa is much cheaper than CF_3SO_2Na. Is it possible to use CF_3COONa instead of CF_3SO_2Na?

Lutz Ackermann responded: Very good point, thank you. We have previously only briefly studied this precursor.

Naoki Shida questioned: In photoelectrocatalysis, the balance of kinetics between electrochemical generation of the active species and the following photochemical process seems to be important. Since your reaction is done under constant current conditions, how did you determine the optimal current for the reaction?

Lutz Ackermann answered: We have experimentally explored different currents under galvanostatic conditions, while at the same time exploring different wavelengths as well.

Kevin Moeller asked: In the Tables provided in the paper (https://doi.org/10.1039/d3fd00076a), for most reactions the organic catalysts perform to a mostly equal extent. This was not the case for the complex natural product substrate. Why is this the case, and what are the key aspects we need to think about in order to make the best selection of a catalyst in those cases?

Lutz Ackermann replied: Excellent question. It is likely due to the chemoselectivity when oxidation-sensitive functional groups are present in the substrate.

Bee Hockin requested: Could you give some more details on considerations for scale up of this technique, given the issues with photon flux and other problems in batch for photocatalysis?

Lutz Ackermann answered: We are exploring flow-conditions to enable scale up.

Toshiki Nokami asked: With the oxidative generation of a radical, is it possible to use that radical in a substitution reaction or some other transformation?

Lutz Ackermann replied: Thank you; excellent question. We are currently looking into that.

Zachary A. Nguyen opened discussion of the paper by David P. Hickey: Often 1,2-NADH and 1,6-NADH are made in detectable quantities in NADH regeneration. Was there any lack of regioselectivity in your regeneration of the NADH?

David P. Hickey responded: The 1,2- and 1,6-NADH derivatives exhibit distinctly different oxidation peaks that can be observed when we poise the electrode to a sufficiently low potential (*e.g.*, -2 V *vs.* SCE) in the presence of $\mathbf{1}^+$ without pyruvate. In the presence of pyruvate, no such peaks were observed suggesting that the reaction is highly selective.

Niklas von Wolff asked: In your work, you chose to use a cyclic voltammetry based screening to find suitable reaction conditions to avoid undesirable side-reactions (dimerization). How did you choose your scan-rate for this screening and what can you say about the generalization of such CV-based screening approaches for finding suitable conditions for preparative electrolysis?

David P. Hickey replied: We chose a scan rate of 800 mV s^{-1} because it was fast enough that the 1D, semi-infinite diffusion assumption for convolution voltammetry held reasonably well. At slower scan rates, natural convection interferes with this assumption, and the kinetically-limited decomposition of water becomes more significant at low potentials. At much faster scan rates, capacitive background current becomes significant, and the CVs have lower resolution (fewer data points per scan).

While CV-based screening approaches may inform preparative electrolysis conditions, caution is advised when extrapolating these conditions to bulk electrolysis. They can provide valuable insight regarding the electrochemical reaction mechanism and should provide a good picture of the idealized electrolysis reaction, but due to the timescale of a CV, you are only measuring the initial reaction kinetics. These kinetics may change throughout the duration of the electrolysis. Additionally, electrode surface composition and morphology have a tendency to change during preparative electrolysis, and this tends to occur on a much longer timescale.

Rokas Gerulskis enquired: You mentioned your data suggesting that the mechanism by which pyruvate addition improves NAD reduction selectivity involves the reduction of pyruvate. I assume this suggests the mechanism is simply separating NAD radicals spatially on the electrode surface, such that they are unable to interact to form dimers before interacting with protons or hydrides to reduce to NADH? Have you investigated the effect of structurally related molecules such as mesoxalate, oxalate, biacetyl? Related polymers?

David P. Hickey answered: Recent experiments suggest that the ox2/ox1 ratio is maximized when the diketone is reduced at the same time/in the same potential window as the mimetic. This suggests that the mechanism is not simply a spatial separation of mNAD radicals. In the paper (https://doi.org/10.1039/d3fd00047h), we listed oxalate as one of the compounds that did not work. We have recently found a variety of structurally similar molecules that give similar (and even significantly better) selectivity and preference for the 1,4-mNADH. We hope to publish these studies soon.

Eniola Sokalu asked: From my understanding there is a hydride transfer reaction happening at the electrode surface (NAD to NADH). What prompted investigation into the supporting electrolyte as opposed to investigating the material used at the working electrode?

David P. Hickey replied: We know that other researchers have found supporting electrolyte to play a large role in product distribution in electrochemistry. Specifically, the eBirch reaction[1] relies on a specific combination of electrolyte, proton source and additive to achieve selective formation of a 1,4-dihydrobenzene. Inspired by this, it seemed reasonable to infer that the right supporting electrolyte/additive could afford the desired 1,4-dihydropyridine.

1 B. K. Peters, K. X. Rodriguez, S. H. Reisberg, S. B. Beil, D. P. Hickey, Y. Kawamata, M. Collins, J. Starr, L. Chen, S. Udyavara, K. Klunder, T. J. Gorey, S. L. Anderson, M. Neurock, S. D. Minteer and P. S. Baran, *Science*, 2019, **363**, 838–845, DOI: 10.1126/science.aav5606.

Dylan G. Boucher enquired: Regarding the structure of pyruvate, in Table 1 in your paper (https://doi.org/10.1039/d3fd00047h) you have several other additives that are structurally similar to pyruvate but do not promote the desired transformation. What is it about the structure/properties of pyruvate that enable this? In other words, what is the mechanism of this selectivity?

David P. Hickey responded: Recent experiments suggest that the active motif is a reducible dicarbonyl group. We are presently working to better understand the mechanism, and we invite the audience (and readers) to stay tuned for updates.

Shelley D. Minteer questioned: The vast majority of dehydrogenases are NAD dependent where NAD is basically a mediator. As electrochemists, we could design the "perfect" mediator, so wouldn't it be better to use the perfect electrochemical mediator and *de novo* design the enzyme around the "perfect" mediator rather mimicking the NAD?

David P. Hickey answered: There are many dehydrogenases that do utilize $NAD^+/$ NADH where $NAD^+/$NADH behaves as purely a redox mediator (*i.e.*, a one- or two-electron mediator). For these enzymes, it makes a lot of sense to design an exogenous redox mediator, and this is something that we are actively pursuing. For dehydrogenases (especially alcohol dehydrogenases) that utilize $NAD^+/$NADH as a hydride transfer catalyst, in which selectivity and activity depend on specific binding of both substrate and cofactor, we think that designing electrochemically regenerable mimetics of NADH may be easier than designing novel enzymes *de novo*. Given the ubiquity of NAD as a cofactor, it is almost as though nature identified it as the natural "perfect" mediator around which many oxidoreductases evolved.

Pim Broersen asked: To see the influence of potential on your reaction, you run cyclic voltammograms (CVs) to different potentials and back. Since this changes the length of time for the reduction, do you think that there might be an influence of intermediate concentration near the electrode, as more intermediate can build up with a longer (further) CV scan? Would it not be better to perform bulk electrolysis at different potentials to deconvolute these effects?

David P. Hickey replied: Bulk electrolysis could, indeed, disentangle the effects of reduction time and switching potential; however, consumption of the radical intermediate generally occurs exceptionally fast (for reference, dimerization of the neutral radical of N-methylnicotinamide occurs at a rate constant of $k = 4.9 \times 10^6$ M^{-1} s^{-1}),[1] nearing its diffusion-limited rate constant. Consequently, we would not expect the concentration of radical intermediate to fluctuate significantly as a function of reduction time. Supporting this notion, variable scan rate cyclic voltammetry of **1**$^+$ in the presence of pyruvate shows no measurable difference in the relative peak heights of ox1 and ox2 (Fig. 1 here).

1 C. O. Schmakel, K. S. V. Santhanam and P. J. Elving, *J. Electrochem. Soc.*, 1974, **121**, 345, DOI: 10.1149/1.2401814.

Alexander Kuhn enquired: More classic NADH regeneration approaches are using for example rhodium catalysts. Is it possible to use your findings about the beneficial effect of pyruvate together with such classic systems in order to obtain a kind of synergy effect, leading to an even more efficient and more selective regeneration?

David P. Hickey answered: This is an interesting concept that may be worth exploring further. We are presently unaware of any synergistic effect reported in the literature, but this is something that we are very interested to investigate in the future.

Toshio Fuchigami asked: It is known that cathodic dimerization of NAD$^+$ can be suppressed by using some kinds of liquid crystal modified electrodes since

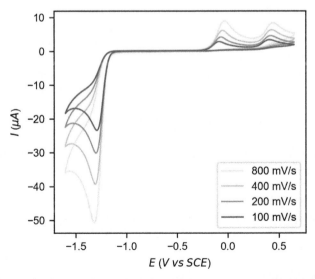

Fig. 1 Cyclic voltammograms of 2 mM **1**$^+$ in an aqueous solution of 100 mM sodium pyruvate at pH 7.23; performed using a boron doped diamond (BDD) working electrode, Pt wire counter electrode, and saturated calomel electrode (SCE) reference at 25 °C.

NAD^+ is adsorbed on the cathode separately. Is your methodology much better than such modified electrodes?

David P. Hickey replied: We have not explored liquid crystal modified electrodes, so it may not be appropriate to compare the two approaches. I do think the idea of physical sequestration of NAD radicals is very interesting and there are several possible approaches to implement such a strategy. From the evidence we currently have, our approach seems to be fundamentally different than physical sequestration, but the designation of better/worse may relate to the specific application.

Belen Batanero questioned: Have you tried other keto acids such as ketoglutaric acid? What is the anodic process? A decarboxylative reaction? Did you observe less dimerization product when R at nitrogen is a hindered group?

David P. Hickey answered: We have observed similar effects with methyl pyruvate, pyruvanilide, and 4-methoxypyruvanilide, among other pyruvate derivatives. The use of excess glyoxalic acid with 1^+ also results in a small ox2 peak; however, the reaction is not as selective and results in multiple additional oxidation peaks.

For the anodic reaction, we are under the assumption that water oxidation is the primary reaction, although we have not studied this extensively. For the studies in our paper (https://doi.org/10.1039/d3fd00047h), the products were inferred from CV analysis. Consequently, we do not expect the counter electrode reaction to impact the product distribution. Regarding the R group, we do have some unpublished data suggesting that sterics of the N-substituent do impact dimerization rates of the corresponding mimetics, but the correlation is not straightforward. We will address this in future work.

Matthew J. Milner said: You stated that you also looked at the influence of the electrode material on the product distribution, and mentioned glassy carbon, as well as the BDD used in the paper. Both of these are typically inert materials, where it would be expected that reduction would cleanly produce solvated neutral radical **1***. It is conceivable that the pyruvate perhaps prevents the radical–radical dimerization either through stabilization of the radical or by sequestration through pyruvate–radical reaction and later release. It might then be possible that a less inert electrode material, on whose surface the **1*** radicals are adsorbed and stabilized, would also disfavour dimerization and allow the second reduction step to occur cleanly. Did you try any less inert electrode materials such as copper or steel?

David P. Hickey responded: This is a very interesting idea! We had considered a version of this idea prior to our work on pyruvate; however, we have not done very much to study the impact of varying electrode materials yet. The advantage of BDD (and to a lesser extent GC) electrodes is their exceptionally high overpotential for proton reduction, which was ideal for clean voltammetry of 1^+. But we are very interested in the idea of using the electrode as an active participant in the overall reaction.

Mickaël Avanthay opened discussion of the paper by Victoria Flexer: When the electrolysis has depleted the middle layer of ions, do you observe any issues with the lowered connectivity?

Victoria Flexer replied: Yes, absolutely. This was indeed observed in our experiments. In Fig. 3 of our article (https://doi.org/10.1039/d3fd00051f) it is shown that the overall cell resistance sharply increases when roughly 92% of the total ionic content has been depleted from the middle compartment. This is attributed to the decrease in conductivity in the middle compartment, following eqn (1) (Section 2.4 of our manuscript).

Emily Carroll asked: You mentioned this work is done in batch experiments, do you envision this system eventually moving to a flow reactor system? What are the limitations and considerations for this?

Victoria Flexer answered: All our experiments carried out so far, including those shown in this paper (https://doi.org/10.1039/d3fd00051f), previous ones, and even our most recent ones were all carried out in batch mode (more precisely, in batch recirculation mode, to increase the volume of solutions we could work with). Indeed, it makes perfect sense to think that we should move towards a flow reactor system if we would like to continue our road towards a technologically viable solution.

The amount of ions that needs to be transferred to cathodic and anodic compartments is huge (above 1 mol cations and 1 mol anions per litre, or higher). With this consideration, possible directions include increasing the applied current density or increasing the residence time. Regarding current density, we still have to try our system at higher values, this is certainly pending. However, there is a limit to the values that we can test, and these are given by the maximum current density that membranes can withstand (provided by the manufacturer). Even if we would change to membranes that can withstand higher current density than the ones we are currently using, it probably would not be enough.

Regarding the possibility of increasing the residence time in the reactor compartment, this is very tricky. This is because residence time is also linked to solution flow in the compartment, which is in turn most often linked to forced mass transferred. As far as I know, in most electrodialysis types of systems, forced mass transfer is provided by the flow of solutions into and out of the compartments. Thus, increasing the residence time but at the same time decreasing the forced convection would result in strong concentration gradients that could be at the very least detrimental in terms of increasing cell voltage.

Eventually, our best hypothesis now is that we should work at as high a current density as possible, and with several continuous flow reactors working in series. In this configuration, the outlet of the middle compartment in the first reactor is the feed of the middle compartment in the second reactor, and so on.

Shelley D. Minteer enquired: Do you have a method for separating toxic metals (*i.e.* arsenic, chromium, *etc.*) from the brines?

Victoria Flexer responded: Toxic metals are not a problem in lithium rich continental brines. Nor in the ones we work with in South America, nor in the ones in the USA or China. The only cations observed in these brines are Li^+, Na^+, K^+, Mg^{2+} and Ca^{2+}. We do acknowledge however that other lithium rich sources, such as selected geothermal brines or produced waters in the oil industry, could display large concentrations of such toxic species.

All arsenic and chromium toxic species are multivalent, the same applies to other toxic elements that could be found in those brines (Ba^{2+}, for example). Thus, my first try would be to attempt at their crystallization as hydroxides, generating hydroxyl groups electrochemically in very similar types of configurations as shown in Fig. 2 and 3 here. Indeed, this is how we deplete Mg^{2+} and Ca^{2+} in our multistage electrolytic approach. See details for Mg^{2+} and Ca^{2+} in ref. 1–3.

I have to acknowledge that while I am quite confident that it is possible to largely deplete those toxic species, I am not sure whether this can be done in such a way that the crystallized solids will not be a mixture of several species, either hydroxides of different cations, or even oxo/hydroxy compounds of the same cation.

1 C. H. Díaz Nieto, N. A. Palacios, K. Verbeeck, A. Prévoteau, K. Rabaey and V. Flexer, *Water Res.*, 2019, **154**, 117–124, DOI: 10.1016/j.watres.2019.01.050.
2 M. L. Vera, C. J. O. Palacios, C. H. Díaz Nieto, N. A. Palacios, N. Di Carlantonio, F. G. Luna, W. R. Torres and V. Flexer, *J. Solid State Electrochem.*, 2022, **26**, 1981–1994, DOI: 10.1007/s10008-022-05219-6.
3 C. H. Díaz Nieto, M. A. Mata, C. J. O. Palacios, N. A. Palacios, W. R. Torres, M. L. Vera and V. Flexer, *Electrochim. Acta*, 2023, **454**, 142401, DOI: 10.1016/j.electacta.2023.142401.

Shelley D. Minteer questioned: Considering the need to separate and use multiple cations from these brines, can you develop a single reactor system (rather than a series of systems) to separate the cations in one reactor system?

Victoria Flexer replied: Considering the system shown in my slide (and shown here as Fig. 4), the reactors labelled as Stages II and III are physically identical. Indeed, the ones we have in our lab are used alternatively for Stage II or Stage III experiments.

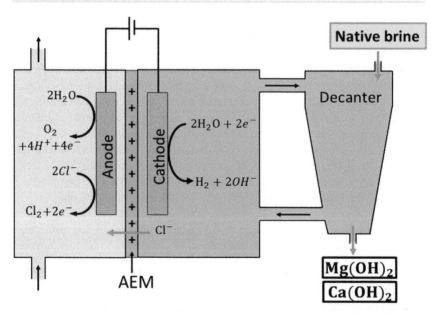

Fig. 2 Schematic representation of configuration to remove Mg^{2+} and Ca^{2+} as crystallised $Mg(OH)_2$ and $Ca(OH)_2$. A similar scheme could be used to attempt the removal of other multivalent species, such as arsenic, chromium, and others. Figure taken from ref. 1 (DOI: 10.1016/j.watres.2019.01.050) cited below.

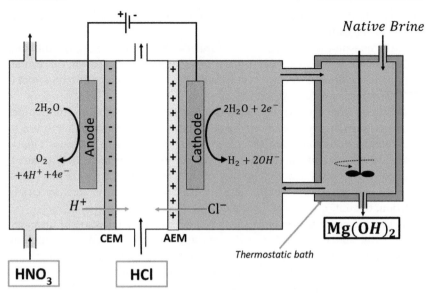

Fig. 3 Schematic representation of an alternative configuration to remove Mg^{2+} as crystallised $Mg(OH)_2$. Such a configuration could also be applied for the removal of other multivalent species, similarly to proposed in Fig. 2 above. Figure taken from ref. 2 (DOI: 10.1007/s10008-022-05219-6) cited below.

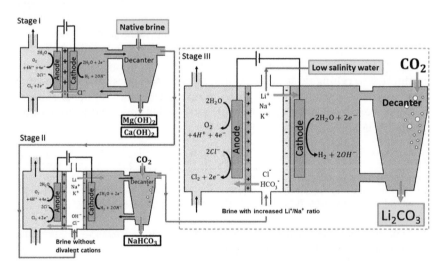

Fig. 4 Schematic representation of full multi-stage electrolytic setup for the joint recovery of several solids and gases together with Li_2CO_3.

The system shown in Fig. 4 and labelled as Stage I is a two compartment reactor, *i.e.* similar but different from the three compartment reactors used for Stages II and III. However, we can deplete Mg^{2+} and Ca^{2+} cations also in a three compartment setup as shown to the left of Fig. 1 in our paper (https://doi.org/10.1039/d3fd00051f). However, this electrolyzer is still not identical to the one used for Stages II and III, since there is an anion exchange membrane (AEM) in contact with the cathodic compartment (while it is a cation exchange membrane (CEM) in Stages II and III).

If the question is more directed towards the idea of recovering the different solids (see left and center of Fig. 1 in our paper) in one single reactor, this cannot in principle be done since Stage I is carried out in a different configuration from II and III, as just explained above.

Switching from $NaHCO_3$ to Li_2CO_3 is however quite straightforward. In order to trigger $NaHCO_3$ crystallization, the pH needs to be kept around 7.5 to 8.0, so that bicarbonate cations are the predominant C species in solution. This is carried out by a careful equilibrium between current density and CO_2 flow rate. It is important to recall here that neither $LiHCO_3$ nor $KHCO_3$ crystallize as solids, they can only exist in solution. Thus, while bicarbonate is the major C species in solution, the only solid formed should be $NaHCO_3$.

In order to promote later on Li_2CO_3 precipitation, the pH needs to be increased and kept at about 11. This is carried out again by simultaneously controlling current density and CO_2 flow rate. When the predominant C species in solution is carbonate, Li_2CO_3 should be the major species to crystallize, since it is about 50 times less soluble (molar solubilities compared) than Na_2CO_3 and K_2CO_3.

We should however recall that electrodes, and particularly cation and anion exchange membranes and solids do not really work well together. We do acknowledge that in the lab, since we were carrying out proof-of-concept type experiments, we allowed for solid formation within the cathodic compartments and in the vicinity of both membrane and electrode. However, this is something that absolutely needs to be corrected in the case of potential large scale implementation of the technology. To the best of my knowledge, there is no electro-membrane process in use today in industry where solids are allowed to be formed in the vicinity of the membrane. Thus, our upgrading of the system should firstly be directed to getting rid of solids within the electrochemical reactors, and only then, we could potentially start thinking about reducing the number of reactors.

Some ideas for avoiding solid formation in the first stage of the process were previously published in ref. 1 and 2.

1 M. L. Vera, C. J. O. Palacios, C. H. Díaz Nieto, N. A. Palacios, N. Di Carlantonio, F. G. Luna, W. R. Torres and V. Flexer, *J. Solid State Electrochem.*, 2022, **26**, 1981–1994, DOI: 10.1007/s10008-022-05219-6.
2 C. H. Díaz Nieto, M. A. Mata, C. J. O. Palacios, N. A. Palacios, W. R. Torres, M. L. Vera and V. Flexer, *Electrochim. Acta*, 2023, **454**, 142401, DOI: 10.1016/j.electacta.2023.142401.

Shahid Rasul asked: What is the selectivity towards hydrogen in the presence of abundant oxygen during electrochemical reactions?

Victoria Flexer responded: We have not assessed quantitatively the H_2 production. However, we cannot imagine which other reduction reaction could be happening. We have worked only with a mixture of chlorides (NaCl + KCl + LiCl), none of which can undergo reduction in aqueous media. Our cathode is composed of stainless steel 316, and we have not observed corrosion in the absence of an applied electric field when we have left our cathode in contact with brine. Thus, we do not expect a partly corroded cathode to be reduced. There are no organics in our solutions either. While we have worked with artificial brines only, native continental brines such as the ones found in South America, the

Nevada Desert in the USA, or those found in China do not contain organics either. The story would be very different if we were to work with geothermal brines, or particularly produced waters from the oil industry that could certainly contain organics that could react at the electrode surface.

Eniola Sokalu enquired: In the Li_2CO_3 crystallisation stage do you find there to be issues with solubility of CO_2? If so could changes such as increasing or decreasing the solution temperature have an effect?

Victoria Flexer replied: We have not yet conducted experiments with varying temperatures in the cathodic compartment or side crystallization vessel. Increasing the temperature should on one side result in a decrease in solubility for CO_2. However, Li_2CO_3 displays inverse solubility with temperature, *i.e.* contrary to most solids, its solubility decreases with temperature. Therefore, I believe that if we think from the perspective of technology implementation it might be best to increase temperature to maximize the amount of Li^+ that is recovered as solid Li_2CO_3. Even though the solubility of CO_2 will decrease, carbonate cations are constantly removed from the aqueous media, and thus more CO_2 will be solubilized.

David P. Hickey asked: Some of the primary challenges associated with electrodialysis for the removal of inorganic salts are high energy consumption, removal of dilute salt solutions, and electrode fouling. Recent work by Chen *et al.*[1] and Beh *et al.*[2] have utilized redox shuttles to continuously refresh the electrode–solution interface and, thereby, address many of these challenges. Is it possible to use a similar approach in your system for the removal of lithium salts?

1 F. Chen, J. Wang, C. Feng, J. Ma and T. D. Waite, *Chem. Eng. J.*, 2020, **401**, 126111, DOI: 10.1016/j.cej.2020.126111.
2 E. S. Beh, M. A. Benedict, D. Desai and J. B. Rivest, *ACS Sustainable Chem. Eng.*, 2019, 7, 13411–13417, DOI: 10.1021/acssuschemeng.9b02720.

Victoria Flexer answered: This is indeed a very clever idea. While I find it very interesting to carry out experiments at lab scale, I have serious doubts of whether such a strategy could potentially be applicable at industrial mining scale. The reason is that Li^+ cations are very diluted in brines, and thus the volumes of brines that need to be processed daily are enormous, in the order of 20 million litres of brine daily in a medium size facility (to see how this estimation was calculated, see ref. 1). Using redox mediators at such large scale seems at the very least challenging. One of the keys of brine processing is searching for ideas that can be implemented at very large scale.

1 M. L. Vera, W. R. Torres, C. I. Galli, A. Chagnes and V. Flexer, *Nat. Rev. Earth Environ.*, 2023, **4**, 149–165, DOI: 10.1038/s43017-022-00387-5.

Conflicts of interest

Bee Hockin is a member of staff at the Royal Society of Chemistry and works in the Editorial Office of Faraday Discussions, and there are no other conflicts to declare.

Faraday Discussions

PAPER

Mechanistic studies of Ni-catalyzed electrochemical homo-coupling reactions of aryl halides†

Jian Luo, [a] Michael T. Davenport, [b] Arianna Carter,[b] Daniel H. Ess *[b] and T. Leo Liu *[a]

Received 17th March 2023, Accepted 23rd May 2023

DOI: 10.1039/d3fd00069a

Ni-catalyzed electrochemical arylation is an attractive, emerging approach for molecular construction as it uses air-stable Ni catalysts and efficiently proceeds at room temperature. However, the homo-coupling of aryl halide substrates is one of the major side reactions. Herein, extensive experimental and computational studies were conducted to examine the mechanism of Ni-catalyzed electrochemical homo-coupling of aryl halides. The results indicate that an unstable $Ni^{II}(Ar)Br$ intermediate formed through oxidative addition of the cathodically generated Ni^{I} species with aryl bromide and a consecutive chemical reduction step. For electron-rich aryl halides, homo-coupling reaction efficiency is limited by the oxidative addition step, which can be improved by negatively shifting the redox potential of the Ni-catalyst. DFT computational studies suggest a $Ni^{III}(Ar)Br_2/Ni^{II}(Ar)Br$ ligand exchange pathway for the formation of a high-valent $Ni^{III}(Ar)_2Br$ intermediate for reductive elimination and production of the biaryl product. This work reveals the reaction mechanism of Ni-catalyzed electrochemical homo-coupling of aryl halides, which may provide valuable information for developing cross-coupling reactions with high selectivity.

Aryl groups widely exist in various chemicals and represent the basic backbone or essential functional groups for many molecules or materials. Therefore, the arylation reaction is one of the core strategies for molecular construction in contemporary synthetic chemistry.[1–4] Ni-catalyzed electrochemical arylation between aryl halides and nucleophiles has attracted great attention due to its merits of using non-precious transition-metal catalysts, mild reaction conditions, controllable reaction process, and using electrical energy to drive chemical transformations, thereby avoiding the use of stoichiometric chemical oxidants or

[a]Department of Chemistry and Biochemistry, Utah State University, Logan, Utah 84322, USA. E-mail: leo.liu@usu.edu

[b]Department of Chemistry and Biochemistry, Brigham Young University, Provo, Utah 84604, USA. E-mail: dhe@chem.byu.edu

† Electronic supplementary information (ESI) available: Additional experimental details and figures and tables. See DOI: https://doi.org/10.1039/d3fd00069a

reductants.[5-8] Literature has witnessed the rapid development of Ni-catalyzed electrochemical arylation reactions, including arylation of alkyl halides,[9-13] carboxylates,[14-16] amines,[8,17,18] alcohols,[19] thiols,[20,21] olefins,[22-24] carbon dioxide,[25,26] *etc.*

However, low yields and side reactions were observed in the Ni-catalyzed electrochemical arylation reactions *via* cross-coupling.[5,12,13,27] One of the major side reactions is the homo-coupling of aryl halides (Scheme 1A).[7,9,27] Under reducing conditions, biaryl products are produced *via* symmetrical coupling of aryl halides, which makes the cross-coupling reaction challenging for nucleophiles with low reactivity.[27] Although the Ni-catalyzed electrochemical homo-coupling of aryl halides has been applied in the synthesis of polypyridine ligands and axially chiral BINOL derivatives,[28-31] the flat symmetric biaryl structure of products limited the broad application of this reaction. To achieve high selectivity for cross-coupling products, suppression of aryl–aryl homo-coupling side reactions is expected, which requires in-depth mechanistic understanding of this competitive reaction. There are several possible pathways for the formation of biaryl products under electrolytic conditions. As shown in Scheme 1B, the oxidative addition of cathodic reduction generated Ni^I species to aryl halide yields a $Ni^{III}(Ar)X_2$ species.[7,17] The $Ni^{III}(Ar)X_2$ species can then decompose to produce the biaryl products *via* a bimolecular pathway. Alternatively, the $Ni^{III}(Ar)X_2$ species

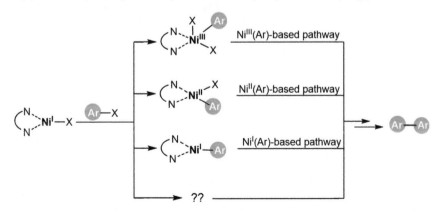

(A) Ni-catalyzed cross-coupling reactions

Desired Cross-Coupling

Undesired Homo-Coupling

(B) Possible pathways for the production of aryl-aryl homo-coupling product

$Ni^{III}(Ar)$-based pathway

$Ni^{II}(Ar)$-based pathway

$Ni^I(Ar)$-based pathway

??

Scheme 1 Ni-catalyzed electrochemical cross-coupling reactions and the undesired homo-coupling side reactions.

can be further reduced (chemically or electrochemically) to either a $Ni^{II}(Ar)X$ or a $Ni^{I}(Ar)$ species and then be converted to the biaryl products.

Herein, we conducted experimental and computational studies to elucidate the mechanism of Ni-catalyzed electrochemical homo-coupling of aryl halides. Our results indicate that an unstable $Ni^{II}(Ar)X$ intermediate formed through the oxidative addition of aryl bromide to the cathodically generated Ni^{I} species and a consecutive chemical reduction step. For electron-rich aryl halides, homo-coupling reaction efficiency is limited by the oxidative addition step, which can be improved by using electron-rich ligands for a Ni-catalyst. Density functional theory (DFT) computational studies suggest a $Ni^{III}(Ar)X_2/Ni^{II}(Ar)X$ ligand exchange pathway for the formation of a high-valent $Ni^{III}(Ar)_2$ intermediate for reductive elimination and production of the biaryl product.

As shown in Fig. 1A, $NiBr_2 \cdot DME$/dtbbpy $(1:1.5)$ catalyst showed a quasi-reversible redox signal with $E_{1/2} = -1.78$ V (vs. ferrocene/ferrocenium ($Fc^{+/0}$)) (dashed curve) standing for the $Ni^{II/I}$ redox couple. After adding 10 eq. aryl bromide, methyl 4-bromobenzoate, the reductive peak current intensity was obviously increased, meanwhile, the return peak disappeared. It indicates an irreversible oxidative addition reaction between cathodically generated (dtbbpy) $Ni^{I}Br$ species and the aryl halide. It is consistent with our previous results.[7] To further confirm the oxidative addition of the aryl halide to the Ni^{I} species, we synthesized the (dtbbpy)$Ni^{I}Br$ species through controlled potential electrolysis (CPE) of (dtbbpy)$Ni^{II}Br_2$ (see ESI† for details). As shown in Fig. 1B, the UV-vis absorption spectrum of the (dtbbpy)$Ni^{I}Br$ species displayed three absorption peaks at 395 nm, 421 nm, and 568 nm (blue curve), which is consistent with the literature.[32] We also tried to prepare the Ni^{I} species chemically by mixing (dtbbpy) Ni^{0} and (dtbbpy)$Ni^{II}Br_2$ $(1:1)$. However, it is interesting that weak absorption peaks were observed at 433 nm, 577 nm, and 621 nm (orange curve) in the UV-vis absorption spectrum of the (dtbbpy)Ni^{0}/(dtbbpy)$Ni^{II}Br_2$ mixture. It indicates that almost no (dtbbpy)$Ni^{I}Br$ species was generated in the comproportionation of (dtbbpy)Ni^{0} and (dtbbpy)$Ni^{II}Br_2$. The main product of this reaction is likely to be a Ni^{I} dimer, $[(dtbbpy)Ni^{I}Br]_2$, which shows no UV-vis absorption peak in the

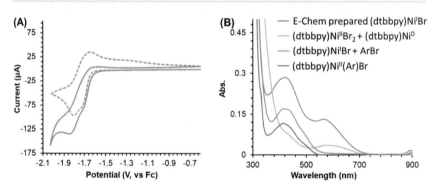

Fig. 1 Electrochemical and UV-vis studies of Ni-catalyzed aryl halide homo-coupling. (A) Cyclic voltammograms (CV) of the Ni-catalyst with (solid line) and without (dash line) methyl 4-bromobenzoate; (B) UV-vis spectra of electrochemically prepared (blue) and chemically prepared (orange) (dtbbpy)$Ni^{I}Br$ species, a mixture of (dtbbpy)$Ni^{I}Br$ and 1 eq. ArBr (green), and chemically synthesized (dtbbpy)$Ni^{II}(Ar)Br$ (purple).

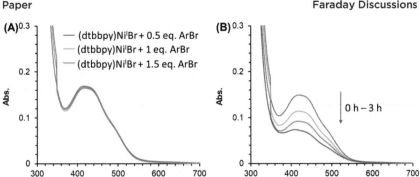

Fig. 2 UV-vis studies of the (dtbbpy)Ni^{II}(Ar)Br intermediate. (A) UV-vis absorption of (dtbbpy)Ni^{I}Br reacted with 0.5 eq., 1 eq., and 1.5 eq. ArBr; (B) stability tests of the (dtbbpy) Ni^{II}(Ar)Br intermediate ((dtbbpy)Ni^{I}Br/ArBr 1 : 1).

visible light range.[32] The Ni^{I} dimer species was reported to be unreactive towards aryl halide oxidative addition.[32] We further reacted aryl bromide with the electrochemically synthesized (dtbbpy)Ni^{I}Br species. A (dtbbpy)Ni^{II}(Ar)Br intermediate quickly formed through oxidative addition and subsequent chemical reduction, as the UV-vis spectrum of the mixture (green curve) is identical to that of independently synthesized (dtbbpy)Ni^{II}(Ar)Br (purple curve). The (dtbbpy) Ni^{II}(Ar)Br intermediate showed a main UV-vis absorption peak at 416 nm and two shoulder peaks at 435 and 495 nm.

To monitor the reaction between Ni^{I} species and aryl bromide, we titrated the (dtbbpy)Ni^{I}Br species with aryl bromide. As shown in Fig. 2A, when the (dtbbpy) Ni^{I}Br species was mixed with 0.5–1.5 equivalent methyl 4-bromobenzoate, UV-vis absorption of the mixtures was nearly identical. It indicates that, as shown in Scheme 2, when (dtbbpy)Ni^{III}(Ar)Br$_2$ is formed *via* oxidative addition between (dtbbpy)Ni^{I}Br species and aryl bromide, it is rapidly converted to (dtbbpy)Ni^{II}(Ar) Br through a comproportionation reaction with (dtbbpy)Ni^{I}Br species. The overall reaction is 2(dtbbpy)Ni^{I}Br + ArBr → (dtbbpy)Ni^{II}(Ar)Br + (dtbbpy)Ni^{II}Br$_2$. Therefore, despite the addition of more than 0.5 equivalent aryl bromide, the concentration of (dtbbpy)Ni^{II}(Ar)Br intermediate in the mixture didn't increase. The (dtbbpy)Ni^{II}(Ar)Br intermediate was found to be unstable in DMAc at room temperature. As shown in Fig. 2B, a significant decrease in UV-vis absorption of the (dtbbpy)Ni^{II}(Ar)Br intermediate was observed, indicating its gradual decomposition over several hours. After 12 h, homo-coupling biaryl product, dimethyl [1,1′-biphenyl]-4,4′-dicarboxylate, was detected in the (dtbbpy)Ni^{I}Br/ArBr (1 : 1) mixture with 62% yield.

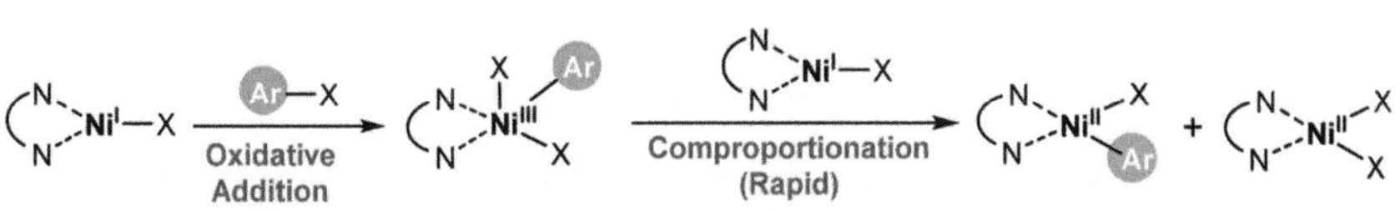

Scheme 2 Oxidative addition and subsequent chemical reduction between Ni^{I} species and aryl bromide.

The above experimental results demonstrate a Ni^I-based oxidative addition pathway for Ar–Br bond activation, which is distinct from the previously proposed Ni^0-based oxidative addition mechanism.[9,28] We further questioned how to accelerate the oxidative addition step of aryl halides to Ni^I species, especially for less reactive electron-rich aryl halides such as 4-bromo-N,N-dimethylaniline. In the oxidative addition reaction, Ni^I species and aryl halides act as a reductant and an oxidant, respectively. Thus, we hypothesized that negatively shifting the reduction potential of the Ni^I species could improve its reactivity towards electron-rich aryl halide. As shown in Fig. 3, when 2,2'-bpy was used as ligand for the Ni-catalyst, a fully reversible redox signal of $Ni(2,2'\text{-bpy})_3^{2+/+}$ couple was observed at $E_{1/2} = -1.60$ V (vs. Fc) (orange dash). After adding 10 eq. 4-bromo-N,N-dimethylaniline, the reduction peak at $E = -1.71$ V was slightly increased, and the oxidation peak at -1.51 V was slightly decreased (orange solid), indicating a very slow oxidative addition reaction. When we replaced the 2,2'-bpy ligand with an electron-rich 4,4'-dimethoxy-2,2'-bipyridyl (dmobpy) ligand, the redox potential of the $Ni(dmobpy)_3^{2+/+}$ redox couple negatively shifted to $E_{1/2} = -1.93$ V (vs. Fc) (green dash). Meanwhile, a stronger current response for the oxidative addition reaction was detected (green solid).

Constant current electrolysis of 4-bromo-N,N-dimethylaniline with a Ni-catalyst was performed by using glassy carbon (RVC) as a cathode and Zn metal as a sacrificial anode. As shown in Scheme 3, in the presence of $Ni(2,2'\text{-bpy})_3Br_2$ catalyst, the homo-coupling product, $N^4,N^4,N^{4'},N^{4'}$-tetramethyl-[1,1'-biphenyl]-4,4'-diamine, was obtained in 54% yield after electrolysis at 4 mA current for 12 hours. 33% of 4-bromo-N,N-dimethylaniline remained in the reaction mixture. When $Ni(dmobpy)_3Br_2$ catalyst was used for the same reaction, 4-bromo-N,N-dimethylaniline was completely converted within 12 hours. The homo-coupling product was obtained with 81% yield. These results suggest that the reaction

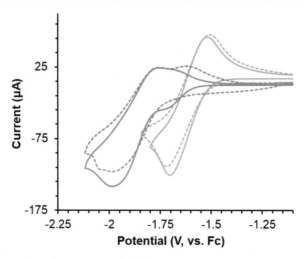

Fig. 3 Cyclic voltammetry (CV) studies of a Ni-catalyzed electrochemical homo-coupling reaction of electron-rich 4-bromo-N,N-dimethylaniline. $Ni(2,2'\text{-bpy})_3Br_2$ (orange dash), $Ni(2,2'\text{-bpy})_3Br_2$ + ArBr (orange solid), $Ni(dmobpy)_3Br_2$ (green dash), $Ni(dmobpy)_3Br_2$ + ArBr (green solid).

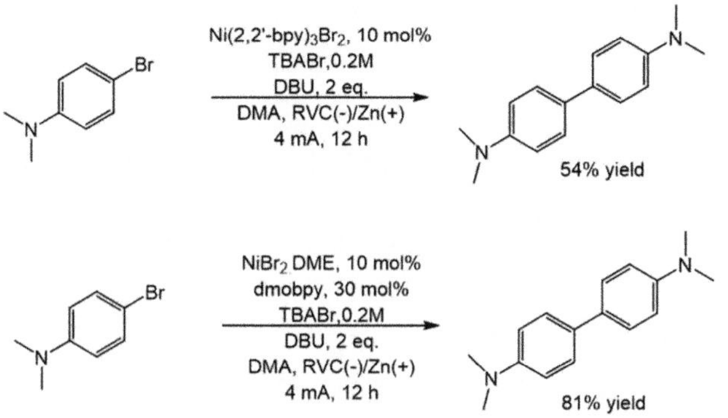

Scheme 3 Ni-catalyzed electrochemical homo-coupling reactions of electron-rich aryl bromide.

rate of the oxidative addition step and reaction efficiency of homo-coupling of electron-rich aryl halides can be enhanced by negatively shifting the redox potential of the Ni-catalyst or generating an electron-rich Ni^I intermediate.

To reveal the details of the reaction mechanism, we used M06-L/6-31G**(LANL2DZ for Ni) DFT calculations to evaluate the energetic impact of the possible reaction pathways.[33–36] Fig. 4A shows the potential energy profile (enthalpies and Gibbs energies and spin states) for reduction of $(dtbbpy)Ni^{II}Br_2$ to generate an open coordination site for oxidative addition with the aryl bromide. The energy barrier from the $(dtbbpy)Ni^IBr$ intermediate (**C**) through **TS 1** to give the $(dtbbpy)Ni^{III}(Ar)Br_2$ intermediate (**D**) is only 6.2 kcal mol^{-1} ($\Delta H^{\ddagger}$) and 7.8 kcal mol^{-1} ($\Delta G^{\ddagger}$). The 3D depictions of this oxidative addition process are displayed in Fig. 5. From **D**, there are several different avenues to generate the homo-coupling biaryl product. We first examined ligand exchange between two **D** (Fig. 4B). It requires a Gibbs energy change of 14.8 kcal mol^{-1} to form (dtbbpy) $Ni^{III}(Ar)_2Br$ (**F**) and $(dtbbpy)Ni^{III}Br_3$ (**G**). After this ligand exchange step, the reductive elimination requires only 3.9 kcal mol^{-1} energy barrier through **TS 2**. While this is a reasonable energy barrier for the room temperature reaction, our experimental studies indicate that **D** is rapidly converted to the $Ni^{II}(Ar)$ intermediate (**E**) (Scheme 2). Therefore, this $Ni^{III}(Ar)$-based homo-coupling pathway is less possible. Our calculations are also consistent with the possibility that **D** is converted to **E**. Fig. 4A shows that reduction and bromide loss of **D** resulting in **E** is exothermic and exergonic by 12.1 kcal mol^{-1}.

Because of the fast $Ni^{III}(Ar)Br_2$ (**D**) to $Ni^{II}(Ar)Br$ (**E**) conversion, we then examined several biaryl forming pathways from **E**. Fig. 4C shows a possible disproportionation route between two **E** to form **F** ready for reductive elimination to produce the biaryl product and a low-valent Ni^I species **H**. The energy barrier for this disproportionation is 18.0 kcal mol^{-1}, which makes this pathway less favorable. Fig. 4D shows a $Ni^I(Ar)$-based homo-coupling pathway. Under electrolytic conditions, **E** can be electrochemically reduced to $Ni^I(Ar)$ species (**I**) at the cathode. The high-valent **F** is formed *via* oxidative addition between **I** and the aryl halide. The energy barrier for the oxidative addition reaction is 8.3 kcal mol^{-1}. Finally, we identified a lower energy pathway that is shown in Fig. 4E. In this pathway, ligand exchange is much more thermodynamically feasible between **E**

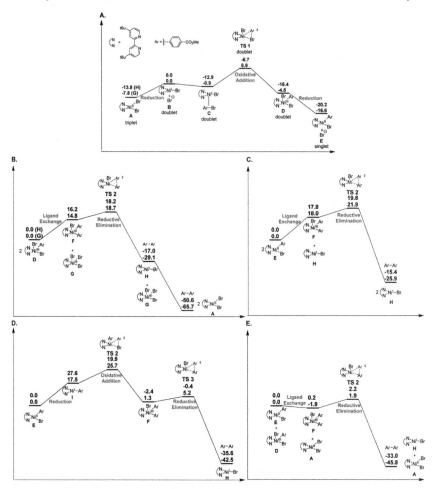

Fig. 4 M06-L/6-31G**[LANL2DZ] energy landscapes for the Ni-catalyzed electro-chemical aryl bromide homo-coupling.

and the transiently generated **D**. The generation of **A** and **F** from **D** and **E** is very close to thermally neutral. Therefore, the subsequent reductive elimination through **TS 2** has an overall barrier of only 1.9 kcal mol^{-1}. This biaryl forming pathway is reasonable considering that under the electrolytic conditions the Ni^{I} species (**B**) is generated slowly and its concentration is much lower than that in the pre-synthesized sample we studied in Fig. 2, which makes the compro-portionation reaction between **D** and **H** to produce **E** slow. Therefore, coexistence of (dtbbpy)Ni^{III}(Ar)Br$_2$ (**D**) and (dtbbpy)Ni^{II}(Ar)Br (**E**) is possible in the reaction mixture, enabling the reaction pathway shown in Fig. 4E.

In summary, we performed experimental and computational studies for mechanistic understanding of the Ni-catalyzed electrochemical homo-coupling reaction of aryl halides. The results indicate that an unstable (dtbbpy)Ni^{II}(Ar)Br intermediate is formed through oxidative addition of aryl bromide to the cathodically generated Ni^{I} species and sequent chemical reduction. For electron-

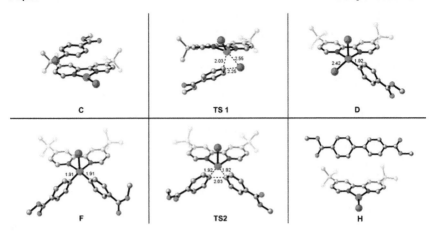

Fig. 5 3D depiction of M06-L structures of the intermediates and transition states for Ni-catalyzed electrochemical aryl bromide homo-coupling.

rich aryl halides, the oxidative addition reaction rate and homo-coupling reaction efficiency can be enhanced by negatively shifting the redox potential of the Ni-catalyst. DFT computational studies suggest a $Ni^{III}(Ar)Br_2/Ni^{II}(Ar)Br$ bimolecular ligand-exchange pathway for the formation of a high-valent $Ni^{III}(Ar)_2$ intermediate for reductive elimination and production of the biaryl product. This work reveals the reaction mechanism of Ni-catalyzed electrochemical homo-coupling of aryl halides, which may be of great significance for developing cross-coupling reactions with high selectivity.

Conflicts of interest

There are no conflicts to declare.

Acknowledgements

We thank the National Institutes of Health (grant no. R15GM143721) and National Science Foundation (grant no. 1847674) for supporting this study. We acknowledge that the NMR studies are supported by NSF's MRI program (award number 1429195). Arianna Carter was supported through the National Science Foundation Chemistry and Biochemistry REU Site to Prepare Students for Graduate School and an Industrial Career (CHE-2050872).

References

1 K. Okamoto, J. Zhang, J. B. Housekeeper, S. R. Marder and C. K. Luscombe, C–H Arylation Reaction: Atom Efficient and Greener Syntheses of π-Conjugated Small Molecules and Macromolecules for Organic Electronic Materials, *Macromolecules*, 2013, **46**, 8059–8078.
2 C. Liu, H. Zhang, W. Shi and A. Lei, Bond Formations between Two Nucleophiles: Transition Metal Catalyzed Oxidative Cross-Coupling Reactions, *Chem. Rev.*, 2011, **111**, 1780–1824.

3 D. Alberico, M. E. Scott and M. Lautens, Aryl–Aryl Bond Formation by Transition-Metal-Catalyzed Direct Arylation, *Chem. Rev.*, 2007, **107**, 174–238.

4 J.-R. Pouliot, F. Grenier, J. T. Blaskovits, S. Beaupré and M. Leclerc, Direct (Hetero)arylation Polymerization: Simplicity for Conjugated Polymer Synthesis, *Chem. Rev.*, 2016, **116**, 14225–14274.

5 S. R. Waldvogel, S. Lips, M. Selt, B. Riehl and C. J. Kampf, Electrochemical Arylation Reaction, *Chem. Rev.*, 2018, **118**, 6706–6765.

6 M. Yan, Y. Kawamata and P. S. Baran, Synthetic Organic Electrochemical Methods Since 2000: On the Verge of a Renaissance, *Chem. Rev.*, 2017, **117**, 13230–13319.

7 J. Luo, B. Hu, W. Wu, M. Hu and T. L. Liu, Nickel-Catalyzed Electrochemical $C(sp^3)$–$C(sp^2)$ Cross-Coupling Reactions of Benzyl Trifluoroborate and Organic Halides, *Angew. Chem., Int. Ed.*, 2021, **60**, 6107–6116.

8 C. Li, Y. Kawamata, H. Nakamura, J. C. Vantourout, Z. Liu, Q. Hou, D. Bao, J. T. Starr, J. Chen, M. Yan and P. S. Baran, Electrochemically Enabled, Nickel-Catalyzed Amination, *Angew. Chem., Int. Ed.*, 2017, **56**, 13088–13093.

9 R. J. Perkins, D. J. Pedro and E. C. Hansen, Electrochemical Nickel Catalysis for Sp^2–Sp^3 Cross-Electrophile Coupling Reactions of Unactivated Alkyl Halides, *Org. Lett.*, 2017, **19**, 3755–3758.

10 K.-J. Jiao, D. Liu, H.-X. Ma, H. Qiu, P. Fang and T.-S. Mei, Nickel-Catalyzed Electrochemical Reductive Relay Cross-Coupling of Alkyl Halides to Aryl Halides, *Angew. Chem., Int. Ed.*, 2020, **59**, 6520–6524.

11 L. Peng, Y. Li, Y. Li, W. Wang, H. Pang and G. Yin, Ligand-Controlled Nickel-Catalyzed Reductive Relay Cross-Coupling of Alkyl Bromides and Aryl Bromides, *ACS Catal.*, 2018, **8**, 310–313.

12 T. B. Hamby, M. J. LaLama and C. S. Sevov, Controlling Ni redox states by dynamic ligand exchange for electroreduction Csp3–Csp2 coupling, *Science*, 2022, **376**, 410–416.

13 B. L. Truesdell, T. B. Hamby and C. S. Sevov, General $C(sp^2)$–$C(sp^3)$ Cross-Electrophile Coupling Reactions Enabled by Overcharge Protection of Homogeneous Electrocatalysts, *J. Am. Chem. Soc.*, 2020, **142**, 5884–5893.

14 Y. Mo, Z. Lu, G. Rughoobur, P. Patil, N. Gershenfeld, A. I. Akinwande, S. L. Buchwald and K. F. Jensen, Microfluidic electrochemistry for single-electron transfer redox-neutral reactions, *Science*, 2020, **368**, 1352–1357.

15 H. Li, C. P. Breen, H. Seo, T. F. Jamison, Y.-Q. Fang and M. M. Bio, Ni-Catalyzed Electrochemical Decarboxylative C–C Couplings in Batch and Continuous Flow, *Org. Lett.*, 2018, **20**, 1338–1341.

16 T. Koyanagi, A. Herath, A. Chong, M. Ratnikov, A. Valiere, J. Chang, V. Molteni and J. Loren, One-Pot Electrochemical Nickel-Catalyzed Decarboxylative Sp^2–Sp^3 Cross-Coupling, *Org. Lett.*, 2019, **21**, 816–820.

17 Y. Kawamata, J. C. Vantourout, D. P. Hickey, P. Bai, L. Chen, Q. Hou, W. Qiao, K. Barman, M. A. Edwards, A. F. Garrido-Castro, J. N. deGruyter, H. Nakamura, K. Knouse, C. Qin, K. J. Clay, D. Bao, C. Li, J. T. Starr, C. Garcia-Irizarry, N. Sach, H. S. White, M. Neurock, S. D. Minteer and P. S. Baran, Electrochemically Driven, Ni-Catalyzed Aryl Amination: Scope, Mechanism, and Applications, *J. Am. Chem. Soc.*, 2019, **141**, 6392–6402.

18 C. Zhu, A. P. Kale, H. Yue and M. Rueping, Redox-Neutral Cross-Coupling Amination with Weak *N*-Nucleophiles: Arylation of Anilines, Sulfonamides,

Sulfoximines, Carbamates, and Imines via Nickelaelectrocatalysis, *JACS Au*, 2021, **1**, 1057–1065.

19 H.-J. Zhang, L. Chen, M. S. Oderinde, J. T. Edwards, Y. Kawamata and P. S. Baran, Chemoselective, Scalable Nickel-Electrocatalytic O-Arylation of Alcohols, *Angew. Chem., Int. Ed.*, 2021, **60**, 20700–20705.

20 D. Liu, H.-X. Ma, P. Fang and T.-S. Mei, Nickel-Catalyzed Thiolation of Aryl Halides and Heteroaryl Halides through Electrochemistry, *Angew. Chem., Int. Ed.*, 2019, **58**, 5033–5037.

21 Y. Wang, L. Deng, X. Wang, Z. Wu, Y. Wang and Y. Pan, Electrochemically Promoted Nickel-Catalyzed Carbon–Sulfur Bond Formation, *ACS Catal.*, 2019, **9**, 1630–1634.

22 S. Condon, D. Dupré, G. Falgayrac and J.-Y. Nédélec, Nickel-Catalyzed Electrochemical Arylation of Activated Olefins, *Eur. J. Org. Chem.*, 2002, **2002**, 105–111.

23 S. Condon-Gueugnot, E. Leonel, J.-Y. Nedelec and J. Perichon, Electrochemical Arylation of Activated Olefins using a Nickel Salt as Catalyst, *J. Org. Chem.*, 1995, **60**, 7684–7686.

24 B. R. Walker and C. S. Sevov, An Electrochemically-Promoted, Nickel-Catalyzed Mizoroki-Heck Reaction, *ACS Catal.*, 2019, **9**, 7197–7203.

25 G.-Q. Sun, W. Zhang, L.-L. Liao, L. Li, Z.-H. Nie, J.-G. Wu, Z. Zhang and D.-G. Yu, Nickel-catalyzed electrochemical carboxylation of unactivated aryl and alkyl halides with CO_2, *Nat. Commun.*, 2021, **12**, 7086.

26 G.-Q. Sun, P. Yu, W. Zhang, W. Zhang, Y. Wang, L.-L. Liao, Z. Zhang, L. Li, Z. Lu, D.-G. Yu and S. Lin, Electrochemical reactor dictates site selectivity in N-heteroarene carboxylations, *Nature*, 2023, **615**, 67–72.

27 G. L. Beutner, E. M. Simmons, S. Ayers, C. Y. Bemis, M. J. Goldfogel, C. L. Joe, J. Marshall and S. R. Wisniewski, A Process Chemistry Benchmark for sp^2–sp^3 Cross Couplings, *J. Org. Chem.*, 2021, **86**, 10380–10396.

28 H. Qiu, B. Shuai, Y.-Z. Wang, D. Liu, Y.-G. Chen, P.-S. Gao, H.-X. Ma, S. Chen and T.-S. Mei, Enantioselective Ni-Catalyzed Electrochemical Synthesis of Biaryl Atropisomers, *J. Am. Chem. Soc.*, 2020, **142**, 9872–9878.

29 K. W. R. de França, M. Navarro, É. Léonel, M. Durandetti and J.-Y. Nédélec, Electrochemical Homocoupling of 2-Bromomethylpyridines Catalyzed by Nickel Complexes, *J. Org. Chem.*, 2002, **67**, 1838–1842.

30 W.-W. Chen, Q. Zhao, M.-H. Xu and G.-Q. Lin, Nickel-Catalyzed Asymmetric Ullmann Coupling for the Synthesis of Axially Chiral Tetra-ortho-Substituted BiarylDials, *Org. Lett.*, 2010, **12**, 1072–1075.

31 J. L. Oliveira, M. J. Silva, T. Florêncio, K. Urgin, S. Sengmany, E. Léonel, J.-Y. Nédélec and M. Navarro, Electrochemical coupling of mono and dihalopyridines catalyzed by nickel complex in undivided cell, *Tetrahedron*, 2012, **68**, 2383–2390.

32 N. A. Till, S. Oh, D. W. C. MacMillan and M. J. Bird, The Application of Pulse Radiolysis to the Study of Ni(I) Intermediates in Ni-Catalyzed Cross-Coupling Reactions, *J. Am. Chem. Soc.*, 2021, **143**, 9332–9337.

33 Y. Zhao and D. G. Truhlar, A new local density functional for main-group thermochemistry, transition metal bonding, thermochemical kinetics, and noncovalent interactions, *J. Chem. Phys.*, 2006, **125**(19), 194101.

34 R. Ditchfield, W. J. Hehre and J. A. Pople, Self-Consistent Molecular-Orbital Methods. IX. An Extended Gaussian-Type Basis for Molecular-Orbital Studies of Organic Molecules, *J. Chem. Phys.*, 2003, **54**, 724–728.

35 F. Weigend and R. Ahlrichs, Balanced basis sets of split valence, triple zeta valence and quadruple zeta valence quality for H to Rn: Design and assessment of accuracy, *Phys. Chem. Chem. Phys.*, 2005, **7**, 3297–3305.

36 M. J. Frisch, G. W. Trucks, H. B. Schlegel, G. E. Scuseria, M. A. Robb, J. R. Cheeseman, G. Scalmani, V. Barone, G. A. Petersson, H. Nakatsuji, X. Li, M. Caricato, A. V. Marenich, J. Bloino, B. G. Janesko, R. Gomperts, B. Mennucci, H. P. Hratchian, J. V. Ortiz, A. F. Izmaylov, J. L. Sonnenberg, D. Williams-Young, F. Ding, F. Lipparini, F. Egidi, J. Goings, B. Peng, A. Petrone, T. Henderson, D. Ranasinghe, V. G. Zakrzewski, J. Gao, N. Rega, G. Zheng, W. Liang, M. Hada, M. Ehara, K. Toyota, R. Fukuda, J. Hasegawa, M. Ishida, T. Nakajima, Y. Honda, O. Kitao, H. Nakai, T. Vreven, K. Throssell, J. A. Montgomery Jr, J. E. Peralta, F. Ogliaro, M. J. Bearpark, J. J. Heyd, E. N. Brothers, K. N. Kudin, V. N. Staroverov, T. A. Keith, R. Kobayashi, J. Normand, K. Raghavachari, A. P. Rendell, J. C. Burant, S. S. Iyengar, J. Tomasi, M. Cossi, J. M. Millam, M. Klene, C. Adamo, R. Cammi, J. W. Ochterski, R. L. Martin, K. Morokuma, O. Farkas, J. B. Foresman and D. J. Fox, *Gaussian 16 Rev. C.01*, Wallingford, CT, 2016.

Faraday Discussions

ROYAL SOCIETY
OF **CHEMISTRY**

PAPER

Exploring electrolyte effects on metal–alkyl bond stability: impact and implications for electrosynthesis

Dylan G. Boucher, [iD] Zachary A. Nguyen and Shelley D. Minteer*

Received 27th February 2023, Accepted 20th March 2023
DOI: 10.1039/d3fd00054k

Transition metal catalysis hinges on the formation of metal–carbon bonds during catalytic cycles. The stability and reactivity of these bonds are what determine product chemo-, regio-, and enantioselectivity. The advent of electrosynthetic methodologies has placed the current understanding of these metal–alkyl bonds into a new environment of charged species and electrochemically induced reactivity. In this paper, we explore the often neglected impact of supporting electrolyte on homogeneous electrocatalytic mechanisms using the catalytic reduction of benzyl chlorides *via* Co and Fe tetraphenylporphyrins as a model reaction. The mechanism of this reaction is confirmed to proceed through the formation of the metal–alkyl intermediates. Critically, the stability of these intermediates, in both the Co and Fe systems, is found to be affected by the hydrodynamic radius of the supporting electrolyte, leading to differences in electrolyte–solvent shell. These studies provide important information for the design of electrosynthetic reactions, and provide a starting point for the rational design of functional supporting electrolytes.

1. Introduction

The formation and reactivity of metal–carbon bonds are foundational to the fields of catalysis and organometallic chemistry.[1] An important part of any catalytic cycle, the formed organometallic intermediates offer a wide array of reactivity and chemical transformations. As such, understanding the molecular and chemical design principles that determine the specific reactivity of these intermediates can unlock the path to new and useful chemical transformations. For example, the Suzuki cross-coupling reaction, which is a palladium-catalyzed method of forming C–C bonds and one of the most used organic methods, proceeds *via* alkyl-metal bond formation from aryl and alkyl halides and boronic acids.[2] Nature utilizes these species as well, with the reversible metal–organic bond homolysis of vitamin B12 [3–5] and radical SAM enzymes.[6] These species use metal–alkyl bond homolysis to provide access to highly unstable organic radicals by providing

Department of Chemistry, University of Utah, Salt Lake City, UT, USA. E-mail: minteer@chem.utah.edu

stability *via* the reversible formation of organometallic intermediates to increase the effective lifetime of the organic radical.[7] Understanding the myriad of ways the intermediate metal–carbon bonds can be stabilized or destabilized is thus crucial to promoting the desired reactivity.

Metal–carbon bonds have also proven central to the burgeoning field of electrosynthesis.[8] As many organometallic catalytic cycles involve changes in the oxidation state of the metal or utilize chemical reductants or oxidants to turn over catalytic cycles,[9] electrochemical methods are uniquely suited to providing this reactivity in a more sustainable and controllable way (Scheme 1a).[10] In fact, electrochemical reduction or oxidation of metal complexes can provide access to high-energy intermediates with unique reactivity not typically accessible by chemical means.[11] *In situ* generation of reduced or oxidized intermediates has been studied *via* electrochemical methods for decades. One of the most well-studied examples is the reduction of M(II) species to M(I) which can act as a potent nucleophile with alkyl halides.[5,12–16] In the case of vitamin B12 mimics like cobalt or iron tetraphenylporphyrin (TPP), an established S_N2 reaction yields a quasi-stable metal–alkyl intermediate that is observable *via* electrochemistry and isolable to study *via* other methods as well (Scheme 1b). Thus, these complexes have provided a wealth of information on the stability and reactivity of organometallic intermediates, information that has enabled a new generation of electrochemical organometallic methodologies.

However, not all of the ways electrochemical methods have been adapted to organic transformations are well understood. One prime example of this is the inclusion of supporting electrolyte. In electrochemical experiments, supporting electrolyte plays the crucial role of increasing solution conductivity by providing a high concentration of charged species.[17] However, many of these supporting electrolyte systems are comprised of common counter ions (coordinating and non-coordinating) for charged organometallic species and, thus, are likely to provide charge stabilization to catalytic intermediates (Scheme 1c). The energetic landscape of these homogeneous electrochemical transformations is likely quite different than their chemical counterparts, with certain intermediates stabilized *versus* others and thereby impacting product chemo-, regio-, and enantioselectivity, depending on the specific reactions. Many studies have noted the impact supporting electrolyte can have on electroorganic transformations. For example,

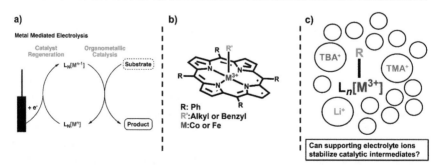

Scheme 1 (a) Illustration of electrocatalysis mediated by a metal complex. (b) Structure of the alkyl–metal porphyrin catalysts used in this study. (c) Illustration of the ion and solvent shell surrounding catalytic intermediates.

the Lin group noted the marked decrease in enantioselectivity of a hydro-cyanation reaction when using a smaller supporting electrolyte (LiClO$_4$) *versus* a larger supporting electrolyte (tetrabutylammonium tetrafluoroborate, TBABF$_4$),[18] citing differences in the polarizability of the electrical double layer at the electrode surface.[19] In another example, the Reisman group showed the clear dependence of yield and enantioselectivity of an alkenyl coupling reaction.[20] NaI was used both as a supporting electrolyte and an iodide source, which is postulated to enhance electron transfer or facilitate the formation of organo-iodide electrophiles.[21] These reactions show the distinct possibilities of rationally designing electrolytes to accomplish a number of tasks beyond solution conductivity, and to actually provide distinct reaction selectivity. Unfortunately, a useful quantitative understanding of the impact of the electrolyte on electro-chemical organic reactions with established electroanalytical techniques has yet to be realized.

Here, we investigate the impact of supporting electrolyte on the stability of electrochemically generated organometallic species using cyclic voltammetry (CV).[22] Monitoring the catalytic reduction of benzyl chloride *via* Co and Fe TPP complexes using CV, we observed the formation of metal–alkyl intermediates. The scan rate dependence of these voltammograms revealed that the metal–organic species decays upon further electrochemical reduction of the organometallic intermediate. We illustrate that this decay is dependent on electrolyte hydrodynamic radius, with enhanced stability in diffuse, weakly coordinating electrolytes (tetrabutylammonium counter ions) and decreased stability (faster decay) with hard, strongly coordinating electrolytes (like Li$^+$ cations). These effects were observed for both Co(II) and Fe(III)Cl TPP complexes, highlighting the generality of the results for this reaction. These results show the critical role the supporting electrolyte can play in electroorganic reactions and provide a new rationale for choosing a particular supporting electrolyte species in screening organic reactions. We anticipate these studies will help enable the rational design of new, functional supporting electrolytes for organic transformations.

2. Experimental

All experiments were carried out in an mBraun glovebox with O$_2$ and moisture levels below 0.5 ppm. Electrochemical techniques were performed on a Biologic VSP potentiostat routed into the glovebox. Electrochemistry was performed using a three-electrode setup with a 0.07 cm^2 glassy carbon working electrode (CH Instruments, CHI 104), a platinum mesh counter electrode (0.5 cm^2, Strem, 99.99%), and a silver wire pseudo-reference electrode (CH Instruments, CHI 112) with a Teflon frit and glass tube filled with electrolyte. In all cases, extra dry dimethylformamide was used as the solvent (DMF, Acros 99.95% Extra Dry) and the supporting electrolytes tetrabutylammonium hexafluorophosphate (TBAPF$_6$, Sigma-Aldrich, >99.0% for electrochemical analysis), tetramethylammonium hexafluorophosphate (TMAPF$_6$, Sigma-Aldrich, >99.0% for electrochemical analysis), and lithium perchlorate (LiClO$_4$, Sigma-Aldrich, >99.0% for electrochemical analysis) were all recrystallized and dried before use. All measurements were referenced to ferrocene immediately after the experiment. For cyclic voltammetry experiments, cobalt(II) tetraphenylporphyrin (Strem, 99%) or iron(III) tetraphenylporphyrin chloride (Strem, 99%) were dissolved in electrolyte to afford a 1 mM solution of the analyte of interest. 2 mL of this solution was added to a glass scintillation vial for measurements. Another 2 mL of this solution was used to dissolve the substrate 1-(chloroethyl)benzene (TCI America, 97+%) to afford a solution that was 1 M in substrate. From this solution, aliquots were taken and added *via* micropipette to the electrochemical cell. Thus substrate was added while keeping the catalyst concentration constant. Concentrations were corrected to account for dilution of the substrate for analysis. Voltammetry was performed at multiple scan rates, 10, 25, 50, and 100 mV s^{-1}, and all peak and current plateau analysis was done with the current of the first scan. Solution was agitated between experiments to ensure a refresh of the electrode surface.

3. Results and discussion

3.1 Cobalt tetraphenylporphyrin

First we investigated the cyclic voltammetry of Co(II)(TPP). The complex exhibited a reversible wave centered at -1.28 V *vs.* Fc$^{+/0}$ associated with the Co(II)/Co(I) redox couple over several scan rates (Fig. 1a and b, blue traces). With the addition of a benzyl chloride substrate, 1-(chloroethyl)benzene (PhEtCl), a second wave appears, with a cathodic peak centered around -1.6 V *vs.* Fc$^{+/0}$, and the return wave of the Co(II)/Co(I) couple begins to disappear, indicating consumption of Co(I) in a chemical process. At moderate scan rates (100 mV s^{-1}), the new peak is quasi-reversible and proportional to the concentration of PhEtCl added (Fig. 1a). At low scan rates (10 mV s^{-1}), this wave loses its peak-like behavior and becomes a plateau, indicative of catalytic behavior. The magnitude of this plateau current is also proportional to the concentration of substrate PhEtCl, increasing dramatically as more was added. The current at -1.7 V *vs.* Fc$^{+/0}$ (roughly where the steady-state plateau current is achieved for all scan rates) as a function of substrate concentration was normalized to the current at this potential when no substrate is present to compare between scan rates. Analysis of the normalized plateau current revealed a linear dependence of the current with respect to square root substrate concentration at all scan rates, indicative of a homogeneous chemical process (Fig. 1c). The current normalization revealed that the greatest

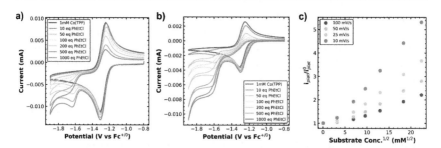

Fig. 1 (a) Cyclic voltammetry of 1 mM Co(TPP) with increasing concentrations of phe-nylethylchloride at 100 mV s^{-1}, (b) at 10 mV s^{-1}. (c) Normalized plateau currents (taken to be around -1.7 V *vs.* Fc$^{+/0}$, normalized to the current at this potential when no substrate was present) for several scan rates, showing that a chemical process contributes to current increases at low scan rates. Working electrode was a 0.07 cm^2 glassy carbon electrode in a solution of 0.1 M TBAPF$_6$ in DMF.

current increases were observed at slow scan rates and decreased steadily as the scan rate was increased, indicative of an EC$'$ catalytic mechanism.[23,24]

These voltammetry studies reveal a number of important features of Co(TPP)'s reaction with benzyl chloride. First, the decrease of the Co(I) return oxidative wave with PhEtCl concentration with no substantial increase in the forward reductive wave is indicative of an EC process, wherein Co(II) is reduced to Co(I) (E step), which reacts subsequently with the benzyl chloride (C step). The chemical step consumes Co(I) at the electrode surface, and thus leaves none for the return wave. In this context, the appearance of a new peak at −1.6 V vs. Fc$^{+/0}$ as the return wave disappears can be attributed to the reduction of the product of the C step, the alkylated metal complex, Co(III)–R(TPP) to Co(II)–R(TPP).[12] At slower scan rates, this quasi-reversible peak changes shape to an irreversible current plateau. This sigmoidal shape is indicative of a catalytic process,[24] presumably the decay of the reduced alkylated Co species to the original Co(II) state and the carbanion. In the kinetic zone diagram formalism developed by Savéant, this transition from quasi-reversible peak to plateau corresponds to the transition from zone KD to KS. In other words, at low concentrations the chemical step is slow relative to the vol-tammetric time-scale, and increasing concentration or decreasing scan rate gives the catalytic response. The normalized peak current was correlated to the square root of substrate concentration, as expected for a catalytic process.[23] The linear correlation is in keeping with the expectation of a homogeneous process under kinetic control (zone K). The overall reaction can be summarized as an ECEC$'$ process as outlined in Scheme 2.

Owing to our ability to resolve the decay of the metal–alkyl species via the current plateau at slow scan rates, we resolved to study this process in a number of supporting electrolytes. We envisioned that the hydrodynamic radius of the dissolved metal complex was indicative of the solvation environment of the metal–alkyl intermediate. Essentially, a smaller hydrodynamic radius is indicative of a harder, more compact coordination environment (Li$^+$), while a softer, more diffuse coordination environment is expected for the larger hydrodynamic radius

$$L_n[Co^{II}] + e^- \overset{k_e}{\rightleftharpoons} L_n[Co^I]^- \tag{1}$$

$$L_n[Co^I]^- + R\text{-}Cl \xrightarrow{k_1} L_n[Co^{III}]\text{-}R + Cl^- \tag{2}$$

$$L_n[Co^{III}]\text{-}R + e^- \overset{k_e}{\rightleftharpoons} L_n[Co^{II}]\text{-}R^- \tag{3}$$

$$L_n[Co^{II}]\text{-}R^- \xrightarrow{k_2} L_n[Co^{II}] + R^- \tag{4}$$

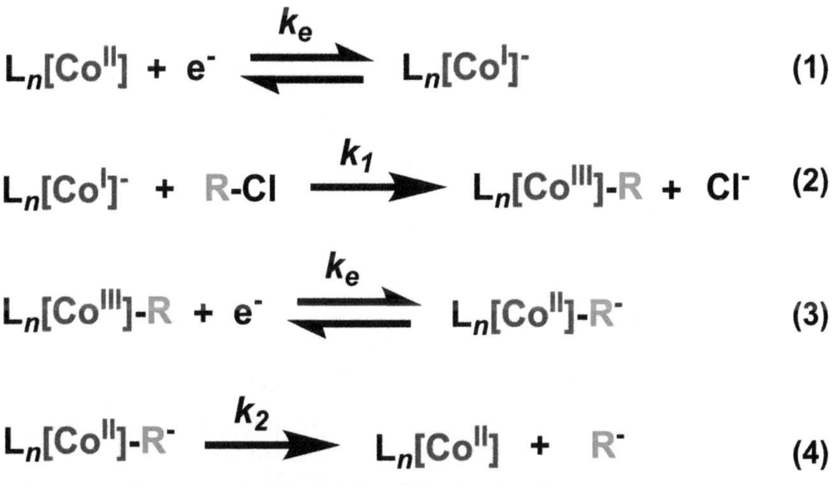

Scheme 2   Overall mechanism of Co(TPP) reduction of a benzyl chloride.

(TBA$^+$). The supporting electrolytes used, ordered from largest to smallest hydrodynamic radius (solvated ion size), were tetrabutylammonium hexafluorophosphate (TBAPF$_6$), tetramethylammonium hexafluorophosphate (TMAPF$_6$), and lithium perchlorate (LiClO$_4$). This was verified from the diffusion constant of Co(TPP) which was calculated in each supporting electrolyte (when no substrate was present) according to the Randles–Sevcik equation.[17] The diffusion constant varied with electrolyte as 3.8×10^{-6} cm^2 s^{-1} for TBAPF$_6$, 6.3×10^{-6} cm^2 s^{-1} for TMAPF$_6$, and 6.9×10^{-6} cm^2 s^{-1} for LiClO$_4$, where a lower diffusion constant indicates a smaller (compacted) hydrodynamic radius in keeping with changes in solution viscosity. The same plateau behavior was observed for all electrolytes at slow scan rates of 10 mV s^{-1}. The normalized current plateau at -1.7 V vs. Fc$^{+/0}$ as a function of square-root PhEtCl concentration is outlined in Fig. 2, with different electrolytes providing different increases in the magnitude of the plateau current. The difference in the current increases is maximized at high concentrations of PhEtCl. The largest current increases were seen for LiClO$_4$, intermediate increases were observed for TMAPF$_6$ and the smallest current increases were observed for TBAPF$_6$.

The differences in current increases between different supporting electrolytes clearly show the impact supporting electrolyte has on the catalytic process. As established above, the current increase of the plateau is attributed to the decay of the metal–alkyl species to return to the original oxidation state of Co(II), where it can undergo the catalytic process again and thereby increasing the current. The differences in current increases between the different electrolytes can be

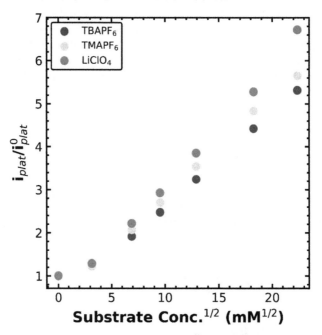

Fig. 2 Normalized plateau currents at -1.7 V vs. Fc$^{+/0}$ for 1 mM Co(TPP) in a DMF solution containing 0.1 M TBAPF$_6$ (blue), TMAPF$_6$ (grey), or LiClO$_4$ (red) for increasing amounts of the substrate PhEtCl. Scan rate was 10 mV s^{-1} with a 0.07 cm^2 glassy carbon working electrode.

interpreted as an increase or decrease in the rate of the catalytic step, decay of the metal alkyl species. Differences in diffusion can also impact the current plateau[24] (and electrolyte does impact diffusion as seen above) but normalization to the current when no substrate is present cancels out this effect. Interference from capacitive charging can be discounted, for the same reason. Importantly, the increase in rate correlates to the hydrodynamic radius of the supporting electrolyte: the largest cationic supporting electrolyte ion is TBA^+ which gives the smallest current increases and the smallest cationic supporting electrolyte is Li^+ which gives the largest current increases. We attribute this to stabilization (or destabilization) of the charged metal–alkyl intermediate $Co(II)$–R^- due to the hardness of the supporting electrolyte solvation shell. In the case of the large, diffuse cationic species, TBA^+, the solvent–ion shell surrounding the intermediate can better stabilize the charged organometallic species, owing to its softer, more organic nature. In contrast, the small cation Li^+ likely destabilizes the organometallic intermediate with its harder ion shell, promoting catalysis *via* the decay of the metal–alkyl bond. These results indicate a method to stabilize or destabilize useful organometallic intermediates during electroorganic reactions *via* supporting electrolyte choice.

3.2   Iron chloride tetraphenylporphyrin

Given the supporting-electrolyte dependence of Co(TPP) on the reduction of benzyl chlorides, we wanted to extend these results to a related, but more complicated system. FeCl(TPP) appeared as an interesting candidate as it is reported to do similar chemistry but requires an extra reduction (as it begins in the Fe(III) state) and the dissociation of the coordinating chloride counter ion. We reasoned these features might reveal more ways that the supporting electrolyte can impact the mechanism of catalysis. Cyclic voltammetry of a 1 mM solution of Fe(III)Cl(TPP) showed two successive reductive waves (Fig. 3a), one centered at −0.65 V *vs.* Fc$^{+/0}$ associated with the reduction of Fe(III) to Fe(II), and another at −1.5 V *vs.* Fc$^{+/0}$ associated with the reduction of Fe(II) to Fe(I). The second of these peaks, the Fe(II)/Fe(I) couple is reversible, but the first peak, the Fe(III)/Fe(II) couple exhibits a scan rate dependent feature on the return wave associated with re-

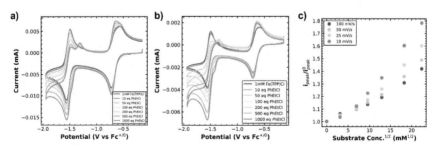

Fig. 3 (a) Cyclic voltammetry of 1 mM FeCl(TPP) with increasing concentrations of phenylethylchloride at 100 mV s^{-1}, (b) at 10 mV s^{-1}. (c) Normalized peak currents (of peak at −1.55 V *vs.* Fc$^{+/0}$, normalized to the current when no substrate was present) for several scan rates, showing that a slow chemical process contributes to current increases at low scan rates. Working electrode was a 0.07 cm^2 glassy carbon electrode in a solution of 0.1 M TBAPF$_6$ in DMF.

coordination of the chloride ion. Upon the addition of PhEtCl, the second peak begins to increase in magnitude and the reverse peak associated with reoxidation of Fe(I) begins to disappear. Further addition at high scan rates shows the emergence of a second oxidative wave in the return scan that lies positive of the reoxidation of Fe(I). At slower scan rates, the increase in the peak current is much greater (Fig. 3b). In addition, the formation of a shoulder wave near the second reductive wave (evident at 100 mV s^{-1} but much clearer at slow scans) becomes observable, and the emergence of the second oxidative wave is much less pronounced. The current increase of the second reductive wave (normalized to the current of the second reductive wave when no substrate is present) revealed a linear correlation with the square root substrate concentration (Fig. 3c). Slower scan rates exhibited larger normalized current increases than the faster scan rates, indicating a slow chemical process taking place.

As with the Co(TPP) complex, this voltammetry reveals critical info about the mechanism of benzyl chloride reduction with FeCl(TPP). First, the current increases upon the addition of PhEtCl are restricted to the Fe(II)/Fe(I) couple, indicating that Fe(I) is the species that reacts with the benzyl chloride, a fact further supported by the decrease in the return oxidation of Fe(I). The emergence of a second oxidative wave on the return scan is indicative of the oxidation of the product of the reaction between Fe(I) and the benzyl chloride, confirmed to be the metal–alkyl intermediate.[16] As such, the formation of the shoulder to the Fe(II) reduction is likely associated with the reduction of the same product, meaning the peaks of the Fe(II)/Fe(I) couple, and the formed Fe(III)–R/Fe(II)–R couple overlap significantly. However, the scan rate dependence suggests that the current increase is not only due to the emergence of a new redox active species. The current increases at slow scan rates are greater than at faster scan rates, which could be due to a slow catalytic process, but could also result from more Fe(III)–R being produced over the longer scan (and subsequently being reduced to produce higher current). Crucially, the new return wave is much less pronounced at slower scan rates, indicating that the reduction of Fe(III)–R to Fe(II)–R is not wholly reversible, and at least some of the reduced Fe(II)–R decays before it can be reoxidized on the return scan. Another interesting note is that the return wave of the Fe(III)/Fe(II) couple is also affected by PhEtCl concentration, presumably because the reaction of Fe(I) with PhEtCl produces chloride ions, which in turn impacts the magnitude of the ion-coupled wave. The overall chemical process can be summarized as EECEC' and is illustrated schematically in Scheme 3.

With this mechanistic assignment, we moved to again assess the stability of the metal–alkyl species in several supporting electrolytes (TBAPF$_6$, TMAPF$_6$, and LiClO$_4$). As with the Co(TPP) system, electrolyte choice impacts the diffusion constant according to changes in electrolyte viscosity: 2.7×10^{-6} cm^2 s^{-1} for TBAPF$_6$, 4.7×10^{-6} cm^2 s^{-1} for TMAPF$_6$, and 4.8×10^{-6} cm^2 s^{-1} for LiClO$_4$ as calculated by the Randles–Sevcik equation. Fig. 4 depicts the normalized peak current (not plateau) as a function of PhEtCl concentration for several different supporting electrolytes. Generally, the smaller supporting electrolyte once again gave higher current increases, though clearly, the dependence is less straightforward than in the case of Co(TPP). Namely, TBAPF$_6$, the largest, provided the smallest current increases until the highest PhEtCl concentration, where it surpassed TMAPF$_6$. The increase in current between PhEtCl concentrations for TBAPF$_6$ looks mostly linear, while it appears that the TMAPF$_6$ increases begin to

$$L_n[Fe^{III}]Cl + e^- \underset{}{\overset{k_e}{\rightleftharpoons}} L_n[Fe^{II}] + Cl^- \qquad (1)$$

$$L_n[Fe^{II}] + e^- \underset{}{\overset{k_e}{\rightleftharpoons}} L_n[Fe^{I}]^- \qquad (2)$$

$$L_n[Fe^{I}]^- + R\text{-}Cl \xrightarrow{k_1} L_n[Fe^{III}]\text{-}R + Cl^- \qquad (3)$$

$$L_n[Fe^{III}]\text{-}R + e^- \underset{}{\overset{k_e}{\rightleftharpoons}} L_n[Fe^{II}]\text{-}R^- \qquad (4)$$

$$L_n[Fe^{II}]\text{-}R^- \xrightarrow{k_2} L_n[Fe^{II}] + R^- \qquad (5)$$

Scheme 3   Overall mechanism of FeCl(TPP) reduction of a benzyl chloride.

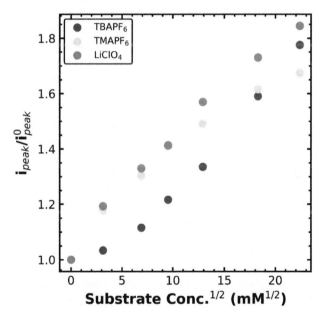

Fig. 4   Normalized peak currents at -1.55 V vs. $Fc^{+/0}$ for 1 mM FeCl(TPP) in a DMF solution containing 0.1 M $TBAPF_6$ (blue), $TMAPF_6$ (grey), or $LiClO_4$ (red) for increasing amounts of the substrate PhEtCl. Scan rate was 10 mV s^{-1} with a 0.07 cm^2 glassy carbon working electrode.

saturate, allowing the $TBAPF_6$ to surpass it. $LiClO_4$ remained the highest in current increases at all concentrations.

As with the Co(TPP) system, the softness or hardness of the cation of the supporting electrolyte appears to have a pronounced effect on the stability of the iron–alkyl bond, with hard (Li^+) supporting electrolytes promoting catalysis *via* decay of the metal alkyl species, presumably for the same reasons as outlined above. These current increases are much smaller than those observed for the Co(TPP) system as well, highlighting the relative stability of the iron–alkyl species *versus* the Co-alkyl species. One important caveat is that at high PhEtCl concentrations, the $TMAPF_6$ begins to saturate, highlighting a more complicated mechanism at play. The saturation may be due to differences in solubility or interference with the ion-coupled movement of the iron chloride species. Regardless, these results indicated the generality of the electrolyte effects for metal porphyrin catalyzed benzyl chloride reduction. Extension of these concepts to other metal complexes with different mechanisms is underway. Systems that undergo reversible bond homolysis similar to vitamin B12 are of particular interest, as this analysis may provide a way to quantify longstanding questions regarding radical cage collapse mechanisms in organometallic catalysis.

4. Conclusions

In this study, we have outlined the impact of supporting electrolyte on metal porphyrin catalyzed benzyl chloride reduction. Cyclic voltammetry studies revealed an ECEC′ mechanism where the final step was the decay of the metal-alkyl intermediate, thus providing a handle on the stability of the metal–alkyl bond stability *via* the current plateau. Plateau current correlation with substrate concentration in different supporting electrolytes revealed a dependence on the hydrodynamic radius of the supporting electrolyte, where the smallest electrolyte $(LiClO_4)$ provided the largest current increase. These results are interpreted as supporting electrolyte altering the stability of the metal–alkyl intermediate *via* the hardness or softness of the electrolyte–solvation shell. However, there is also a possibility that current increases are driven by the stabilization of the carb-anion, in particular the case of Li^+ cations which are known to form organoalkali compounds with aryl and benzylic carbanions.[25,26] Studies identifying the chemical intermediates of these reactions are ongoing, and will no doubt inform electrolyte choice and design in the future. A similar mechanism was explored for FeCl(TPP), albeit with an additional reduction and ion-coupled movement of the chloride. Analysis showed that the same general trend in the current increase and supporting electrolyte hydrodynamic radius could be observed for the Fe–alkyl intermediate. We anticipate these supporting electrolyte effects can be extended to a number of electrochemical reactions. While the precise dependence will depend on the exact mechanism, there are likely a number of unexplored avenues for electrolyte-mechanism interactions. Most importantly, these results have major implications for the design of electrosynthetic reactions. The impact of electrolyte on metal–carbon bond stability means that electrolytes can be chosen rationally, if the mechanism is well understood, to stabilize or destabilize particular intermediates and promote product specificity. Thus, these results are an important first step in the realization of designed supporting electrolytes, that may perform multiple functions beyond the traditional role.

Data availability

All data is made available in the main text.

Conflicts of interest

The authors declare no conflict of interest.

Acknowledgements

This work was supported by the NSF CCI Center for Synthetic Organic Electrochemistry (CHE-2002158).

References

1 J. D. Atwood, *Inorganic and Organometallic Reaction Mechanisms*, VCH Publishers, New York, 2nd edn, 1997.
2 N. Miyaura and A. Suzuki, Palladium-Catalyzed Cross-Coupling Reactions of Organoboron Compounds, *Chem. Rev.*, 1995, **95**(7), 2457–2483.
3 R. G. Finke and B. D. Martin, Coenzyme AdoB$_{12}$ vs. AdoB$_{12}$.$^-$ homolytic Co-C cleavage following electron transfer: a rate enhancement $\geq 10^{12}$, *J. Inorg. Biochem.*, 1990, **40**(1), 19–22.
4 J. Halpern, Mechanisms of coenzyme B12-dependent rearrangements, *Science*, 1985, **227**(4689), 869–875.
5 D. Lexa and J. M. Saveant, Electrochemistry of vitamin B12 I. Role of the base-on/base-off reaction in the oxidoreduction mechanism of the B12r–B12s system, *J. Am. Chem. Soc.*, 1976, **98**(9), 2652–2658.
6 J. B. Broderick, *et al.*, Radical S-adenosylmethionine enzymes, *Chem. Rev.*, 2014, **114**(8), 4229–4317.
7 H. Fischer, The persistent radical effect: a principle for selective radical reactions and living radical polymerizations, *Chem. Rev.*, 2001, **101**(12), 3581–3610.
8 C. A. Malapit, *et al.*, Advances on the Merger of Electrochemistry and Transition Metal Catalysis for Organic Synthesis, *Chem. Rev.*, 2022, **122**(3), 3180–3218.
9 N. G. Connelly and W. E. Geiger, Chemical Redox Agents for Organometallic Chemistry, *Chem. Rev.*, 1996, **96**(2), 877–910.
10 M. Yan, Y. Kawamata and P. S. Baran, Synthetic Organic Electrochemical Methods Since 2000: On the Verge of a Renaissance, *Chem. Rev.*, 2017, **117**(21), 13230–13319.
11 R. Francke and R. D. Little, Redox catalysis in organic electrosynthesis: basic principles and recent developments, *Chem. Soc. Rev.*, 2014, **43**(8), 2492–2521.
12 D. Lexa, J. M. Savéant and J. P. Soufflet, Chemical catalysis of the electrochemical reduction of alkyl halides, *J. Electroanal. Chem. Interfacial Electrochem.*, 1979, **100**(1–2), 159–172.
13 G. B. Maiya, B. C. Han and K. M. Kadish, Electrochemical studies of cobalt–carbon bond formation. A kinetic investigation of the reaction between (tetraphenylporphinato)cobalt(ı) and alkyl halides, *Langmuir*, 1989, **5**(3), 645–650.
14 R. L. Birke, *et al.*, Electroreduction of a series of alkylcobalamins: mechanism of stepwise reductive cleavage of the Co-C bond, *J. Am. Chem. Soc.*, 2006, **128**(6), 1922–1936.
15 C. Costentin, *et al.*, Concertedness in proton-coupled electron transfer cleavages of carbon–metal bonds illustrated by the reduction of an alkyl cobalt porphyrin, *Chem. Sci.*, 2013, **4**(2), 819–823.

16 D. Lexa, J. Mispelter and J. M. Saveant, Electroreductive alkylation of iron in porphyrin complexes. Electrochemical and spectral characteristics of σ-alkylironporphyrins, *J. Am. Chem. Soc.*, 1981, **103**(23), 6806–6812.

17 A. J. Bard, L. F. Faulkner and H. S. White, *Electrochemical Methods: Fundamentals and Applications*, Wiley, 3rd edn, 2022, p. 1104.

18 L. Song, *et al.*, Dual electrocatalysis enables enantioselective hydrocyanation of conjugated alkenes, *Nat. Chem.*, 2020, **12**(8), 747–754.

19 A. Redden and K. D. Moeller, Anodic coupling reactions: exploring the generality of Curtin-Hammett controlled reactions, *Org. Lett.*, 2011, **13**(7), 1678–1681.

20 T. J. DeLano and S. E. Reisman, Enantioselective Electroreductive Coupling of Alkenyl and Benzyl Halides via Nickel Catalysis, *ACS Catal.*, 2019, **9**(8), 6751–6754.

21 A. H. Cherney and S. E. Reisman, Nickel-catalyzed asymmetric reductive cross-coupling between vinyl and benzyl electrophiles, *J. Am. Chem. Soc.*, 2014, **136**(41), 14365–14368.

22 J. M. Savéant and C. Costentin, *Elements of Molecular and Biomolecular Electrochemistry*, John Wiley & Sons Inc, 2nd edn, 2019.

23 C. Costentin and J.-M. Savéant, Multielectron, Multistep Molecular Catalysis of Electrochemical Reactions: Benchmarking of Homogeneous Catalysts, *ChemElectroChem*, 2014, **1**(7), 1226–1236.

24 J. M. Saveant and E. Vianello, Potential-sweep chronoamperometry: Kinetic currents for first-order chemical reaction parallel to electron-transfer process (catalytic currents), *Electrochim. Acta*, 1965, **10**(9), 905–920.

25 H. Gilman and G. L. Schwebke, Improved Method for the Preparation of Benzyllithium, *J. Org. Chem.*, 1962, **27**(12), 4259–4261.

26 R. Waack, L. D. McKeever and M. A. Doran, The nature of carbon–lithium bonding in benzyl-lithium and its variation with solvent, *J. Chem. Soc. D*, 1969, (3), 117b–118.

Faraday Discussions

PAPER

Selective electrosynthesis of platform chemicals from the electrocatalytic reforming of biomass-derived hexanediol†

Yun-Ju Liao,[a] Shih-Ching Huang[a] and Chia-Yu Lin ⓘ *[abc]

Received 25th March 2023, Accepted 16th May 2023

DOI: 10.1039/d3fd00073g

6-Hydroxyhexanoic acid and adipic acid are platform chemicals and are widely used as building blocks for the synthesis of important polymers. Nevertheless, the industrial syntheses of these two chemicals are fossil fuel-based and involve the use of corrosive acid and emission of the NO_x greenhouse gas. In this study, the electrosynthesis of 6-hydroxyhexanoic acid and adipic acid from the electrochemical oxidation of hexanediol at the nanoporous nickel oxyhydroxide modified electrode was explored as an environmentally-benign alternative to the industrial syntheses of 6-hydroxyhexanoic acid and adipic acid. The effects of electrolysis conditions, including the electrolyte pH and applied potentials, on faradaic efficiency and product distribution of the electrochemical oxidation of hexanediol, were thoroughly examined through a series of controlled-potential electrolyses. In addition, the scale-up electrosynthesis of 6-hydroxyhexanoic acid and adipic acid using a flow-type electrolyzer was also demonstrated.

Introduction

6-Hydroxyhexanoic acid (6HA) and adipic acid (AA) are chemicals of industrial importance and mainly used as the building blocks for the industrial production of polycaprolactone and nylon 6,6.[1] However, the industrial production of these two chemicals involves the use of fossil fuel-based feedstocks, multiple and energy-intensive reaction steps, concentrated corrosive nitric acid, and emission of NO_x greenhouse gas, which results in catastrophic environmental impacts (*e.g.*, global warming and ozone depletion).[2] Consequently, the development of other

[a]*Department of Chemical Engineering, National Cheng Kung University, Tainan City 70101, Taiwan. E-mail: CYL44@mail.ncku.edu.tw*

[b]*Hierarchical Green-Energy Materials (Hi-GEM) Research Center, National Cheng Kung University, Tainan 70101, Taiwan*

[c]*Program on Key Materials & Program on Smart and Sustainable Manufacturing, Academy of Innovative Semiconductor and Sustainable Manufacturing, National Cheng Kung University, Tainan 70101, Taiwan*

† Electronic supplementary information (ESI) available. See DOI: https://doi.org/10.1039/d3fd00073g

synthetic routes for the sustainable synthesis of 6HA and AA with fossil-free feedstock and minimal environmental footprint is of great importance.

The electrochemical 1,6-hexanediol oxidation reaction (e-HOR) emerges as a viable, clean, and sustainable approach for the production of 6HA and AA as it provides several unique advantages. To begin with, 1,6-hexanediol, derived from renewable biomass,[1,3–5] is a fossil-free feedstock. In addition, e-HOR uses water as the oxygen source and can be operated at room temperature and ambient pressure. Finally, sustainable e-HOR can be realized by integrating with renewable energy sources, such as solar energy.

No study on the production of 6HA and AA from e-HOR is available currently, but the production of small acids, such as formic acid and acetic acid, from the electrochemical oxidation of alcohols at nickel-based electrocatalysts, has been intensively investigated since the 1970s.[6–15] For instance, the highly selective production of formic acid, in terms of faradaic efficiency of ~100%, from the electrochemical oxidation of ethylene glycol, has been demonstrated using nickel oxyhydroxide-modified nickel phosphide in our previous work.[6] Therefore, it can be expected that nickel-based electrocatalysts may have electrocatalytic activity towards e-HOR. The conditions for both the electrocatalyst preparation and electrolysis of alcohols were found to play an important role in determining the overall activity of these nickel-based electrocatalysts.[6,15–17] For example, the activation of nickel-based electrocatalysts was found to be necessary to achieve high stability and high yield of carboxylic acid from the electrochemical oxidation of longer-chain alcohols.[15,16] Moreover, the activation of nickel phosphide and nickel nanoparticles by potential cycling resulted in the formation of nickel oxyhydroxide (NiOOH) layers of different β-NiOOH contents.[6] As β-NiOOH was the main active species involved in the electrocatalytic oxidation of ethylene glycol, the activated nickel phosphide with higher β-NiOOH content exhibited significantly higher activity than the activated metallic nickel nanoparticles.[6] Huang et al. report that incorporating suitable amounts of iron species into nickel oxyhydroxide can significantly improve the selectivity and activity towards the generation of formate from electrochemical oxidation of methanol at near neutral pH.[7] Consequently, the development of a facile synthetic methodology to control and maximize the content of active β-NiOOH is of great importance for the establishment of a high-performance nickel-based electrocatalytic platform for the production of acids from the oxidation of alcohols.

We previously developed a surfactant-free and facile pulse-current electrodeposition (PED) method for the preparation of the nanoporous nickel oxyhydroxide-borate film (nanoNiOOH-borate).[17] We found that nickel oxyhydroxide film prepared using PED in the presence of borate buffer had high β-NiOOH content and high-performance towards the production of formic acid from the electrochemical oxidation of methanol.[17] Encouraged by the high activity and selectivity of nanoNiOOH–borate, the application of the nanoNiOOH-Bi for the electrosynthesis of 6HA and AA from e-HOR was further explored and investigated in the present study. Through systematic assessments of the effects of electrolysis conditions, the optimal electrolyte pH and applied potential for the efficient production of 6HA and AA have been discovered. The scale-up electrosynthesis of 6HA and AA using a flow-type electrolyzer was also discussed.

Experimental

Chemicals and reagents

All the chemicals used in this work, including adipic acid (AA; 97%, Acros Organics), 1,6 hexanediol (99%, Acros Organics), 6-hydroxyhexanoic acid (6HA; 95%, Alfa Aesar), ε-caprolactone (99.0%, Alfa Aesar), sodium tetraborate decahydrate ($\geq$99.5%, Sigma-Aldrich), nickel(II) sulfate hexahydrate (98%, J.T.Baker), sodium hydroxide ($\geq$99.0%, Sigma-Aldrich), hydrogen peroxide solution ($\geq$30.0%, Sigma-Aldrich), ammonium hydroxide (29%, J.T.Baker), H_2SO_4 solution (1.0 M, Sigma-Aldrich), were used as received from the commercial suppliers without further purification. Fluorine-doped tin oxide coated glass substrate with a sheet resistance of 7 Ω sq^{-1} (FTO; TEC GlassTM 7) was used as substrate and cleaned with NH_4OH–H_2O_2–H_2O solution mixture (volume ratio: 1 : 1 : 5) at 70 °C for 40 min prior to the further modification. Deionized water (18.2 MΩ cm; DIW) was used for the electrode rinsing and electrolyte preparation throughout the work.

Preparation of nickel oxyhydroxide modified electrodes

The nanoporous nickel oxyhydroxide-borate film modified FTO electrode (nanoNiOOH-borate) was prepared using an electrochemical deposition method reported previously.[17,18] Briefly, nanoNiOOH-borate film was deposited onto the cleaned FTO substrate by pulse-current electrodeposition (PED) in a customized three-electrode cell, filled with the borate buffer solution (0.1 M, pH 9.2) containing nickel sulfate (0.5 mM), using a multipalmSens4 workstation (PalmSens B.V., Netherlands) with Hg/HgO (1 M NaOH) as the reference electrode and Pt foil (exposed area: ~4.0 cm^2) as the counter electrode. The overall charge passage, applied pulse current density, pulse frequency, and duty cycle used for the PED process were set at 0.54 C cm^{-2}, 100 µA cm^{-2}, 0.1 Hz, and 0.5, respectively. The typical potential transient recorded during the PED preparation of the nanoNiOOH-borate electrode is shown in Fig. S1a.† For comparison, a NiOOH-coated Ni foil (Ni|NiOOH) was prepared by subjecting the nickel foil (99.95%) to potential cycling in 0.1 M NaOH at a scan rate of 50 mV s^{-1} between 0.8 and 1.8 V vs. RHE for 225 cycles.[17] The typical cyclic voltammetry recorded during the preparation of the Ni|NiOOH electrode is shown in Fig. S1b.†

Physical characterization

The surface morphology of the prepared electrodes was characterized using scanning electron microscopy (SEM) with a Hitachi SU-8010 microscope. Raman spectra of the prepared electrodes were acquired using an exposure time 1 s per scan (20 scans in total) with a 532 nm excitation laser provided by a RAMaker Raman spectrometer (BEII).

Electrochemical characterization

The electrocatalytic properties of the prepared nanoNiOOH-borate electrode towards the electrochemical oxidation of hexanediol were characterized in a customized H-cell or flow-type electrolyzer (Scheme S1†) that is connected to a Ivium-n-Stat workstation (Ivium Technologies B.V., Netherlands). The anodic and cathodic compartments of the H-cell and flow-type electrolyzer were separated with a Neosepta ASE anion exchange membrane (ASTOM Corporation, Tokyo, Japan). The nanoNiOOH-borate working electrode and Hg/HgO reference electrode were placed in the anodic compartment, whereas the Pt counter electrode was placed in the cathodic compartment. All potentials were reported against the reversible hydrogen electrode (RHE) scale using eqn (1).

$$E(\text{V } vs. \text{ RHE}) = E(\text{V } vs. \text{ Hg/HgO}) + 0.12 + 0.059 \times \text{pH} \tag{1}$$

The electrocatalytic properties of the prepared nanoNiOOH-borate electrodes in different electrolyte solutions were characterized by cyclic voltammetry (CV) at a scan rate of 10 mV s^{-1}, controlled-potential electrolyses (CPEs) at various applied potentials, and controlled-current electrolysis. 95% IR compensation was applied for the CV and CPE experiments. The amount of electrochemically available nickel species for the electrocatalysis (N_{Ni}) was determined using linear sweep voltammetry (LSV), at a scan rate of 10 mV s^{-1}, in NaOH solution (1.0 M); the charge (Q) responsible for the oxidation of Ni(OH)$_2$ to NiOOH was quantified by integrating the area under the anodic wave, and converted to N_{Ni} using eqn (2):

$$N_{Ni} = \frac{Q}{n_e F} \tag{2}$$

where F and n_e are, respectively, faradaic constant (96 485 C mol^{-1}) and the number of electrons transferred for the Ni(OH)$_2$/NiOOH redox reaction. The typical values of N_{Ni} for the nanoNiOOH-borate and Ni|NiOOH electrodes were ~0.39 and 0.06 μmol cm^{-2}, respectively. Note that the n_e value used in this work was 1 to provide a lower bound for the determination of turnover frequency (*vide infra*).

Product analysis

The liquid products, *i.e.*, adipic acid (AA) and 6-hydroxyhexanoic acid (6HA), generated from the CPEs were analyzed and quantified using a high-performance liquid chromatograph (Nexera-i LC2040C 3D Plus), equipped with a Shodex SUGAR SH1821 column, with diluted H$_2$SO$_4$ solution (0.5 mM) as eluent at a flow rate of 0.6 mL min^{-1} at 60 °C. The detectors used for the quantification of AA ad 6HA were a photodiode array ($\lambda = 207$ nm) and a refractive index detector, respectively. The obtained HPLC (high-performance liquid chromatography) signals were converted to the moles of AA (N_{AA}) and 6HA (N_{6HA}) with routinely updated calibration curves. The faradaic efficiencies (FE) for the generation of AA (FE$_{AA}$) and 6HA (FE$_{6HA}$) were determined using eqn (3) and (4):

$$\text{FE}_{AA}(\%) = \frac{N_{AA} \times 8F}{Q_{total}} \times 100\% \tag{3}$$

$$\text{FE}_{6HA}(\%) = \frac{N_{6HA} \times 4F}{Q_{total}} \times 100\% \tag{4}$$

where Q_{total} is the total charge passage for each 2 h CPE. The total current density (J_{total}), AA partial current density (J_{AA}), and 6HA partial current density (J_{6HA}) were derived from Q_{total} using eqn (5)–(7), respectively:

$$J_{total} = \frac{Q_{total}}{7200} \tag{5}$$

$$J_{AA} = \frac{Q_{total} \times FE_{AA}}{7200} \tag{6}$$

$$J_{6HA} = \frac{Q_{total} \times FE_{6HA}}{7200} \tag{7}$$

Turnover frequencies (TOF) for the generation of AA (TOF_{AA}) and 6HA (TOF_{6HA}) were determined using eqn (8) and (9):

$$TOF_{AA} = \frac{R_{AA}}{N_{Ni}} = \frac{N_{AA}}{N_{Ni} \times 7200} \tag{8}$$

$$TOF_{6HA} = \frac{R_{6HA}}{N_{Ni}} = \frac{N_{6HA}}{N_{Ni} \times 7200} \tag{9}$$

where R_{AA} and R_{6HA} are the average production rate of AA and 6HA, respectively, determined from each 2 h CPE.

Results and discussion

The nanoNiOOH-borate electrode was prepared using pulse-current deposition (see the Experimental section), and its SEM and Raman analyses are shown in Fig. 1a and 2, respectively. The SEM image (Fig. 1a) reveals that the nanoNiOOH-borate electrode contained closely-packed particles with dendritic features. On the other hand, two well-defined peaks at 477 cm^{-1} and 556 cm^{-1}, assignable to the Ni–O vibrations in NiOOH,[19–21] were observed in the Raman spectrum of the nanoNiOOH-borate electrode (Fig. 2). These results confirm the successful reproduction of the nanoNiOOH-borate electrode reported previously.[17,18]

Fig. 3 shows the cyclic voltammetry (CV) of the nanoNiOOH-borate electrode in the blank NaOH solution (1.0 M) and NaOH (1.0 M) solution containing 1,6-hexanediol (0.4 M). For comparison, a NiOOH-coated nickel foil (Ni|NiOOH) was also prepared, and its CV analysis was included in the analysis. It can be found that the nanoNiOOH-borate electrode showed CV features containing a couple of redox peaks (anodic o_1 peak and cathodic r_1 peak) with the formal potential of

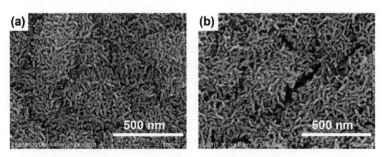

Fig. 1 SEM images of the nanoNiOOH-borate electrode (a) before and (b) after 2 h CPE at 1.55 V *vs.* RHE in NaOH solution containing 1,6-hexanediol (0.4 M).

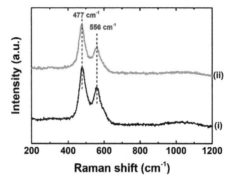

Fig. 2   Raman spectra of the nanoNiOOH-borate electrode (i) before and (ii) after 2 h CPE at 1.55 V *vs.* RHE in NaOH solution containing 1,6-hexanediol (0.4 M).

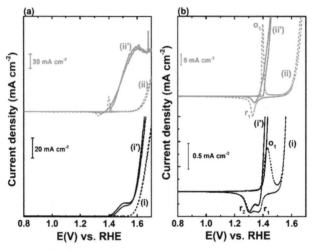

Fig. 3   (a) Original and (b) enlarged CVs of (i and i′) Ni|NiOOH and (ii and ii′) nanoNiOOH-borate, recorded at a scan rate of 10 mV s^{-1}, in the blank NaOH solution (1.0 M; (i) and (ii)) and the NaOH solution (1.0 M) containing 1,6-hexanediol (0.4 M; (i′) and (ii′)).

~1.38 V *vs.* RHE and an anodic wave onset at ~1.53 V *vs.* RHE in the absence of 1,6-hexanediol. The former can be ascribed to the electrochemical inter-conversion of Ni(OH)$_2$ and NiOOH electroactive species (eqn (10)), whereas the latter can be assigned to the oxygen evolution reaction (OER). The Ni|NiOOH electrode exhibited similar CV features to the nanoNiOOH-borate electrode except the appearance of two cathodic peaks (r$_1$ and r$_2$ peaks) during the reverse scan. These r$_1$ and r$_2$ peaks are attributable, respectively, to the cathodic transformation of β-NiOOH and γ-NiOOH to β-Ni(OH)$_2$,[22,23] and, therefore, suggest that the Ni|NiOOH electrode contained a significantly higher amount of γ-NiOOH than the nanoNiOOH-borate electrode. On the other hand, additional CV features were observed for both electrodes in the presence of 1,6-hexanediol, including a remarkable increase in the current density of the o$_1$ peak and a decrease in the current density of the r$_1$ peak, which indicates that the e-HOR at these two electrodes proceeded through an electrocatalytic scheme involving the electro-chemical generation of NiOOH species (eqn (10)) and a subsequent chemical reaction between electrochemically formed NiOOH species and 1,6-hexanediol (eqn (11)). Besides, the nanoNiOOH-borate electrode was found to exhibit more significant current response to 1,6-hexanediol than the Ni|NiOOH electrode. Notably, the nanoNiOOH-borate electrode exhibited a current response that is $\sim$5.5 times higher than the Ni|NiOOH electrode (50.8 mA cm^{-2} vs. 9.3 mA cm^{-2}) at 1.47 V vs. RHE. It is important to note that the decrement in the cathodic current of the r_2 peak in the presence of 1,6-hexanediol was negligible as compared to that of r_1 peak, which suggests that γ-NiOOH is unlikely to be involved in the oxidation of 1,6-hexanediol (eqn (11)). In other words, β-NiOOH would be the active species responsible for the electrocatalytic oxidation of 1,6-hexanediol. This explains why the nanoNiOOH-borate electrode showed a remarkably higher catalytic current response towards the electrocatalytic oxidation of 1,6-hexanediol than the Ni|NiOOH electrode. It is also interesting to note that the applied potential to achieve specific current density in the 1,6-hexanediol-containing NaOH solution was lower than that in the blank NaOH solution. Specifically, a current density of 100 mA cm^{-2} can be achieved at 1.52 V vs. RHE in the presence of 1,6-hexanediol (0.4 M), but was not achievable at potentials < 1.7 V vs. RHE in the blank NaOH solution. This finding suggests that the e-HOR can serve as a lower-energy alter-native anode reaction for hydrogen generation from overall water splitting.

$$Ni(OH)_2 + OH^- \rightleftharpoons NiOOH + e^- + H_2O \tag{10}$$

$$NiOOH + HO(CH_2)_6OH \rightleftharpoons Ni(OH)_2 + products \tag{11}$$

The electrocatalytic properties of the nanoNiOOH-borate electrode towards e-HOR were further investigated by performing a series of 2 h controlled-potential electrolyses (CPEs) at various applied potentials in NaOH solution (1.0 M) containing 1,6-hexanediol (0.4 M), and the results are summarized in Fig. S2† and 4. As can be seen from Fig. 4, the products generated from e-HOR were mainly 6HA and AA, and the generation of these two products began at $\sim$1.40 V vs. RHE, which is higher than the formal potential of the NiOOH/Ni(OH)$_2$ redox couple. This finding suggests that the e-HOR at the nanoNiOOH-borate electrode requires the generation of NiOOH species. In addition, the applied potential was found to have a great influence on the electrocatalytic indexes, including faradaic efficiency, turnover frequency, and generation rate, of the nanoNiOOH-borate electrode towards the electrosynthesis of 6HA and AA. Regarding the electrosynthesis of 6HA, FE$_{6HA}$ decreased from 71.10 $\pm$ 4.04% to 7.99 $\pm$ 0.23% when the applied potential was increased from 1.40 V to 1.65 V vs. RHE, whereas R_{6HA} and TOF$_{6HA}$ exhibited a volcano-like behaviour in the potential range of interest and reached their maximal values, i.e., 0.27 $\pm$ 0.01 mmol cm^{-2} h^{-1} and 0.19 $\pm$ 0.01 s^{-1}, respectively, at 1.45 V vs. RHE. In the case of the electrosynthesis of AA, FE$_{AA}$, R_{AA}, and TOF$_{AA}$ also showed a volcano-like behaviour in the potential range of interest, and all of them reached their maximal values, i.e., 56.74 $\pm$ 5.63%, 0.36 $\pm$ 0.03 mmol cm^{-2} h^{-1}, and 0.26 $\pm$ 0.02 s^{-1}, respectively, at 1.55 V vs. RHE. These potential-dependent electrocatalytic properties of the nanoNiOOH-borate electrode also resulted in potential-

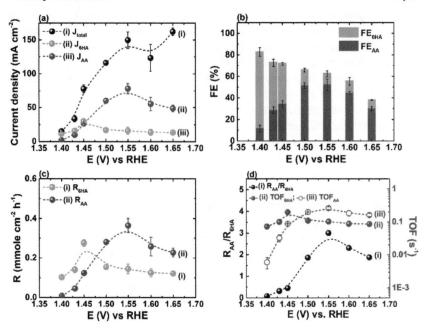

Fig. 4 (a) J_{total}, J_{AA}, and J_{6HA}, (b) FE_{AA} and FE_{6HA}, (c) R_{AA} and R_{6HA}, (d) R_{AA}/R_{6HA} ratio, TOF_{AA}, and TOF_{6HA}, obtained from 2 h CPEs of the nanoNiOOH-borate electrode at various applied potentials in NaOH (1.0 M) solution containing 1,6-hexanediol (0.4 M).

dependent product selectivity. Specifically, the selectivity for AA production, in terms of R_{AA}/R_{6HA}, increased from its lowest value (0.08) to the maximal value (3.01) when the applied potential was increased from 1.40 to 1.55 V *vs.* RHE. In other words, the selectivity for the production of 6HA or AA can be controlled by adjusting the applied potential. The SEM (Fig. 1b) and Raman (Fig. 2) analyses of the nanoNiOOH-borate electrode after 2 h CPE at 1.55 V *vs.* RHE revealed that no obvious change in the surface morphology and oxidation state of the nanoNiOOH-borate electrode was observed after electrolysis, which is indicative of the robustness of the nanoNiOOH-borate electrode. Fig. S3 and Table S1† show the results of 2 h CPEs of the nanoNiOOH-borate electrode at 1.55 V *vs.* RHE in NaOH solution (1.0 M) containing ε-caprolactone (0.4 M). It can be found that AA can be selectively (FE_{AA} = 86.35 ± 0.10%) generated from the electrochemical oxidation of ε-caprolactone. As ε-caprolactone can be fully hydrolyzed into 6-HA in alkaline aqueous solutions,[24,25] this finding suggests that AA is generated through the step-wise oxidation of 1,6-hexanediol with 6-HA as the intermediate. It is important to note that several unknown HPLC signals were found after CPEs (Fig. S4†), and their intensities increased when CPEs were performed at increasing applied potentials, which implies that some byproducts, presumably from oligomerization between the generated products,[9] were generated during the CPEs, especially at elevated potentials, and explains why both FE_{6HA} and FE_{AA} decreased at the applied potentials ≥1.60 V *vs.* RHE. The effects of electrode preparation method on the product selectivity were also investigated by subjecting the Ni|NiOOH electrode to 2 h CPEs at 1.55 V *vs.* RHE, and the results are shown in Table S2†. As revealed, the products generated from e-HOR at the Ni|NiOOH were mainly 6HA and AA. Nevertheless, the overall faradaic efficiency for the generation of these two products by the Ni|NiOOH electrode, *i.e.*, the sum of FE_{6HA} and FE_{AA}, was much less than that by the nanoNiOOH-borate electrode ($35.92 \pm 4.19\%$ *vs.* $62.79 \pm 2.45\%$), which indicates that e-HOR at the Ni|NiOOH electrode was not efficient as compared with the nanoNiOOH-borate electrode. In addition, the Ni|NiOOH electrode showed significantly lower selectivity, in terms of FE_{AA} ($17.67 \pm 4.48\%$ *vs.* $56.74 \pm 5.63\%$) and R_{AA}/R_{6HA} (0.53 *vs.* 3.01), towards the production of AA than the nanoNiOOH-borate electrode. Moreover, the Ni|NiOOH electrode exhibited lower electro-catalytic activity, in terms of TOF_{6HA} (0.06 ± 0.01 s^{-1} *vs.* 0.10 ± 0.02 s^{-1}) and TOF_{AA} (0.03 ± 0.01 s^{-1} *vs.* 0.26 ± 0.02 s^{-1}), than the nanoNiOOH-borate electrode. The inferior electrocatalytic activity of the Ni|NiOOH electrode could be attributed to the lower amount of active β-NiOOH sites available for 1,6-hex-anediol oxidation. These findings suggest that the preparation method/electrode composition would play an important role in determining the amount of active species and, thus the electrocatalytic activity.

As the hydroxide ion is involved in the electrochemical generation of NiOOH species (eqn (10)), the effects of electrolyte pH on the e-HOR activity of the nanoNiOOH-borate electrode were also investigated. Fig. S5† shows the CVs of the nanoNiOOH-borate electrode in the blank and 1,6-hexanediol (0.4 M)-containing electrolyte solutions. As can be seen from Fig. 3 and S5,† the nanoNiOOH-borate electrode showed increasing current response to 1,6-hexanediol with increasing electrolyte pH up to ∼13.9 (1.0 M NaOH). The formation of a second organic-rich phase, presumably due to the limited solubility of 1,6-hexanediol, was observed in the highly concentrated NaOH solution (5 M). In other words, the amount of 1,6-hexanediol available for e-HOR became limited, which resulted in a significant decrease in the current response of the nanoNiOOH-borate electrode in the highly concentrated NaOH solution (5 M). The nanoNiOOH-borate electrode was further subjected to a series of 2 h CPEs in 1,6-hexanediol (0.4 M)-containing borate buffer (0.1 M, pH 9.2) and NaOH solution (0.1 M, pH 13.0), and the results are shown in Fig. S6, S7† and 5. As revealed here, the sum of FE_{6HA} and FE_{AA} increased from $20.06 \pm 2.29\%$ to $55.41 \pm 4.81\%$ as the electrolyte pH was increased from pH 9.2 (0.1 M borate buffer) to pH 13.6 (1.0 M NaOH), which indicates the e-HOR became more efficient at alkaline pH. In addition, the generation of 6HA was more favourable than the generation of AA in 0.1 M NaOH solution regardless of the applied potentials. Notably, the maximal values of R_{6HA} and TOF_{6HA}, achieved at 1.50 V *vs.* RHE, were 0.38 ± 0.09 mmol cm^{-2} h^{-1} and 0.23 ± 0.06 s^{-1}, respectively, whereas the values of R_{AA} and TOF_{AA} achieved at the same applied potential were only 0.04 ± 0.00 mmol cm^{-2} h^{-1} and 0.03 ± 0.00 s^{-1}. In other words, the R_{AA}/R_{6HA} value achieved was only 0.11. These findings indicate that the product distribution was governed by both electrolyte pH and applied potentials.

The generation and dissociation of 6HA and AA would cause a pronounced pH drop within the concentration boundary layer, especially in the prolonged CPEs at higher applied potentials in the batch system. This pronounced pH drop would anodically shift the redox potential for the formation of the active NiOOH species and e-HOR, and subsequently decrease the overpotential available for e-HOR. Both the drops in local pH and overpotential available for HOR would impede the complete oxidation of 1,6-hexanediol into AA. On the other hand, the

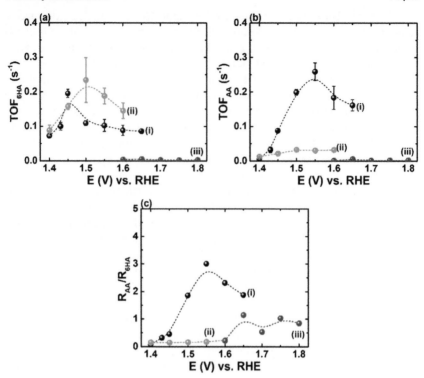

Fig. 5 (a) TOF_{6HA}, (b) TOF_{AA}, and (c) R_{AA}/R_{6HA} of the nanoNiOOH-borate electrode obtained from the 2 h electrolysis in various 1,6-hexanediol (0.4 M)-containing electrolyte solutions ((i) 1 M NaOH (pH 13.6); (ii) 0.1 M NaOH (pH 12.6); (iii) 0.1 M borate buffer (pH 9.2)).

prolonged electrolysis at high applied potential in the batch system would also increase the thickness of the concentration boundary layer and decrease the concentration gradient as well as the mass transfer rate of 1,6-hexanediol, which results in the lower availability of 1,6-hexanediol and enhances the progress of side reactions, such as OER. To minimize these mass-transfer-induced adverse impacts, a flow-type electrochemical cell (Scheme S1†) was established and used for the constant-current electrolyses. In addition, to highlight the benefits of a flow-type electrolyzer, we employed the nanoNiOOH-borate electrode with a significantly higher working area than that used for the electrolyses in the H-cell (5.0 cm^2 *vs.* 0.0706 cm^2). Fig. S8† and Table 1 show the results obtained from the constant-current electrolyses using the flow-type electrolyzer at 25 mA cm^{-2} and 50 mA cm^{-2} in NaOH solution (1.0 M, pH 13.6) containing 1,6-hexanediol (0.4 M). It can be found that the applied current density for the constant-current electrolysis played an important role in determining the product distribution. Specifically, when the applied current was increased from 25 mA cm^{-2} to 50 mA cm^{-2}, FE_{6HA} decreased from 51.24 ± 7.84% to 37.49 ± 1.98%, whereas FE_{AA} increased from 25.46 ± 1.40% to 37.01 ± 0.25%. In addition, the R_{AA}/R_{6HA} and conversion of 1,6-hexanediol were found to increase from 0.26 ± 0.06 to 0.49 ± 0.02 and 13.19 ± 2.39% to 17.29 ± 3.85%, respectively when the applied current density was increased from 25 mA cm^{-2} to 50 mA cm^{-2}. These findings indicate

Table 1 Results of the constant current-electrolyses using a flow-type electrolyzer in 1.0 M NaOH solution containing 0.4 M 1,6-hexanediol

Parameters	Total charge passage (C)	FE_{6HA} (%)	FE_{AA} (%)	R_{6HA} (μmol h^{-1})	R_{AA} (μmol h^{-1})	R_{AA}/R_{6HA}	Conversion (%)
25 mA cm^{-2}	900	51.24 ± 7.84	25.46 ± 1.40	597.49 ± 91.48	148.41 ± 8.16	0.26 ± 0.06	13.19 ± 2.39
50 mA cm^{-2}	900	37.49 ± 1.98	37.01 ± 0.25	874.14 ± 23.03	431.53 ± 2.91	0.49 ± 0.02	17.29 ± 3.85

that the selectivity for the production of AA conversion can be improved by increasing the applied current density. Furthermore, as compared with the batch electrolysis (Fig. 4), the sum of FE_{6HA} and FE_{AA} remain similar ($\sim$75%) regardless of the applied current densities, which suggests that the scale-up of electrode surface area didn't induce obvious mass transfer issues and side reactions. Note that the large working area of the nanoNiOOH-borate electrode (5 cm^2) and the semiconducting nature of the FTO substrate induced significant IR drop, which necessitated a significantly high potential to maintain a specific current density. For instance, an IR-drop-uncorrected potential of $\sim$5.5 V $vs.$ RHE was required to drive a current density of 50 mA cm^{-2} (Fig. S8†). Electrolyses at applied current densities $\geq$ 100 mA cm^{-2} were attempted but failed due to the voltage limitation of the potentiostat used in this study. The effects of the electrode substrate, such as use of an electrode substrate with higher conductivity ($e.g.$, carbon paper), the configuration of the flow-type electrolyzer ($e.g.$, the gap between anode and cathode), and the operation conditions ($e.g.$, applied current density and the adjustment of the flow rate of the electrolyte) are under investigation for further improvement of the performance of the electrolyzer.

Conclusions

The applicability of the nanoNiOOH-borate electrode for the efficient electro-synthesis of platform chemicals, $i.e.$, 6HA and AA, from e-HOR was explored and investigated in this study. The applied potential and the pH of the electrolyte used in the electrolysis of 1,6 hexanediol were found to play a crucial role in determining the overall electrocatalytic activity and selectivity of the nanoNiOOH-borate electrode. The nanoNiOOH-borate electrode exhibited significantly high selectivity towards the generation of 6HA in 0.1 M NaOH regardless of the applied potential. Nevertheless, the generation of AA became more favourable at moderate potentials (1.55 V $vs.$ RHE) in 1.0 M NaOH. The scale-up production of 6HA and AA using a flow-type electrolyzer was also successfully demonstrated.

Author contributions

Concept and design of the study: C.-Y. Lin; experiments and data collection: Y.-J. Liao; analysis and interpretation of the results: Y.-J. Liao and S.-C. Huang; draft manuscript preparation: Y.-J. Liao and S.-C. Huang. All authors reviewed and approved the final version of the manuscript.

Conflicts of interest

There are no conflicts to declare.

Acknowledgements

The authors gratefully acknowledge the financial support from the National Science and Technology Council of Taiwan (110-2221-E-006-018-MY3 and 111-2221-E-006-019-MY3). This research was also supported in part by Higher Education Sprout Project, Ministry of Education to the Headquarters of University Advancement at National Cheng Kung University (NCKU).

Notes and references

1 S. H. Pyo, J. H. Park, V. Srebny and R. Hatti-Kaul, *Green Chem.*, 2020, **22**, 4450–4455.
2 H. A. Wittcoff, B. G. Reuben and J. S. Plotkin, *Industrial Organic Chemicals*, John Wiley & Sons, Hoboken, NJ, 2012.
3 J. Y. He, K. F. Huang, K. J. Barnett, S. H. Krishna, D. M. Alonso, Z. J. Brentzel, S. P. Burt, T. Walker, W. F. Banholzer, C. T. Maravelias, I. Hermans, J. A. Dumesic and G. W. Huber, *Faraday Discuss.*, 2017, **202**, 247–267.
4 X. Kong, Y. F. Zhu, Z. Fang, J. A. Kozinski, I. S. Butler, L. J. Xu, H. Song and X. J. Wei, *Green Chem.*, 2018, **20**, 3657–3682.
5 W. Wang, Y. Wang, Z. X. Zhan, T. Tan, W. P. Deng, Q. H. Zhang and Y. Wang, *Acta Phys.-Chim. Sin.*, 2022, **38**, 2205032.
6 C. Y. Lin, S. C. Huang, Y. G. Lin, L. C. Hsu and C. T. Yi, *Appl. Catal., B*, 2021, **296**, 120351.
7 S. C. Huang, C. C. Cheng, Y. H. Lai and C. Y. Lin, *Chem. Eng. J.*, 2020, **395**, 125176.
8 W. B. Wang, Y. B. Zhu, Q. L. Wen, Y. T. Wang, J. Xia, C. C. Li, M. W. Chen, Y. W. Liu, H. Q. Li, H. A. Wu and T. Y. Zhai, *Adv. Mater.*, 2019, **31**, 1900528.
9 S. E. Michaud, M. M. Barber, K. E. R. Cruz and C. C. L. McCrory, *ACS Catal.*, 2023, **13**, 515–529.
10 V. Mairanovsky, I. Gitlin and S. Rosanov, *J. Solid State Electrochem.*, 2012, **16**, 2399–2404.
11 J. Kaulen and H. J. Schafer, *Tetrahedron*, 1982, **38**, 3299–3308.
12 M. Fleischmann, K. Korinek and D. Pletcher, *J. Electroanal. Chem. Interfacial Electrochem.*, 1971, **31**, 39–49.
13 H. Ruholl and H. J. Schafer, *Synthesis*, 1988, 54–56.
14 B. V. Lyalin and V. A. Petrosyan, *Russ. J. Electrochem.*, 2010, **46**, 1199–1214.
15 A. S. Vaze, S. B. Sawant and V. G. Pangarkar, *J. Appl. Electrochem.*, 1997, **27**, 584–588.
16 H. J. Schafer, *Top. Curr. Chem.*, 1987, **142**, 101–129.
17 C. Y. Lin, Y. C. Chueh and C. H. Wu, *Chem. Commun.*, 2017, **53**, 7345–7348.
18 S.-C. Huang and C.-Y. Lin, *Electrochim. Acta*, 2019, **321**, 134667.
19 R. Kostecki and F. McLarnon, *J. Electrochem. Soc.*, 1997, **144**, 485–493.
20 B. S. Yeo and A. T. Bell, *J. Phys. Chem. C*, 2012, **116**, 8394–8400.
21 M. W. Louie and A. T. Bell, *J. Am. Chem. Soc.*, 2013, **135**, 12329–12337.
22 A. Van der Ven, D. Morgan, Y. S. Meng and G. Ceder, *J. Electrochem. Soc.*, 2006, **153**, A210–A215.
23 Y. H. Chung, I. Jang, J. H. Jang, H. S. Park, H. C. Ham, J. H. Jang, Y. K. Lee and S. J. Yoo, *Sci. Rep.*, 2017, **7**, 13792.
24 H. K. Hall, M. K. Brandt and R. M. Mason, *J. Am. Chem. Soc.*, 1958, **80**, 6420–6427.
25 Y. Yang, W. Du, G. Qian, X. Z. Duan, X. Y. Gu, X. G. Zhou, Z. R. Yang and J. Zhang, *AIChE J.*, 2023, **69**, e17867.

DISCUSSIONS

Understanding and controlling organic electrosynthesis mechanism: general discussion

Mickaël Avanthay, Belen Batanero, Christoph Bondue, Dylan G. Boucher, Pim Broersen, Richard C. D. Brown, Luke Chen, Anthony Choi, Ching Wai Fong, Toshio Fuchigami, David P. Hickey, Alexander Kuhn, Kevin Lam, Yun-Ju Liao, T. Leo Liu, Shelley D. Minteer, Kevin Moeller, Zachary A. Nguyen and Naoki Shida

DOI: 10.1039/d3fd90038j

Shelley D. Minteer opened discussion of the paper by T. Leo Liu: Based on your results, it seems imperative that in the field of electrosynthesis we need to focus more on *in situ* and *operando* techniques to fundamentally understand our systems. What *in situ* and *operando* techniques do you think are best for studying molecular electrocatalysis systems and why?

T. Leo Liu answered: It is a great question. Generally, mechanistic studies are critical to gaining an in-depth understanding of a reaction mechanism, catalytic or non-catalytic. Characterization of catalytic intermediates is highly informative to understand reaction mechanisms. Elemental reaction studies of catalytic intermediates can further gain information about reaction selectivity and kinetics and thus inspire improved catalyst designs and reaction conditions. In this regard, *in situ* and *operando* spectroelectrochemistry (UV-Vis and IR based) are very suitable to investigate reaction intermediates for electrosynthesis. Both UV-Vis and IR with their fast relaxation and spectroscopic features can effectively track short-life, transient species generated by electrochemical reduction or oxidation *in situ* and provide both characterization and kinetic information about a reaction intermediate.

Kevin Lam asked: Why did you choose a trifluoroborate as the coupling partner?

T. Leo Liu replied: Organic trifluoroborate is redox active and used as a nucleophile to generate an organic radical by anodic oxidation. In our case, we used benzyl trifluoroborate. The benzyl radical is relatively stable and can be trapped by a Ni(II)Ar intermediate for cross-coupling. Please see our previous publication.[1] We have identified other redox active nucleophiles for our Ni e-couplings and will publish our new results soon.

1 J. Luo, B. Hu, W. Wu, M. Hu and T. L. Liu, *Angew. Chem., Int. Ed.*, 2021, **60**(11), 6107–6116, DOI: 10.1002/anie.202014244.

David P. Hickey enquired: The amount of bipyridine ligand used in nickel-catalyzed electroreductive cross-couplings was shown previously to be important for product yield when reactions are performed galvanostatically.[1] Did you investigate the effect of ligand concentration relative to amount of NiBr$_2$?

1 Y. Kawamata, J. C. Vantourout, D. P. Hickey, P. Bai, L. Chen, Q. Hou, W. Qiao, K. Barman, M. A. Edwards, A. F. Garrido-Castro, J. N. deGruyter, H. Nakamura, K. Knouse, C. Qin, K. J. Clay, D. Bao, C. Li, J. T. Starr, C. Garcia-Irizarry, N. Sach, H. S. White, M. Neurock, S. D. Minteer and P. S. Baran, *J. Am. Chem. Soc.*, 2019, **141**, 6392–6402, DOI: 10.1021/jacs.9b01886.

T. Leo Liu answered: We did ligand screening and found dtbbpy is the ligand to give the highest cathodic current. We also optimized the ratio of dtbbpy and Ni and found the ratio of 1.5 gives the highest cathodic current. Then we used NiBr$_2$/1.5dtbbpy as the standard catalyst and achieved pretty reasonable yields. We did not test other ratios for electrolysis. Please see our previous publication.[1]

1 J. Luo, B. Hu, W. Wu, M. Hu and T. L. Liu, *Angew. Chem., Int. Ed.*, 2021, **60**(11), 6107–6116, DOI: 10.1002/anie.202014244.

David P. Hickey asked: Under galvanostatic operation, electrochemical nickel-catalyzed cross-coupling reactions can suffer from nucleation of underligated Ni(0) at low ligand concentration. Using your approach, did you observe any formation of insoluble Ni(0) aggregates?

T. Leo Liu replied: It is a great question. With an optimized Ni catalyst which has a ratio of 1.5 between the bis-*tert*-butyl bipyridine ligand and nickel, the redox neutral cross-coupling of benzyl borate and aryl halide takes place at the redox couple of Ni$^{(2+/1+)}$ of *ca.* −1.9 V *vs.* Fc. We used a three electrode cell to monitor the voltage profile of the reaction. The voltage profile stayed around Ni$^{(2+/1+)}$ and did not reach the Ni(0) redox state (−2.5 V *vs.* Fc). We also did other control experiments such as SEM studies and did not observe any deposited Ni on the electrode.

Toshio Fuchigami remarked: You can generate radical species or cations selectively by anodic oxidation of the corresponding trifluoroborate without use of a nickel complex. Please try.

T. Leo Liu responded: Thanks your advice. We will look at one and two electron oxidation of benzyl trifluoroborate without use of a nickel complex.

Toshio Fuchigami queried: Is the cation also stable enough?

T. Leo Liu replied: In our catalysis, the benzyl borate selectively undergoes one electron oxidation to form a benzyl radical. And then the benzyl radical is efficiently trapped by a Ni catalyst without further oxidation to a benzyl cation.

Belen Batanero enquired: Have you got a problem with stability of the nickel catalyst under oxygen? Does the catalyst decompose at room temperature or in protic solvents?

T. Leo Liu responded: I think nickel catalysts at the oxidation state of Ni(+1) are not stable under oxygen. In fact, our Ni-catalyzed cross-coupling reactions are operated under a nitrogen atmosphere. Ni catalysts are stable at room temperature but are sensitive to protic solvents like water as it causes side reactions such as hydrodebromination of aryl bromide.

Belen Batanero asked: What about the use of a manganese complex as the catalyst? Or other similar metallic complexes, do you think any of them can provide the same results?

T. Leo Liu answered: I am not sure about manganese as it is not common to use Mn as a metal for cross-coupling reactions. Based on literature, I think Fe and Co with a set of appropriate ligands can have potential for electrochemical cross-coupling reactions.

Ching Wai Fong enquired: For the amido nickel complex mentioned in the talk, since it forms a stable intermediate that discourages the formation of the homo-coupling product, does that affect the formation of the cross-coupling product as well? What is the difference in selectivity, yield and reaction rate with the introduction of the amido ligand, and does that limit the possible electrophiles used in the cross-electrophile coupling?

T. Leo Liu responded: Thanks for asking the important questions. The nickel amido complex can avoid homo-coupling and thus favor cross-coupling. With an amido ligand, the reaction selectivity for cross-electrophile coupling is significantly improved by 5 times with a good to high reaction yield. We did not pay attention to the rate but we need to do that. The reaction scope is good.

Kevin Moeller asked: Now that you know the role of the Ni(ɪɪ) intermediate in forming the dimer, is there a way to slow down dissociation of the phenyl anion so that it does not transfer to your key Ni(ɪɪɪ) intermediate? What I am interested in is a pathway to the cross-electrophile coupling reaction.

T. Leo Liu replied: It is a great question. As discussed in my presentation, we found that a Ni(ɪɪ)(amino)(Ar) intermediate is not only reluctant to undergo reductive elimination but is also highly stable against Ar–Ar homo-coupling. Our ongoing research revealed that this Ni(ɪɪ)(amino)(Ar) design can be utilized to promote highly selective cross-electrophile couplings.

Mickaël Avanthay opened discussion of the paper by Dylan G. Boucher: Why is the impact on the current intensity higher as the concentration of the substrate increases (Fig. 2 in the paper; https://doi.org/10.1039/d3fd00054k)? Increasing the substrate concentration lowers the cation/substrate ratio, so one could expect an erosion of the effect of the nature of the cation.

Dylan G. Boucher responded: Our interpretation is that the impact of the electrolyte is magnified due to higher turnover of the catalyst at higher substrate concentrations. Essentially, the more frequent turnover means a higher current and more probability of seeing the impact of the electrolyte. It is an interesting point, it would be quite interesting to see how electrolyte concentrations impact this effect, both at higher and lower electrolyte concentrations.

Luke Chen asked: What do you hope to achieve with these results in terms of practical applications and are you able to use the current data to develop a predictive model, especially together with AI and machine learning?

Dylan G. Boucher answered: We are not yet at the predictive stage, though it is clear that the exact dependence or effect of the electrolyte depends on the type of reaction. For example in Prof. Kevin Moeller's work,[1] the anodic cyclization, selectivity was driven by methanol inclusion at the double layer and so benefits from smaller, harder electrolyte ions. In contrast, in Song Lin's hydrocyanation work[2] and in the work presented here (https://doi.org/10.1039/d3fd00054k), where the reaction is driven by a mediated electrochemical reaction, selectivity benefits from a larger, softer electrolyte ion. A general observation might be that formation of a organometallic intermediate is favored by the inclusion of soft alkyl ammonium ions; extending this kind of general insight to other classes of reactions or other intermediates presents an exciting, and really very exciting avenue for electrosynthesis.

1 A. Redden and K. D. Moeller, *Org. Lett.*, 2011, **13**, 1678–1681.
2 L. Song, N. Fu, B. G. Ernst, W. H. Lee, M. O. Frederick, R. A. DiStasio Jr and S. Lin, *Nat. Chem.*, 2020, **12**, 747–754.

Pim Broersen enquired: You take great care in your experimental setup to work in dry conditions; what method do you use to dry your electrolyte salt?

Zachary A. Nguyen replied: These salts were double recrystallized and dried in a vacuum oven at 50 °C for 48+ hours.

Pim Broersen asked: Did you test the final water concentrations of your system *via* Karl-Fischer?

Dylan G. Boucher responded: We did not, but this is a great suggestion for the future.

Zachary A. Nguyen replied: The water concentrations were not determined by Karl-Fischer, but background scans detected no water in the solution.

Pim Broersen commented: In our group we have similar issues where we have to work in very dry conditions; as a tip I would suggest drying electrolyte salts *in vacuo* over P_2O_5.

Dylan G. Boucher responded: This is a great suggestion! We will absolutely look into this; moisture from the electrolyte is a typical source of confusion for these types of experiments.

Pim Broersen remarked: The hardness of the salt can also impact how much water it shuttles to your substrate. In your mechanism you show the final reductive cleavage of the metal–organic bond. In the presence of water this could turn into a proton-coupled electron transfer (PCET) step, thereby reducing the activation barrier necessary for the reaction, giving you an extra unknown variable to your system. It is therefore really important to know that you aren't introducing any water into the system (or if you are, that it is at least constant between different electrolyte salts). Since you are already working under very dry conditions I guess you agree with that statement, but the electrolyte salt remains an often overlooked source of water in these systems, especially since the nature of most salts is that they are hygroscopic.

Dylan G. Boucher responded: You are absolutely correct that proton content of the solution can introduce a more complicated PCET step (as studied by Professors Costentin and Savéant)[1] that can be much faster than the decay in the absence of protons. This is precisely why these experiments were conducted in an inert atmosphere and with dried electrolytes.

1 C. Costentin, G. Passard, M. Robert and J.-M. Savéant, *Chem. Sci.*, 2013, **4**, 819–823.

T. Leo Liu enquired: Have you look at the solvent effect on the structure of an active species in the double layer and thus reaction selectivity?

Dylan G. Boucher replied: Not yet but we are very interested in this! The double layer is full of interesting effects often incongruous with bulk behavior of molecular species. It is reasonable to expect that in the altered electric field and dielectric (see the recent work of Gerald Meyer)[1,2] of the double layer reactivity of molecular electrocatalysts becomes quite different. While we have not explored this aspect yet, we are very interested in the possibility of *operando* and *in situ* techniques for looking at reactions within the double layer.

1 R. E. Bangle, J. Schneider, E. J. Piechota, L. Troian-Gautier and G. J. Meyer, *J. Am. Chem. Soc.*, 2020, **142**, 674–679.
2 R. E. Bangle, J. Schneider, D. T. Conroy, B. M. Aramburu-Troselj and G. J. Meyer, *J. Am. Chem. Soc.*, 2020, **142**, 14940–14946.

T. Leo Liu said: The structure of a redox active species and conditions of an electrical double layer are very intriguing. Gaining such information is very valuable in controlling electrochemical reaction selectivity.

Dylan G. Boucher responded: I agree completely. Selectivity from interfacial engineering is a neglected aspect of the present literature.

Belen Batanero commented: The influence of electrolyte in an electrochemical process is not new (Monsanto adiponitrile electrosynthesis is an example). The nature and distribution of products are sometimes strongly modified by changing the supporting electrolyte. The solvation effect of the solvent on the ion-pairs formed influences the stability of anion radicals and therefore their behaviour. However in your case there are so many parameters influencing at one time that it could be convenient to study step by step in a separated manner.

Dylan G. Boucher responded: I absolutely agree that these effects are not new, and I do not mean to suggest that they are. What is lacking in the present literature, however, is the synthetic logic of choosing one supporting electrolyte species over another for a particular reaction, especially in mediated electrochemical reactions which take place further away from the electrode surface and are less impacted by double layer contributions. Given that electrosynthesis has become quite popular in the past decade, including with many that are new to the field with no electrochemical background, it seems prudent to explicitly outline the "why?" of choosing electrolytes for reaction design.

Considering the complexity of our system, I take the opposing view: it is the simplicity of it that makes it a good model system. The voltammetry of the metal porphyrin species is well understood and their reactions with benzyl halides outlined in detail elsewhere. Furthermore, it involves the formation of a stable organometallic intermediate that decays upon subsequent electron transfer, which makes it perfect for observing electrolyte induced changes in the rate of decay.

David P. Hickey asked: Porphyrins are notorious for adsorbing onto carbon electrodes in an aqueous environment. While you are working in non-aqueous conditions, I wonder whether adsorption may be occurring in the presence of different electrolytes. Did you perform variable scan rate analysis (*i.e.*, peak current *vs.* scan rate or peak current *vs.* square root of scan rate) to determine whether any of the Co or Fe porphyrin complexes are partially adsorbed to the electrode surface?

Dylan G. Boucher replied: This is an excellent point; we do not anticipate any adsorption in organic solvent (DMF) as noted in previous studies of Co and Fe

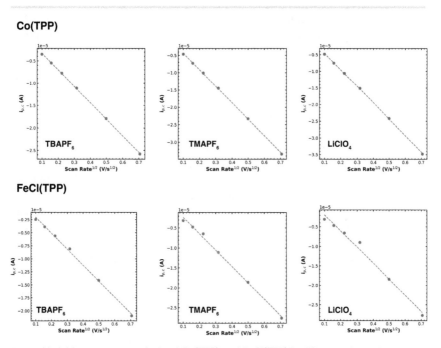

Fig. 1 Variable scan rate analysis of Co(TPP) and FeCl(TPP) in different electrolytes.

tetraphenylporphyrins.[1,2] This was further confirmed by variable scan rate analysis as you mention (see Fig. 1 here). Both Co and Fe TPP catalysts showed a square-root dependence on scan rate, indicating a freely diffusing species. We note a slight deviation from linearity in the case of FeCl(TPP) in LiClO$_4$, but a square root dependence remained a better fit than linear dependence.

1 D. Lexa, J. M. Savéant and J. P. Soufflet, *J. Electroanal. Chem. Interfacial Electrochem.*, 1979, **100**, 159–172.
2 D. Lexa, J. Mispelter and J. M. Saveant, *J. Am. Chem. Soc.*, 1981, **103**, 6806–6812.

Naoki Shida requested: The authors rationalized the electrolyte effect based on the change in the surrounding environment of the molecule of interest *via* the softness and hardness of electrolyte-derived cation species. It seems also possible to think about the ion pairing of electrolyte-derived cations and electrochemically generated species, based on parameters such as Lewis acidity of cations. Is there any comment on this point?

Dylan G. Boucher responded: Yes this is an excellent point! It is our viewpoint that these rationalizations are compatible, and both involve the same operative explanation: it is the strength of ionic interaction that stabilizes (or destabilizes) the intermediate. We chose to describe it in terms of the softness and hardness of the solvation shell for two reasons: (i) in high dielectric solvents (like DMF) and high supporting electrolyte concentrations formation of discrete ion pairs seemed less likely, and (ii) the changes to the hydrodynamic radius seen from the diffusion constant in different electrolytes implied changes to the solvation structure. Again, we think Lewis acidity and hard–soft acid–base interactions are compatible viewpoints here: they are both ways to rationalize the stabilization of charge.

Christoph Bondue said: This comment relates to the statement by Dr Boucher that he has trouble to see how ionic interactions play a role when everything is surrounded by a sea of electrolyte.

The fact that we have activity coefficients means that we have ionic interactions even in solutions. This must be more severe in the electrochemical double layer. In this context one should also consider the primary kinetic salt effect that acts *via* an effect on the activity coefficient.

Dylan G. Boucher replied: I think this is an excellent point. The intention of my comment was to highlight the complicated interplay of solvation shell, solvent dielectric, and high salt concentrations. I did not mean to discount ionic interactions (indeed I think they are actually at the heart of this study), only that, given the relatively high dielectric of DMF and abundance of charge there will be other factors at play as well. After reading more into the subject, I do believe you are correct about the interactions being more severe in the double layer. Considering the dielectric of solvents (mostly studied for water, but perhaps can be extended to other solvents) is dramatically lowered at polarized interfaces, one can expect screening of charge to be drastically lowered as well. This is an interesting point of view that we had not considered! Thank you for bringing it to our attention.

Christoph Bondue remarked: After Dr Boucher still questions the relevance of ionic interactions, I reaffirm that the existence of activity coefficients means that ionic interactions exists. This should be seen in context with the work of Marc Koper on the double layer capacity:[1,2] the reduced Debye length suggests the existence of ion pairing in the electrochemical double layer.

1 K. Doblhoff-Dier and M. T. M. Koper, Modeling the Gouy–Chapman Diffuse Capacitance with Attractive Ion–Surface Interaction, *J. Phys. Chem. C*, 2021, **125**(30), 16664–16673, DOI: 10.1021/acs.jpcc.1c02381.
2 K. Ojha, K. Doblhoff-Dier and M. T. M. Koper, Double-layer structure of the Pt(111)–aqueous electrolyte interface, *Proc. Natl. Acad. Sci. U. S. A.*, 2022, **119**(3), e2116016119, DOI: 10.1073/pnas.2116016119.

Dylan G. Boucher responded: Again, I did not mean to imply that ionic interactions are not relevant or that they are not present: indeed ionic stabilization of the intermediates is the operative mechanism we propose in the paper (https://doi.org/10.1039/d3fd00054k). My comments were meant to imply that discrete ion pairing is likely more complicated in light of high dielectric solvent and high electrolyte concentrations. Marc Koper's work on the double layer capacitance is an excellent example of this complication: the effects of the surface attraction (and reduced Debye length) are only seen at low electrolyte concentrations where the inner Helmholtz capacitance does not dominate.

I think, in the end we are really saying the same thing: interaction of the intermediate with electrolyte is one of electrostatic stabilization.

Alexander Kuhn opened discussion of the paper by Yun-Ju Liao: During your reaction you form also 6-hydroxyhexanoic acid as a bifunctional product and I wonder whether you observed the formation of (poly)condensation products such as polyester type oligomers?

Yun-Ju Liao answered: There is a possibility that there are polyester type oligomers. We have done a mass spectrum on the reactant for those unknown peaks found in the HPLC analysis and found out that there is a nearly double charge-to-mass ratio than 1,6-hexanediol.

Kevin Lam asked: What about scaling up your reaction? Do you think you could easily scale it up to g or kg scales?

Yun-Ju Liao replied: We haven't thought about scaling up our current setup, since the current electrolyte conditions are very alkaline. Therefore, we are now focusing on the optimization of the electrocatalyst so it can conduct under neutral conditions. If we can find the appropriate electrocatalyst, then it is time for scaling up our setup.

Anthony Choi enquired: In your talk (https://doi.org/10.1039/d3fd00073g) you described how you incorporated your coated electrodes into a flow setup. I was wondering after performing your reaction, do you see any degradation of your coated electrodes or any fouling?

Yun-Ju Liao responded: After 2 hours of electrolysis in the flow system, the nano-NiOOH borate remains in the electrode surface.

Anthony Choi asked: I was wondering with the flow setup you described and your reaction, have you thought about how you are going to optimise the reaction if you have to scale it up in the future? Can you make any comments on any issues that may arise?

Yun-Ju Liao answered: We are facing a massive iR drop in our system, so we are now considering two cases. Since the conductivity of the FTO (fluorine-doped tin oxide) causes large resistance, we are trying to use carbon paper to solve this problem. On the other hand, the membrane will also cause the resistance, so a removal can also be done.

Richard C. D. Brown requested: Could you explain whether a reference electrode is included in the flow cell, and if so, how it is located within the electrolyte chamber.

Yun-Ju Liao replied: The location of the reference electrode is shown in Fig. 2 here.

Richard C. D. Brown enquired: Following on from the previous question, how large is the gap between the electrode and membrane in the flow cell?

Yun-Ju Liao answered: The distance between the reference and the working electrode is 0.7 cm.

Toshio Fuchigami asked: What is the difference between a nickel oxide anode and your anode? Regarding the mechanism, I think that nickel oxyhydroxide on the surface of the anode works as an oxidant just like an oxidation mediator. Is that right?

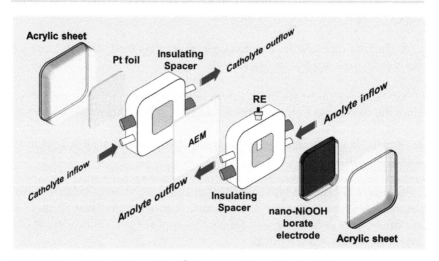

Fig. 2 Schematic illustration of 5 cm^2 flow-type electrochemical cell.

Yun-Ju Liao replied: Yes, we also think that the nickel oxyhydroxide works as the oxidant in this reaction. However, if the nickel oxyhydroxide is the only effect, the reaction should be a potential-independent reaction. In our case the electrooxidation of 1,6-hexanediol is a potential-dependent reaction, so nickel oxyhydroxide may not only act as a chemical oxidant.

Toshio Fuchigami questioned: Your nanoporous anode seems to have a much larger surface area than conventional nickel anodes. Is that right?

Yun-Ju Liao answered: Yes, as we can see in the SEM image, the electrocatalyst is formed of nanoparticles and can have a larger surface area than the nickel-plate anode.

Pim Broersen addressed T. Leo Liu and Dylan G. Boucher: In a broader sense, when looking at electrosynthesis you have two main problems, the fact you have to work in a heterogeneous system, where you have mass transfer limitations, and the fact there is an electrolyte salt that later needs to be purified out. In homogeneous catalysis, you have the same purification issue of the catalyst after the reaction, but since it's not a heterogeneous system there is a benefit. In combining the two, don't you get a little bit of the worst of both worlds? When moving to industrial application, how can these challenges be solved?

Dylan G. Boucher responded: My half joke, half serious answer to this is that every catalyst or reaction should aspire to "put it on the electrode surface and do it in water". Obviously this is an oversimplification of the requirements of organic electrosynthesis, but it is worth considering ways to streamline purification and sustainability of these reactions. At batch scale, the vast majority of pharmaceutical companies utilize homogeneous catalysts or mediators in organic solvent for their electrochemical reactions. Finding ways to achieve the same selectivity that these companies require while also curtailing the use of toxic solvents and streamlining separations is an admirable goal.

T. Leo Liu replied: Regarding mass transport, it can be regulated in a flow synthesis cell where a flow rate can be adjusted as needed.

As for electrolyte salts, they are polar and can be easily removed using column chromatography and even can be recycled with additional work-up. In addition, the cost of electrolyte salts can be negligible compared to valuable organic products.

Conflicts of interest

There are no conflicts to declare.

Faraday Discussions

PAPER

Dihydrolevoglucosenone (Cyrene™), a new possibility of an environmentally compatible solvent in synthetic organic electrochemistry†

Jose Manuel Ramos-Villaseñor,[ab] Jessica Sotelo-Gil,[a] Sandra E. Rodil [iD][c] and Bernardo Antonio Frontana-Uribe [iD] *[ab]

Received 11th March 2023, Accepted 23rd March 2023

DOI: 10.1039/d3fd00064h

Dihydrolevoglucosenone (DLG or Cyrene™) solvent is a green dipolar solvent produced from cellulose waste. Different studies have demonstrated that it can successfully replace dipolar solvents, such as N,N-dimethylformamide (DMF), N,N-dimethylacetamide (DMA) and N-methylpyrrolidinone (NMP), in a variety of chemical reactions. In this paper, the first application of DLG in organic electrosynthesis is described, with results of its use in the electrochemical reduction of benzophenone derivatives (ca. E = −1.75 V vs. Ag/AgCl), as a greener alternative to other dipolar solvents with environmental concerns. Conductivity measurements show that the solvent presents conductivity and viscosity limitations that can be overcome by using EtOH as a cosolvent. The DLG/EtOH mixture resulted in a convenient solvent to carry out galvanostatic electroreductions of starting materials that exhibit high potential value. Furthermore, the reaction pathway (1e⁻ or 2e⁻) was found to be dependent on the supporting electrolyte used; TBABF$_4$ favored 2e⁻ reduction to the corresponding alcohol (52–85%), whereas LiClO$_4$ promoted C–C pinacolic coupling (47–70%).

Introduction

In organic synthesis, the solvent solubilizes organic compounds and promotes organic reactions, but in organic electrosynthesis (OES), besides this, the solvent should also solubilize the corresponding supporting electrolyte to allow conductivity and carry out the electrochemical reaction. Although OES is considered an

[a] Centro Conjunto de Investigación en Química Sustentable UNAEM-UNAM, Toluca, 50200, Estado de México, Mexico. E-mail: bafrontu@unm.mx

[b] Instituto de Química, Universidad Nacional Autónoma de México, Circuito Exterior. Ciudad Universitaria, Coyoacán, 04510 CDMX, Mexico

[c] Instituto de Investigación en Materiales, Universidad Nacional Autónoma de México, Circuito Exterior, Ciudad Universitaria, Coyoacán, 04510 CDMX, Mexico

† Electronic supplementary information (ESI) available. See DOI: https://doi.org/10.1039/d3fd00064h

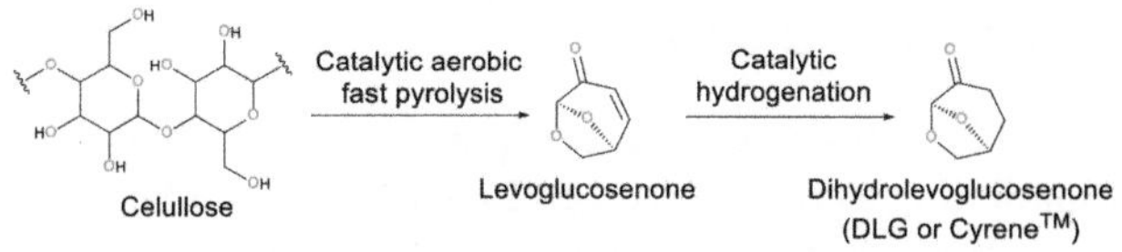

Scheme 1 Production of DLG from cellulose mass.

Table 1 Physicochemical properties of selected dipolar aprotic solvents (DMSO = dimethylsulfoxide)[8,21]

Properties	DLG	NMP	DMF	DMSO	DMA
Bp (°C)	227	202	153	189	165
Density (g mL^{-1})	1.25	1.03	0.948	1.10	0.937
π^*	0.93	0.90	0.88	1	0.85
ε	37.3 (ref. 22)	33	36.7	46.7	37.8
Viscosity (cP)	14.5 (ref. 23)	1.65	0.9	1.99 (ref. 24)	2.14

environmentally friendly methodology that uses electrons as a reagent instead of using toxic redox compounds,[1,2] this is not completely true,[3] since one of the biggest concerns during scaling-up is the use of aprotic solvents such as N,N-dimethylformamide (DMF), N,N-dimethylacetamide (DMA) and N-methylpyrrolidinone (NMP), which are considered toxic to both the environment and human health.[4–6] Nowadays, to increase the green factor of OES reactions, safer alternatives should be explored to replace these traditional dipolar aprotic solvents. Ideally, water should be the chosen solvent due to the excellent conductivity of ionic aqueous solutions, but the possibility to carry out organic electrosynthetic reactions in aqueous media or water–organic-solvent mixtures is limited by the solubility of the organic starting compounds.[7] This environmental concern can be overcome with the use of green solvent alternatives, such as 6,8-dioxabicyclo[3.2.1]octan-2-one, known as dihydrolevoglucosenone (DLG), which is a solvent commercially available as Cyrene™.[8] It is a biodegradable and biorenewable solvent derived from cellulose mass, and manufactured on a large scale[8] (Scheme 1). It is a biorenewable non-toxic solvent, and a green alternative to traditional dipolar solvents. According to Table 1, DLG has some similar physicochemical properties to DMF, DMA and NMP, such as boiling point (bp), polarity and dipolar moment; these features could make the use of DLG in electrosynthesis possible, particularly to substitute DMF.[9] Several groups worldwide have shown the possibility of using DLG successfully in organic synthesis. For example, it has been used in palladium-catalysed C–C cross-coupling reactions,[10,11] the Baylis–Hillman reaction,[12] the Menschutkin reaction,[13] the synthesis of amides,[14] ureas[15] and isothiocyanates,[16] biocatalysis[17] and the fabrication of value-added materials.[18–20] However, the use of DLG as a solvent in electrochemistry, in particular OES, has not been explored yet.

In this article, the first use of DLG as a solvent in the electrochemical reduction of benzophenone and its derivatives is reported, demonstrating its potential to successfully replace DMF. Some limitations were found, but its mixture with EtOH allowed them to be successfully overcome.

Results and discussion

Given that DLG has a similar polarity to dipolar aprotic solvents, the dissolution of classical supporting electrolytes (Bu_4NBF_4, Bu_4NClO_4 and Bu_4NPF_6) was tested at up to 1 M; Bu_4NPF_6 reaches oversaturation at concentration (c) > 0.8 M, but the first two dissolved completely. Salts with a smaller ammonium cation, *e.g.*, Et_4N^+ or Me_4N^+, or with Li^+ did not show apparent dissolution at room temperature (25°) at $c = 0.01$ M. Nonetheless, to achieve useful electrolyte solutions, solubility is not enough, and ionic dissociation is required. This process can be monitored by conductivity (σ) measurements of the electrolyte solution in DLG at different concentrations (Fig. 1).

In pure DLG (Fig. 1a), at low concentrations the conductivity remains similar for the three salts, but at $c > 0.2$ M the conductivity increases slightly in the

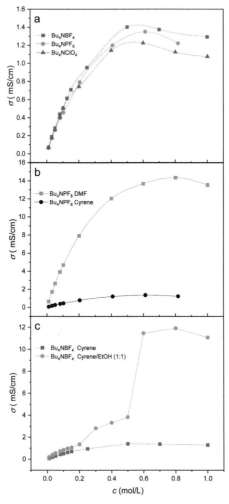

Fig. 1 (a): Plot of σ *vs.* c of tetrabutylammonium salts in DLG. (b) Plot of σ *vs.* c of Bu_4NPF_6 in DLG (black line) in comparison with DMF (red line). (c) Plot of σ *vs.* c of Bu_4NBF_4 in DLG (black line) in comparison with a DLG/EtOH (1 : 1) mixture. $T = 25\ °C$.

following order: $Bu_4NBF_4 > Bu_4NPF_6 > Bu_4NClO_4$, indicating that the first has better conductivity than the others in this solvent. Interestingly, a maximum in conductivity is reached for the three electrolytes in the range of 0.5–0.6 M (Bu_4NBF_4 1.4 mS cm^{-1}, Bu_4NPF_6 1.35 mS cm^{-1} and Bu_4NClO_4 1.22 mS cm^{-1}). At higher concentrations the conductivity decreases slowly due to the formation of ion pairs. It was not possible to obtain the conductivity of Bu_4NPF_6 at $c > 0.8$ M due to oversaturation. Fig. 1b shows that the conductivity of BuN_4PF_6 in DLG is approximately one order of magnitude smaller than that of Bu_4NPF_6 in DMF, a fact explained by the viscosity of DLG, which is 16 times higher than that of DMF (Table 1). Thus, although DLG can dissolve common supporting electrolytes used in OES, its high viscosity has a negative impact on the conductivity of the electrolyte solution, a property necessary to carry out electrochemical transformations on a large scale. As shown in Fig. 1a, conductivities in this medium are low, resembling electrolyte solutions in low-dielectric-constant solvents; for example, the conductivity of Bu_4NBF_4 at 1 M in THF is 4.4 mS cm^{-1} (ref. 25) and that in DME is 4.4 mS cm^{-1} (ref. 25), which are even higher than in DLG (1.29 mS cm^{-1}). Viscosity problems can be avoided by adding a green cosolvent with the purpose of reducing the viscosity and increasing the conductivity.[26] Ethanol was chosen as a green cosolvent in a 1 : 1 proportion of DLG/EtOH and as shown in Fig. 1c, this resulted in an 8.5-fold increase in conductivity at Bu_4NBF_4 concentrations of 0.7–0.8 M. In the range of low concentrations, $c < 0.2$ M, the conductivity shows a similar behaviour, but a similar behaviour, but showing almost two times higher values. From 0.2 M, the conductivity undergoes a major increment until 0.5 M. At higher concentrations, an unexpected rise in conductivity occurs, reaching a similar value to that of Bu_4NPF_6 in DMF as a traditional dipolar aprotic solvent. This unexpected behaviour could be due to changes in the intermolecular interactions of the DLG/EtOH blend, promoting an efficient dissociation of the supporting electrolyte at these concentration levels. The greatest conductivity value is reached at a concentration of 0.8 M and then the conductivity decreases due to formation of ionic pairs. Further studies about the physicochemistry of this electrolyte system is beyond the objectives of this article.

With the purpose of exploring the electrochemical behaviour of DLG on a glassy carbon (GC) electrode, cyclic voltammetry was carried out with Bu_4NBF_4 and Bu_4NPF_6 as supporting electrolytes at $c = 0.15$ M. The potential window of Cyrene™ obtained by cyclic voltammetry is depicted in Fig. 2. These experiments showed that DLG is electrochemically stable in the potential range between -1.5 to 1.75 V. When EtOH is added in 50% proportion, and using Bu_4NBF_4, the potential window of the DLG/EtOH 1 : 1 mixture surprisingly showed a cleaner and larger potential window of -1.8 V to 2.5 V on the glassy carbon electrode (Fig. 3). The use of Bu_4NPF_6 (not shown) instead of Bu_4NBF_4 did not significantly change the potential window or electrochemical behaviour in this solvent mixture.

The large anodic window of the DLG/EtOH mixture indicates that this electrolytic system is stable under electrochemical oxidation conditions until 2.5 V, opening the door to the anodic oxidation of many organic compounds in this green solvent mixture. Although its cathodic window is slightly smaller than that of DMF, it is sufficiently wide for high-potential-value electroreductions like carbonyl reductions, with the reduction potential (E_{red}) generally higher than -1.5 V vs. Ag/Ag$^+$. Furthermore, a good response was obtained at $c = 0.15$ M, even

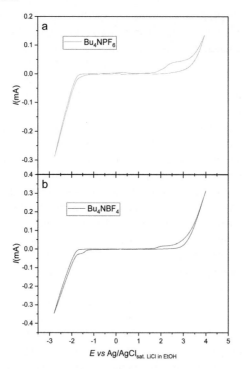

Fig. 2 Cyclic voltammograms of Bu_4NPF_6 (a) and Bu_4NBF_4 (b) at c = 0.15 M in DLG. Working electrode (WE): glassy carbon, counter electrode (CE): Pt, reference electrode (RE): Ag/AgCl, sat. LiCl(s) in EtOH. Scan rate: 0.1 V s^{-1}, second cycle is shown.

if the conductivity was only double that with pure DLG (Fig. 1c). The supporting electrolyte is also a major concern in OES; its cost and recyclability sometimes limit reactions on large scales.[27] Thus, this concentration was chosen to pursue our electrosynthetic study. With the electrochemical stability results in hand, it was decided to explore the use of the DLG/EtOH mixture in OES in this electrolytic system.

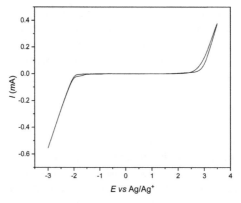

Fig. 3 Representative cyclic voltammogram of the potential window of 0.15 M Bu_4NBF_4 in a DLG/EtOH (1 : 1) mixture. WE: glassy carbon, CE: Pt, RE: Ag/AgCl, sat. LiCl in EtOH. Scan rate: 0.1 V s^{-1}.

Xia and collaborators have reported the preparation of alcohols and diols from electrochemical reduction of carbonyl compounds, such as aldehydes and ketones, in which DMF is used as the solvent.[28] Cyclic voltammetry experiments in 1 : 1 DLG/EtOH with Bu$_4$NBF$_4$ revealed that the reduction of benzophenone and derivatives occurs irreversibly at cathodic potentials higher than −1.3 V vs. Ag/AgCl (ESI†); thus, their reductive transformation was investigated. Using benzo-phenone (1) as a model substrate under Xia's conditions, and replacing DMF with the green solvent mixture, the cathodic reduction was carried out under galva-nostatic conditions (5 mA cm^{-2}) in an undivided cell equipped with a glassy carbon electrode as the cathode and Pt as the anode; 1,4-diazabicyclo[2.2.2]octane (DABCO) was used as a sacrificial reagent. Under these conditions, the corre-sponding alcohol (2) was obtained in a yield of 85% (Table 2, entry 1). This yield was higher than that reported with DMF in Xia's report (79%). Attempts to decrease the amount of DABCO or change it to a cheaper amine resulted in lower yields of the reduced compound (entries 2–6). Increasing the amount of EtOH and decreasing the amount of DABCO to 1.5 equiv. generated the alcohol in 70% yield (entry 7); varying the current intensity to lower and higher values was not bene-ficial (entries 8 and 9), nor even increasing the amount of DABCO to 2 eq. (entry 10). Using LiClO$_4$ instead of Bu$_4$NBF$_4$ resulted in a 62% yield, with the presence of

Table 2 Optimization of the conditions for the cathodic reduction of benzophenone using Cyrene™/EtOH (1 : 1) as a green solvent mixture[a]

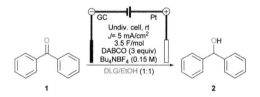

Entry	Variation from standard reaction conditions	Yield[b] (%)
1	None	85
2	2 equiv. of DABCO	30
3	1.5 equiv. of DABCO	20
4	Et$_3$N instead of DABCO	20
5	DIPEA instead of DABCO	10
6	Py instead of DABCO	10
7	1.5 equiv. of DABCO, DLG/EtOH(1 : 2)	70
8	1.5 equiv. of DABCO, DLG/EtOH(1 : 2), 2.5 mA cm^{-2}	60
9	1.5 equiv. of DABCO, DLG/EtOH(1 : 2), 7.5 mA cm^{-2}	55
10	2 equiv. of DABCO, DLG/EtOH(1 : 2)	50
11	LiClO$_4$ instead of Bu$_4$NBF$_4$	62 + 25[c]
12	DLG/EtOH (1 : 2)	70
13	Water instead of EtOH	10
14	MeOH instead of EtOH	20
15	EtOH as solvent	50[c]
16	No constant current	NR[d]

[a] Standard reaction conditions: **1** (1 mmol), DABCO (3 mmol, 3.0 equiv.), Bu$_4$NBF$_4$ (0.15 M), 1 : 1 DLG/EtOH (8 mL), constant current (10 mA), under air, rt, 3.5 F mol^{-1}. [b] Yields of isolated product. [c] Pinacol coupling product was isolated. [d] No reaction.

the dimeric compound (**3**), generated by the C–C pinacolic coupling, in 25% yield (entry 11).

The solvent composition was very important during the cathodic reduction of benzophenone, wherein the use of a 1 : 2 DLG/EtOH mixture gave **2** in 70% yield (entry 12), which is interesting considering that DLG is still an expensive solvent. The change to H_2O or MeOH instead of EtOH did not work well (entries 13 and 14). When the reaction was carried out in the absence of DLG, only using EtOH, the selective formation of the dimeric compound (**3**) was observed in 50% yield (entry 15), indicating a clear effect of the solvent mixture on the reaction pathway. The presence of DLG was essential for the selective reduction of benzophenone to the diphenylmethanol derivative (**2**). This unexpected behaviour was rationalized, considering that during the cathodic reduction of benzophenone, the corresponding anion radical is formed at the electrode surface, and it is stabilized by the Bu_4N^+ cation present at the electric double layer when DLG is present in high concentration. This process facilitates the second electron transfer near the electrode surface, yielding the diphenylmethanol (**2**) derivative. DLG bears an internal cyclic ketal and ketone group, both of which can coordinate with cations; thus, the change of the solvent concentration and electrolyte cation might affect the radical-anion stability. For example, in Table 2 entry 11, the use of $LiClO_4$ promoted the partial transformation through the $1e^-$ pathway to the pinacol derivative (**3**), and the change of solvent inhibited the $2e^-$ reduction route. These results indicate that changes in the solvent–cation–radical-anion interactions can control the reduction of benzophenone to obtain **2** or **3** derivatives selectively, slightly modifying the reduction conditions. To test this hypothesis, a series of experiments using $LiClO_4$ were carried out to favour the C–C pinacolic coupling (Table 3). Mainly, the current intensity was explored using an EtOH-enriched solvent mixture containing 1 : 2 DLG/EtOH, because previous experiments demonstrated that the presence of this alcohol favoured the $1e^-$ reduction reaction. When the electrolysis was carried out at a constant current density of 5 mA cm^{-2}, as in the previous experiments, **3** was

Table 3 Optimization of conditions for the electrochemical pinacol C–C coupling of benzophenone (**1**)[a]

Entry	Variation from standard reaction conditions	Yield[b] (%)
1	None	62
2	2.5 mA cm^{-2}	60
3	7.5 mA cm^{-2}	85
4	Bi[c] (−) instead of GC (−)	56

[a] Standard reaction conditions: **1** (1 mmol), DABCO (1.5 mmol, 1.5 equiv.), $LiClO_4$ (0.15 M), 1 : 2 DLG/EtOH (8 mL), constant current density (5 mA cm^{-2}), under air, rt, 3.5 F mol^{-1}. [b] Yields of isolated product. [c] Bi–In alloy electrode was used.

obtained in 60% yield (entry 1); using 2.5 mA cm^{-2} did not result in a significant difference (entry 2). When a constant current density of 7.5 mA cm^{-2} was applied, the yield of **3** increased to 85% (entry 3).

Interestingly, a Bi–In electrode (entry 4) resulted in a 56% yield of the pinacolic compound. Bi electrodes have been proposed as substitutes for Hg electrodes, which have been largely used for high-potential cathodic reactions in electrochemistry.

With the optimized electrochemical conditions in hand for the transformation of **1** to **2** (2e$^-$ reduction) or **3** (1e$^-$ reduction), the scope of both methodologies was explored using a series of benzophenone derivatives. First, benzophenone derivatives containing the electron-donating group OMe and electron-withdrawing groups Br and COOMe were successfully reduced to alcohols **2a**, **2b** and **2d**, respectively, in moderate yields (48–65%) (Scheme 2). The metallocene derivative benzoylferrocene was also tolerated, affording the corresponding alcohol (**2c**) in 60% yield. Interestingly, under these conditions the organometallic compound tolerates the cathodic conditions applied.

For electrochemical pinacol C–C coupling, the same substrate scope was explored with useful results (Scheme 3). With the methoxy derivative, a good yield of **3a** (70%) was obtained. The bromo derivative transformation failed, giving compound **2b** in 47% yield, while with the compound bearing the methoxycarbonyl group, the corresponding diol (**3d**) was obtained in a moderate yield (49%). Attempts to obtain pinacol product **3c** failed; instead alcohol **2c** was obtained in 44% yield. This result is probably due to the steric hindrance of the ferrocene moiety, which hindered the approach of the radical to bring about the C–C coupling. Additionally, this electrochemical methodology was applied to produce the jet-fuel precursor hydrofuroin (**3f**) in 62% yield from 2-furaldehyde. Classical methodologies to produce aromatic pinacols generally involve two steps,[29] (i) benzoin condensation and (ii) catalytic hydrogenation, making the electrochemical method in an environmentally friendly solvent mixture a convenient route to prepare these compounds.

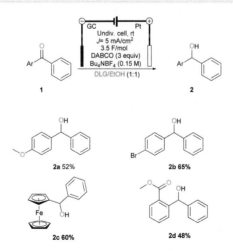

Scheme 2 Substrate scope for the 2e$^-$ electroreduction of benzophenone derivatives: **1** (1 mmol), DABCO (3 mmol, 3.0 equiv.), Bu$_4$NBF$_4$ (0.15 M), 1 : 1 DLG/EtOH (8 mL), constant current (10 mA), under air, rt, 3.5 F mol^{-1}. Yields of isolated products are reported.

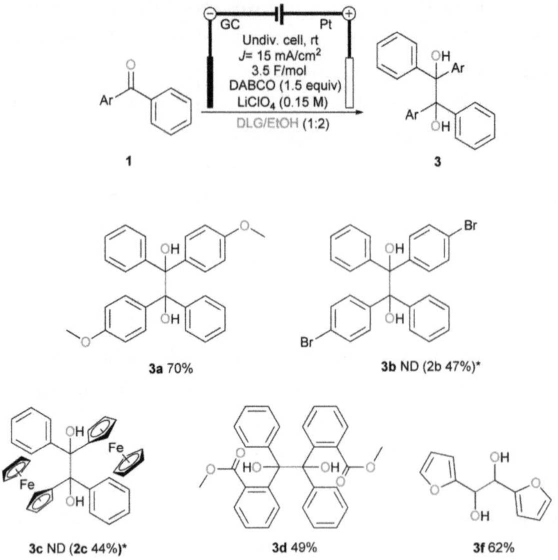

Scheme 3 Substrate scope for pinacolic C–C coupling. **1** (1 mmol), DABCO (1.5 mmol, 1.5 equiv.), LiClO$_4$ (0.15 M), 1 : 2 DLG/EtOH (8 mL), constant current (10 mA), under air, rt, 3.5 F mol^{-1}. Yields of isolated products are reported. ND = not detected. *Ferrocenylphenylmethanol and (4-bromophenyl)phenylmethanol were isolated instead of **3b** and **3c**.

Conclusions

The conductivity measurements of common supporting-electrolyte Bu$_4$NBF$_4$, Bu$_4$NClO$_4$ and Bu$_4$NPF$_6$ solutions using DLG as a solvent showed that they have low conductivity values due to the high viscosity of the solvent. These solutions are electrochemically stable in the potential range between −1.5 and 1.75 V. Viscosity limitations can be overcome using a green cosolvent, such as EtOH. Using the solvent mixture 1 : 1 DLG/EtOH, similar conductivity values to DMF solutions were obtained and the cyclic voltammetry studies demonstrated a larger potential window of −1.8 V to 2.5 V. Using this environmentally friendly solvent mixture, electrosynthetic reactions that use DMF as a solvent were successfully replaced. The use of 1 : 1 DLG/EtOH and Bu$_4$NBF$_4$ favoured the 2e$^-$ reduction reaction, generating diphenylmethanol derivatives, whereas using 1 : 2 DLG/EtOH and LiClO$_4$, the reaction gave mainly the C–C pinacolic coupling products. This proof of concept using DLG showed the interesting potential of its cosolvent mixtures in OES reactions to render the electrochemical methodology safer and greener. The use of other environmentally safe, low-viscosity and aprotic cosolvents, besides EtOH, is now envisaged and the results will be reported in due time.

Experimental

Chemicals

The starting materials (**1**) were synthesized following synthetic procedures re-ported in the literature.[30,31] Absolute ethanol was purchased from Tecsiquim and was used without prior treatment. Dihydrolevoglucosenone (DLG or Cyrene™) was purchased from Sigma-Aldrich and was used without prior treatment for electrosynthesis. For electrochemical measurements, it was dried over magnesium sulphate and distilled under vacuum at 62 °C and 4 mmHg, followed by a silica gel (230–400 mesh) filtration, prior to its use.

Conductivity measurements

Conductivity measurements were carried out with a Mettler Toledo EasyPlus P30 conductimeter utilizing an INLAB760 model dip probe with a cell constant of 0.805 cm^{-1}. Typically, 10 mL of solution was added to the conductivity cell and then stirred, and the temperature of the solution was controlled with a thermostatic bath (25 °C $\pm$ 0.3) throughout the experiment. The dip probe contains a thermistor, and the temperature of the solution was recorded during each experiment.

Electrochemistry

Cyclic voltammetry experiments were carried out using a Metrohm Autolab PGSTAT101 potentiostat connected to a PC and controlled using the NOVA 2.1 software package. A three-electrode setup was used, consisting of a Pt counter electrode and a 1.5 mm-diameter glassy carbon working electrode. The reference used was an Ag/AgCl (LiCl$_{(sat)}$ in EtOH) reference electrode separated from the bulk solution using a porosity frit. The ferrocene $E°$ determined in Bu$_4$NBF$_4$ (0.15 M) in 1 : 1 DLG/EtOH and LiClO$_4$ (0.15 M) in 1 : 2 DLG/EtOH was 0.63 V for both systems. Prior to each experiment, the working electrode was polished using alumina paste (0.03 μm) on a wet polishing cloth followed by rinsing with distilled water and drying.

Electrosynthesis

General procedure for electrochemical reduction of carbonyl compounds to alcohols. The carbonyl compound (1 mmol), DABCO (3 mmol) and Bu$_4$NBF$_4$ (0.15 M) dissolved in 8 mL of DLG/EtOH (1 : 1) mixture were transferred into an undivided electrolysis cell (ElectraSyn®) equipped with a glassy carbon cathode and a platinum anode. Constant-current electrolysis with a current density of 2.5–7.5 mA cm^{-2} was carried out at room temperature. After application of 3.5 F per mol, the electrolysis was stopped, and 80 mL of water was added into the electrolysis mixture and the mixture was stirred for 30 min. The product was extracted with EtOAc (25 mL × 3) and the combined extracts were evaporated to minimal volume. KOH (0.2 g) was added, and the mixture was heated to 50 °C for 30 minutes to hydrolyse and eliminate the DLG remnants in the organic phase. 20 mL of EtOAc was added and the mixture was filtered, and the organic phase was dried over Na$_2$SO$_4$, filtered, and concentrated under vacuum rotary evaporation. The desired alcohol derivatives were purified by column chromatography over silica gel (eluent: hexane : EtOAc = 9 : 1).

General procedure for electrochemical pinacol C–C coupling of carbonyl compounds. The carbonyl compound (1 mmol), DABCO (1.5 mmol) and LiClO$_4$ (0.15 M) in 8 mL of DLG/EtOH (1 : 2) mixture were transferred into an undivided electrolysis cell (ElectraSyn®) equipped with a glassy carbon cathode and a platinum anode. Constant-current electrolysis with a current density of 2.5–7.5 mA

cm^{-2} was carried out at room temperature. After application of 3.5 F per mol, the electrolysis was stopped, and 80 mL of water was added into the electrolysis mixture and the mixture was stirred for 30 min. The product was extracted with EtOAc (25 mL × 3) and the combined extracts were evaporated to minimal volume. KOH (0.2 g) was added, and the mixture was heated to 50 °C for 30 minutes to hydrolyse and eliminate the DLG remnants in the organic phase. 20 mL of EtOAc was added and the mixture was filtered, and the organic phase was dried over Na_2SO_4, filtered, and concentrated under vacuum rotary evaporation. The desired diol derivatives were purified by column chromatography over silica gel (eluent: hexane : EtOAc = 9.5 : 0.5).

Bismuth–indium electrode preparation

The Bi–In films were prepared using a magnetron sputtering technique using a homemade Bi–In target. The diameter of the target was $2''$ and it was produced by melting together 95 wt% Bi (99.95% purity) and 5 wt% In (99.99% purity) in a carbon-coated stainless-steel mold. After cooling, the target was polished to remove surface oxide impurities and attached to the magnetron using a colloidal silver paste, to ensure good conductivity. Magnetron and glass substrates were placed in a vacuum chamber achieving a base pressure of 6.7×10^{-4} Pa. The deposition was carried out under an Ar atmosphere using a flux of 18 standard cubic centimetres per minute (sccm) and a working pressure of 2 Pa. The power applied to the Bi–In target was 20 W of direct current. The substrate holder was rotated at 10 rpm to achieve good film uniformity. The deposition time was 10 min. The thickness was measured using a mechanical profilometer, obtaining an average value of 457.4 ± 3.6 nm. The morphology of the films was analysed using an optical profilometer and scanning electron microscopy. The average roughness was 1.58 nm.

Compound characterization

4-Methoxyphenyl-phenylmethanol (**2a**):[32] white solid; ^{1}H NMR (300 MHz, CDCl$_3$): δ 7.42–7.30 (m, 7H), 6.8 (d, J = 9 Hz, 2H), 5.82 (s, 1H), 3.82 (s, 3H), 2.36 (s, 1H).

4-Bromophenyl-phenylmethanol (**2b**):[32] white solid; ^{1}H NMR (300 MHz CDCl$_3$): δ 7.41–7.09 (m, 9H), 5.8 (s, 1H).

Ferrocenylphenylmethanol (**2c**):[33] yellow solid; ^{1}H NMR (300 MHz, CDCl$_3$): δ 7.32–7.30 (m, 2H), 7.27–7.22 (m, 2H), 7.19–7.17 (m, 1H), 5.38 (s, 1H), 4.14–4.08 (m, 9H), 2.42 (s, 1H).

Methyl-2-(hydroxy(phenyl)methyl)benzoate (**2d**):[34] white solid; ^{1}H NMR (300 MHz, CDCl$_3$): δ 8.06–7.94 (m, 1H), 7.67–7.52 (m, 3H), 7.45–7.30 (m, 5H), 6.40 (s, 1H), 3.61 (s, 3H), 2.35 (s, 1H).

1,2-Bis(4-methoxyphenyl)-1,2-diphenylethane-1,2-diol (**3a**):[35] white solid; ^{1}H NMR (300 MHz, CDCl$_3$): δ 7.26–7.19 (m, 5H), 7.11–7.03 (m, 9H), 6.63–6.56 (m, 4H), 3.65 (s, 3H), 3.63 (s, 3H), 2.97 (s, 2H).

Dimethyl-2,2′-(1,2-dihydroxy-1,2-diphenylethane-1,2-diyl)dibenzoate (**3d**): white solid; ^{1}H NMR (300 MHz, CDCl$_3$): δ 7.96 (d, J = 6 Hz, 2H), 7.67–7.45 (m, 11H), 7.37–7.25 (m, 5H), 3.53 (s, 6H), 2.89 (s, 2H).

1,2-Di(furan-2-yl)ethane-1,2-diol (**3f**):[36] white solid; ^{1}H NMR (300 MHz, CDCl$_3$): δ 7.52 (s, 2H), 6.63 (d, J = 3 Hz, 2H), 6.47 (dd, J = 3 Hz, J = 6 Hz, 2H), 5.25 (s, 2H), 2.99 (s, 2H).

Conflicts of interest

There are no conflicts to declare.

Acknowledgements

The support of CONACYT (through project A1-S-18230) and PAPIIT-DGAPA UNAM IV200222 is recognized. The technical support of Citlalit Martínez, Ma. Nieves Zavala-Segovia, Triana Cruz, Alejandra Nuñez, Javier Pérez and Ma. Carmen García is also acknowledged.

Notes and references

1 B. A. Frontana-Uribe, R. D. Little, J. G. Ibanez, A. Palma and R. Vasquez-Medrano, *Green Chem.*, 2010, **12**, 2099.

2 E. J. Horn, B. R. Rosen and P. S. Baran, *ACS Cent. Sci.*, 2016, **2**, 302–308.

3 Y. Yuan and A. Lei, *Nat. Commun.*, 2020, **11**, 802.

4 F. P. Byrne, S. Jin, G. Paggiola, T. H. M. Petchey, J. H. Clark, T. J. Farmer, A. J. Hunt, C. Robert McElroy and J. Sherwood, *Sustainable Chem. Processes*, 2016, **4**, 7.

5 A. Jordan, C. G. J. Hall, L. R. Thorp and H. F. Sneddon, *Chem. Rev.*, 2022, **122**, 6749–6794.

6 R. K. Henderson, C. Jiménez-González, D. J. C. Constable, S. R. Alston, G. G. A. Inglis, G. Fisher, J. Sherwood, S. P. Binks and A. D. Curzons, *Green Chem.*, 2011, **13**, 854.

7 F. Harnisch and U. Schröder, *ChemElectroChem*, 2019, **6**, 4126–4133.

8 J. E. Camp, *ChemSusChem*, 2018, **11**, 3048–3055.

9 N. A. Stini, P. L. Gkizis and C. G. Kokotos, *Green Chem.*, 2022, **24**, 6435–6449.

10 K. Wilson, J. Murray, C. Jamieson and A. Watson, *Synlett*, 2018, **29**, 650–654.

11 K. L. Wilson, A. R. Kennedy, J. Murray, B. Greatrex, C. Jamieson and A. J. B. Watson, *Beilstein J. Org. Chem.*, 2016, **12**, 2005–2011.

12 S. Sangon, N. Supanchaiyamat, J. Sherwood, C. R. McElroy and A. J. Hunt, *React. Chem. Eng.*, 2020, **5**, 1798–1804.

13 J. Sherwood, M. De bruyn, A. Constantinou, L. Moity, C. R. McElroy, T. J. Farmer, T. Duncan, W. Raverty, A. J. Hunt and J. H. Clark, *Chem. Commun.*, 2014, **50**, 9650–9652.

14 K. L. Wilson, J. Murray, C. Jamieson and A. J. B. Watson, *Org. Biomol. Chem.*, 2018, **16**, 2851–2854.

15 L. Mistry, K. Mapesa, T. W. Bousfield and J. E. Camp, *Green Chem.*, 2017, **19**, 2123–2128.

16 R. Nickisch, P. Conen, S. M. Gabrielsen and M. A. R. Meier, *RSC Adv.*, 2021, **11**, 3134–3142.

17 N. Guajardo and P. Domínguez de María, *Mol. Catal.*, 2020, **485**, 110813.

18 C. Grune, J. Thamm, O. Werz and D. Fischer, *J. Pharm. Sci.*, 2021, **110**, 959–964.

19 S. Borova and R. Luxenhofer, *Beilstein J. Org. Chem.*, 2023, **19**, 217–230.

20 R. A. Milescu, C. R. McElroy, T. J. Farmer, P. M. Williams, M. J. Walters and J. H. Clark, *Adv. Polym.*, 2019, **2019**, 1–15.

21 T. Marino, F. Galiano, A. Molino and A. Figoli, *J. Membr. Sci.*, 2019, **580**, 224–234.

22 L. Cseri and G. Szekely, *Green Chem.*, 2019, **21**, 4178–4188.

23 H. J. Salavagione, J. Sherwood, M. De bruyn, V. L. Budarin, G. J. Ellis, J. H. Clark and P. S. Shuttleworth, *Green Chem.*, 2017, **19**, 2550–2560.

24 R. G. LeBel and D. A. I. Goring, *J. Chem. Eng. Data*, 1962, **7**, 100–101.

25 K. Izutsu, *Electrochemistry of Nonaqueous Solutions*. Wiley, Germany, 2002.

26 C. Sullivan, Y. Zhang, G. Xu, L. Christianson, F. Luengo, T. Halkoski and P. Gao, *Green Chem.*, 2022, **24**, 7184–7193.

27 B. Schille, N. O. Giltzau and R. Francke, *Angew. Chem., Int. Ed.*, 2018, **57**, 422–426.

28 L. Wang, X. Zhang, R. Y. Xia, C. Yang, L. Guo and W. Xia, *Synlett*, 2022, **33**, 1302–1308.

29 A. Kabro, E. C. Escudero-Adán, V. V. Grushin and P. W. N. M. van Leeuwen, *Org. Lett.*, 2012, **14**, 4014–4017.

30 R. H. Vekariya and J. Aubé, *Org. Lett.*, 2016, **18**, 3534–3537.

31 A. Salvo, V. Campisciano, H. Beejapur, F. Giacalone and M. Gruttadauria, *Synlett*, 2015, **26**, 1179–1184.

32 J. Yang, X. Chen and Z. Wang, *Tetrahedron Lett.*, 2015, **56**, 5673–5675.

33 B. Lu, Q. Wang, M. Zhao, X. Xie and Z. Zhang, *J. Org. Chem.*, 2015, **80**, 9563–9569.

34 P. B. Jones, M. P. Pollastri and N. A. Porter, *J. Org. Chem.*, 1996, **61**, 9455–9461.

35 Z. Qiu, H. D. M. Pham, J. Li, C.-C. Li, D. J. Castillo-Pazos, R. Z. Khaliullin and C.-J. Li, *Chem. Sci. J.*, 2019, **10**, 10937–10943.

36 X. Shang, Y. Yang and Y. Sun, *Green Chem.*, 2020, **22**, 5395–5401.

Faraday Discussions

PAPER

Electroorganic synthesis in aqueous solution *via* generation of strongly oxidizing and reducing intermediates

Seyyedamirhossein Hosseini, [iD] [a] Joshua A. Beeler, [iD] [a] Melanie S. Sanford [iD] *[b] and Henry S. White [iD] *[a]

Received 14th March 2023, Accepted 6th April 2023

DOI: 10.1039/d3fd00067b

Water is the ideal green solvent for organic electrosynthesis. However, a majority of electroorganic processes require potentials that lie beyond the electrochemical window for water. In general, water oxidation and reduction lead to poor synthetic yields and selectivity or altogether prohibit carrying out a desired reaction. Herein, we report several electroorganic reactions in water using synthetic strategies referred to as reductive oxidation and oxidative reduction. Reductive oxidation involves the homogeneous reduction of peroxydisulfate ($S_2O_8^{2-}$) *via* electrogenerated $Ru(NH_3)_6^{2+}$ at potential of -0.2 V *vs.* Ag/AgCl (3.5 M KCl) to form the highly oxidizing sulfate radical anion ($E^{0\prime}$ ($SO_4^{\cdot-}/SO_4^{2-}$) = 2.21 V *vs.* Ag/AgCl), which is capable of oxidizing species beyond the water oxidation potential. Electrochemically generated $SO_4^{\cdot-}$ then efficiently abstracts a hydrogen atom from a variety of organic compounds such as benzyl alcohol and toluene to yield product in water. The reverse analogue of reductive oxidation is oxidative reduction. In this case, the homogeneous oxidation of oxalate ($C_2O_4^{2-}$) by electrochemically generated $Ru(bpy)_3^{3+}$ produces the strongly reducing carbon dioxide radical anion ($E^{0\prime}$ ($CO_2^{\cdot-}/CO_2$) = -2.1 V *vs.* Ag/AgCl), which can reduce species at potential beyond the water or proton reduction potential. In preliminary studies, the $CO_2^{\cdot-}$ has been used to homogeneously reduce the C–Br moiety belonging to benzyl bromide at an oxidizing potential in aqueous solution.

Introduction

A goal of chemical industries is to develop processes that can be performed in aqueous solution. Besides intrinsic greenness, several features such as abundance and minimal solvent handling make water a very appealing solvent. In this context, developing electroorganic reactions that can be carried out in an aqueous

[a]*Department of Chemistry, University of Utah, 315 S 1400 E Salt Lake City, Utah 84112, USA. E-mail: white@chemistry.utah.edu*

[b]*Department of Chemistry, University of Michigan, 930 North University Avenue, Ann Arbor, Michigan 48109, USA. E-mail: mssanford@umich.edu*

solution is a very attractive idea. In comparison to popular organic solvents such as dimethyl formamide (DMF) and acetonitrile (MeCN), water has a large dielectric constant, thus allowing for the use of common inorganic supporting electrolytes and reagents at high concentrations. However, in general, electro-organic reactions in an aqueous solution are limited by the small potential window of water, which is defined as the potentials beyond which water undergoes direct reduction and oxidation (*i.e.*, eqn (1) and (2)).[1,2]

$$2H_2O + 2e^- \rightleftharpoons H_2 + 2OH^-, E^0 = -1.1 \text{ V } vs.$$
$$\text{Ag/AgCl (3.5 M KCl) (alkaline solution)} \quad (1)$$

$$2H_2O \rightleftharpoons 4H^+ + O_2 + 4e^-, E^0 = 1.03 \text{ V } vs. \text{ Ag/AgCl (3.5 M KCl) (acidic solution)} (2)$$

The majority of the electroorganic reactions, such as the electrochemical reduction of C–X (X = Cl, Br, I) bonds or electrochemical C–H bond oxidation, require the application of potentials beyond the water potential window.[3] Therefore, water reduction/oxidation interferes with electroorganic reactions, yielding undesirable side products and low reaction yields.

In a recent report, we have proposed that the electrogeneration of highly oxidizing/reducing intermediates from water soluble reagents, and interception of these transient species by organic compounds, provides a promising approach to perform electroorganic reactions that would normally be prohibited by the oxidation or reduction of water.[12] Electrogeneration of highly oxidizing and reducing intermediates was developed by Bard and co-workers to create excited state species in aqueous solutions that emitted light, a process referred to as electrogenerated chemiluminescence or ECL.[4] In particular, two approaches for generating ECL are based on: (1) $S_2O_8^{2-}$ reduction to generate the strong oxidant $SO_4^{\cdot-}$ and (2) $C_2O_4^{2-}$ oxidation to generate the strong reductant $CO_2^{\cdot-}$.[4-6] In the present work, we employed $S_2O_8^{2-}$ reduction and $C_2O_4^{2-}$ oxidation as the initial steps in performing synthetic organic reactions in aqueous solutions.

Fig. 1 shows the cyclic voltammogram for the direct reduction of $S_2O_8^{2-}$ (red trace) and oxidation of $C_2O_4^{2-}$ (blue trace) at a glassy carbon (GC) electrode. Briefly, the one-electron reduction of $S_2O_8^{2-}$ generates $S_2O_8^{3\cdot-}$, which rapidly dissociates to form SO_4^{2-} and the highly oxidizing sulfate radical anion ($E^{0'}$ ($SO_4^{\cdot-}/SO_4^{2-}$) = 2.21 V *vs.* Ag/AgCl (3.5 M KCl)).[5-7] Electrogenerated $SO_4^{\cdot-}$ can be further reduced at the working electrode. Alternatively, $SO_4^{\cdot-}$ can accept an electron from a molecular species in solution, resulting in oxidation of that species (*e.g.*, halides, organometallic complexes, arenes).[8] Because the reduction of $S_2O_8^{2-}$ results in the oxidation of a molecule in solution, we refer to this electrochemical reaction sequence as reductive oxidation. In contrast to $S_2O_8^{2-}$, the one-electron oxidation of $C_2O_4^{2-}$ yields the transient $C_2O_4^{\cdot-}$ that rapidly dissociates to CO_2 and the highly reducing carbon dioxide radical anion ($E^{0'}$ ($CO_2^{\cdot-}/CO_2$) = -2.1 V *vs.* Ag/AgCl (3.5 M KCl)).[9,10] Electrogenerated $CO_2^{\cdot-}$ can undergo a subsequent 1e oxidation at the working electrode or can be used to reduce a solution species. Hence, the electrochemical reaction sequence is referred to as oxidative reduction.

The short-lived intermediates $S_2O_8^{3\cdot-}$ ($\tau_{1/2} \sim 5$ ps)[12] and $C_2O_4^{\cdot-}$ ($\tau_{1/2} \sim 1$ μs)[11] decompose to $SO_4^{\cdot-}$ and $CO_2^{\cdot-}$, respectively, in close vicinity to the electrode surface and can be rapidly reduced to SO_4^{2-} or oxidized to CO_2, respectively.

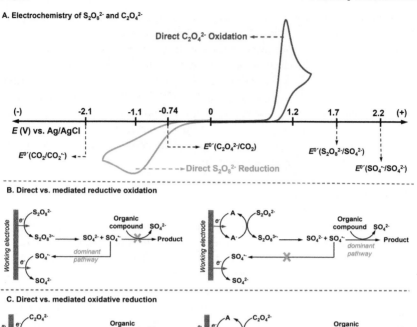

A. Electrochemistry of $S_2O_8^{2-}$ and $C_2O_4^{2-}$

B. Direct vs. mediated reductive oxidation

C. Direct vs. mediated oxidative reduction

Fig. 1 (A) Cyclic voltammetry in an O_2-free aqueous solution containing 5.0 mM $Na_2S_2O_8$ (red trace), 5.0 mM $Na_2C_2O_4$ (blue trace) and 0.1 M Na_2SO_4 (pH = 6.8). Schematic of direct vs. mediated mechanisms for the (B) reductive oxidation and (C) oxidative reduction processes. Voltammograms were recorded at a scan rate (v) of 100 mV s^{-1} using a 0.07 cm^2 GC working electrode, a Pt mesh auxiliary electrode, and a Ag/AgCl (3.5 M KCl) reference electrode.

Consequently, direct oxidation of $C_2O_4^{2-}$ or reduction of $S_2O_8^{2-}$ generates low amounts of $CO_2^{\bullet-}$ and $SO_4^{\bullet-}$ in solution, respectively, that can be used to carry out synthetic transformations (left side of Fig. 1B and C). To avoid the oxidation of $CO_2^{\bullet-}$ and reduction of $SO_4^{\bullet-}$ at the electrode surface, an outer-sphere redox mediator (i.e., species A in Fig. 1B and C) is required to reduce $S_2O_8^{2-}$ and oxidize $C_2O_4^{2-}$. In contrast to the direct reactions of these species at the electrode, the use of a mediator yields $SO_4^{\bullet-}$ and $CO_2^{\bullet-}$ several tens of micrometers from the working electrode,[12] where they can be intercepted by organic compounds for synthetic purposes.

$S_2O_8^{2-}$ is widely employed in (non-electrochemical) organic synthesis. $SO_4^{\bullet-}$, generated by heat or light, is an efficient reagent for hydrogen atom abstraction from organic compounds, and there are several reports where $S_2O_8^{2-}$ is utilized for C–H activation reactions.[13–15] In contrast, to the best of our knowledge, $C_2O_4^{2-}$ has never been utilized for synthetic purposes. Rather, the $CO_2^{\bullet-}$ is typically generated from formate oxidation.[16–18] Recent reports show that photochemically generated $CO_2^{\bullet-}$ can be used to reductively cleave the C–Cl moiety in electron-deficient chloroarenes[16,17] or in hydrocarboxylation reactions.[18] In both cases, however, an organic solvent was employed to carry out the chemical reaction, and the $SO_4^{\bullet-}$ and $CO_2^{\bullet-}$ were generated using light or heat.

Herein, we describe our recent efforts to apply the mediated electrochemical reduction of $S_2O_8^{2-}$ by electrogenerated hexaammineruthenium(II) $(Ru(NH_3)_6^{2+})$ to carry out C–H oxidation of benzyl alcohol and toluene in aqueous solution. Additionally, we describe preliminary studies of the mediated electrooxidation of oxalate using electrogenerated tris(2,2′-bipyridine)ruthenium(III) $(Ru(bpy)_3^{3+})$ to form $CO_2^{\bullet-}$ in an aqueous solution. Electrogenerated $CO_2^{\bullet-}$ is used for the homogeneous reduction of the C–Br moiety in benzyl bromide and for the reduction of Zn^{2+} in an aqueous solution.

Results and discussion

The direct reduction of $S_2O_8^{2-}$ and rapid dissociation of $S_2O_8^{3\bullet-}$ produces $SO_4^{\bullet-}$ within ~10 nm of the GC electrode. Therefore, the direct reduction of the $SO_4^{\bullet-}$ at the GC electrode occurs prior to being intercepted by an organic compound. Several outer-sphere mediators were explored for the homogeneous reduction of $S_2O_8^{2-}$. Fig. 2 shows the cyclic voltammetry (CV) of the GC electrode in solutions containing nitrobenzene (NB), hexaammineruthenium(III) chloride $(Ru(NH_3)_6^{3+})$, potassium hexacyanoferrate(III) $(Fe(CN)_6^{3-})$, and potassium hexachloroiridate(III) $([IrCl_6]^{3-})$ in the presence and absence of $S_2O_8^{2-}$. All four mediators show chemically and thermodynamically reversible 1e reduction behavior in the absence of $S_2O_8^{2-}$. Upon addition of $S_2O_8^{2-}$, the voltammetric responses of $[IrCl_6]^{3-}$ and $Fe(CN)_6^{3-}$ remain unchanged. On the other hand, the CVs of $Ru(NH_3)_6^{3+}$ and NB in the presence of $S_2O_8^{2-}$ show an increase in the cathodic peak current, concurrent with the disappearance of the reverse anodic peak. These results indicate that the electrogenerated $Ru(NH_3)_6^{2+}$ and NB radical anion are both capable of reducing $S_2O_8^{2-}$, presumably by a 1e transfer. In comparison with nitrobenzene, $Ru(NH_3)_6^{2+}$ reduces $S_2O_8^{2-}$ at a less negative potential and generates a larger electrocatalytic current. Therefore, $Ru(NH_3)_6^{3+}$ was selected as the electrocatalyst to homogeneously generate $SO_4^{\bullet-}$ for the subsequent electro-organic reactions. We also note that $Ru(NH_3)_6^{2+}$ reduces $S_2O_8^{2-}$ at potentials approximately 0.5 V more positive than where $S_2O_8^{2-}$ is directly reduced at GC. Thus, $Ru(NH_3)_6^{2+}$ acts as an efficient electrocatalyst for $S_2O_8^{2-}$ reduction, generating $SO_4^{\bullet-}$ in the solution at potentials where $S_2O_8^{2-}$ is not directly reduced at

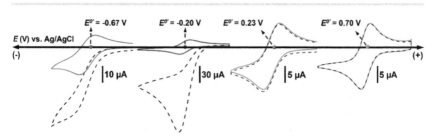

Fig. 2 Cyclic voltammetry of 0.5 mM (magenta) NB, (blue) $Ru(NH_3)_6^{3+}$, (green) $Fe(CN)_6^{3-}$, and (red) $[IrCl_6]^{3-}$ in an O_2-free aqueous solution, in the absence and presence of 5 mM $Na_2S_2O_8$ (dashed black trace). Voltammograms were recorded at a scan rate (ν) of 100 mV s^{-1} using a 0.07 cm^2 GC working electrode, a Pt mesh auxiliary electrode, and a Ag/AgCl (3.5 M KCl) reference electrode. The solution contained 0.1 M Na_2SO_4 (pH = 6.8).

the GC electrode. This combination of properties ensures that $SO_4^{\bullet-}$ is available for synthetic purposes, as shown on the right-side column of Fig. 1B.

Fig. 3 shows the variation of the height of the cathodic peak upon addition of 2.0 mM benzyl alcohol (**BA**) to a solution containing 0.5 mM $Ru(NH_3)_6^{3+}$ and 4.5 mM $S_2O_8^{2-}$.[12] The dependence of the cathodic peak current as a function of **BA** concentration is shown in Fig. 3B. We have previously proposed a five-step mechanism for the electrocatalytic reduction of $S_2O_8^{2-}$ *via* $Ru(NH_3)_6^{3+}$ (eqn (3)–(7)) based on CV analysis. The 1e reduction of $Ru(NH_3)_6^{3+}$ yields $Ru(NH_3)_6^{2+}$ (eqn (3)), which homogeneously reduces $S_2O_8^{2-}$ resulting in the regeneration of $Ru(NH_3)_6^{3+}$ and the generation of $S_2O_8^{3\bullet-}$ (eqn (4)). Electrogenerated $S_2O_8^{3\bullet-}$ dissociates rapidly (*i.e.*, $\tau_{1/2} < 5$ ps) forming SO_4^{2-} and $SO_4^{\bullet-}$ (eqn (5)). Finally, $SO_4^{\bullet-}$ reduces by 1e either *via* homogeneous reduction by $Ru(NH_3)_6^{2+}$ (eqn (6)) or directly at the working electrode (eqn (7)).

$$Ru(NH_3)_6^{3+} + e^- \rightleftharpoons Ru(NH_3)_6^{2+} \tag{3}$$

$$Ru(NH_3)_6^{2+} + S_2O_8^{2-} \rightleftharpoons Ru(NH_3)_6^{3+} + S_2O_8^{3\bullet-} \tag{4}$$

$$S_2O_8^{3\bullet-} \rightleftharpoons SO_4^{2-} + SO_4^{\bullet-} \tag{5}$$

$$Ru(NH_3)_6^{2+} + SO_4^{\bullet-} \rightleftharpoons Ru(NH_3)_6^{3+} + SO_4^{2-} \tag{6}$$

$$SO_4^{\bullet-} + e^- \rightleftharpoons SO_4^{2-} \tag{7}$$

The decrease in the cathodic peak height (*i.e.*, i_{pc}) in the presence of **BA** results from $SO_4^{\bullet-}$ abstracting a hydrogen atom from benzyl alcohol to form the corresponding benzyl alcohol radical (**BAR**) (eqn (8)). The **BAR** can then be oxidized to yield benzaldehyde (**BAL**) *via* reaction with $SO_4^{\bullet-}$ (eqn (9)) or $Ru(NH_3)_6^{3+}$ (eqn

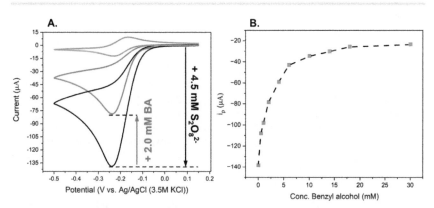

Fig. 3 (A) Cyclic voltammetric response in an O_2-free H_2O–MeCN solution (80% H_2O v/v) containing (red) 0.50 mM $Ru(NH_3)_6^{3+}$, (black) 0.50 mM $Ru(NH_3)_6^{3+}$ and 4.5 mM $S_2O_8^{2-}$, and (green) 0.50 mM $Ru(NH_3)_6^{3+}$, 4.5 mM $S_2O_8^{2-}$, and 2.0 mM BA. (B) Cathodic peak current (i_{pc}) *versus* the concentration of benzyl alcohol. Voltammograms were recorded at a scan rate (ν) of 100 mV s^{-1} using a 0.07 cm^2 GC working electrode, a Pt mesh auxiliary electrode, and a Ag/AgCl (3.5 M KCl) reference electrode. The solution contained 0.1 M Na_2SO_4 (pH = 6.8).

(10)). A direct consequence of eqn (8) and (9) is the homogeneous conversion of $SO_4^{\cdot-}$ to SO_4^{2-} in solution rather than at the electrode. Thus, i_{pc} decreases with increasing concentration of **BA**, as shown in Fig. 3B.

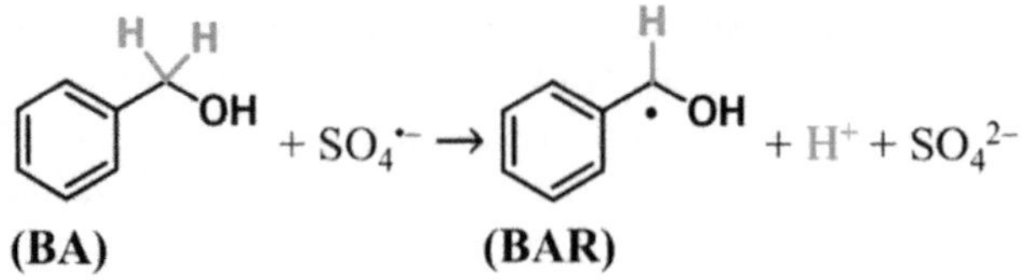

$$\text{(BA)} + SO_4^{\cdot-} \rightarrow \text{(BAR)} + H^+ + SO_4^{2-} \qquad (8)$$

$$\text{(BAR)} + SO_4^{\cdot-} \rightarrow \text{(BAL)} + H^+ + SO_4^{2-} \qquad (9)$$

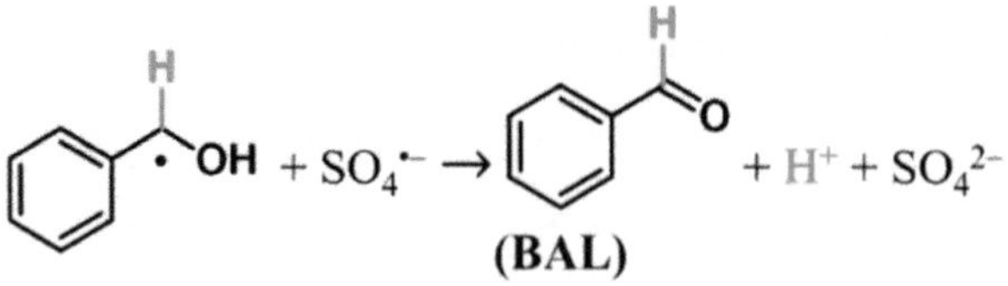

$$\text{Ru(NH}_3)_6^{3+} + \textbf{BAR} \rightarrow \text{Ru(NH}_3)_6^{2+} + H^+ + \textbf{BAL} \qquad (10)$$

The utility of the $\text{Ru(NH}_3)_6^{3+}/S_2O_8^{2-}$ system for the electrochemical oxidation of **BA** was examined *via* constant potential electrolysis at -0.5 V *vs.* Ag/AgCl (3.5 M KCl) in a divided cell, using a reticulated vitreous carbon (RVC) disk as the working electrode.[12] The scope of the aliphatic and benzylic alcohol oxidation *via* mediated reductive oxidation is shown in Fig. 4A. Mediated reductive oxidation of

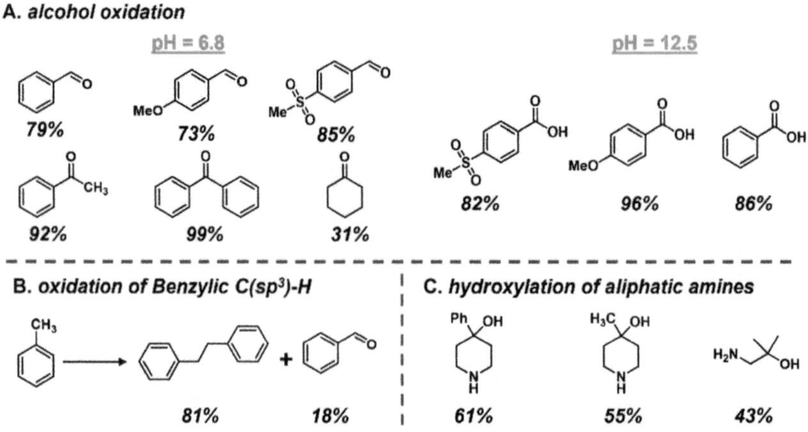

A. alcohol oxidation

pH = 6.8

79%   73%   85%

92%   99%   31%

pH = 12.5

82%   96%   86%

B. oxidation of Benzylic C(sp³)-H

81%   18%

C. hydroxylation of aliphatic amines

61%   55%   43%

Fig. 4   Utility of reductive oxidation for: (A) selective oxidation of aliphatic and benzylic alcohols; (B) oxidation of benzylic $C(sp^3)$–H; and (C) hydroxylation of aliphatic amines. Constant potential electrolysis was performed at -0.5 V *vs.* Ag/AgCl (3.5 M KCl) in an O_2-free H_2O–MeCN solution (80% H_2O v/v) containing 0.1 M Na_2SO_4 using a divided cell. In (A) and (C), a RVC disk was employed as the working electrode, whereas in (B) a graphite rod working electrode was used.

various alcohols in neutral pH (*i.e.*, pH = 6.8) yields aldehydes/ketones whereas electrolysis in basic pH (*i.e.*, pH = 12.5) predominately yields carboxylic acids.

Electrochemical reductive oxidation of **BA** involves the reaction of $SO_4^{\cdot-}$ with a relatively weak benzylic C–H bond (BDE ~ 80 kcal mol^{-1}).[19] In preliminary experiments, we next examined the utility of reductive oxidation toward activation of the significantly stronger benzylic $C(sp^3)$–H bond of toluene (BDE ~ 88 kcal mol^{-1}).[20] The benzylic C–H activation of toluene is typically carried out using photochemical methods.[21,22] As shown in Fig. 5A, the peak potential for the direct oxidation of toluene overlaps with the onset of water oxidation. Therefore, any attempt to directly oxidize toluene at $E > 1.95$ V *vs.* Ag/AgCl (3.5 M KCl) in an aqueous solution results in water oxidation. However, reductive oxidation *via* reduction of $S_2O_8^{2-}$ at negative potentials allows for toluene oxidation. Upon addition of 5.0 mM toluene to an aqueous solution containing 0.5 mM $Ru(NH_3)_6^{3+}$ and 5.0 mM $S_2O_8^{2-}$, the height of the cathodic peak decreases (Fig. 5B). In the absence of toluene, electrogenerated $SO_4^{\cdot-}$ is reduced by $Ru(NH_3)_6^{2+}$ (*i.e.*, eqn (6)) or directly at the working electrode (*i.e.*, eqn (7)). However, electrogenerated $SO_4^{\cdot-}$ homogeneously oxidizes toluene, which manifests itself as a decrease in the cathodic peak current, suggesting that reductive oxidation enables toluene oxidation in aqueous solution. Electrolysis of 5.0 mM toluene was carried out in the presence of 1.0 mM $Ru(NH_3)_6^{3+}$ and 25 mM $S_2O_8^{2-}$ at a constant potential of −0.5 V *vs.* Ag/AgCl (3.5 M KCl). Electrolysis was performed in a divided cell containing an O_2-free solution of H_2O–MeCN (80% v/v H_2O) using a graphite rod as the working electrode. Analysis of the product distribution revealed that ~81% of the toluene was oxidized to bibenzyl, and ~18% of the toluene was converted to benzaldehyde as the result of overoxidation *via* $SO_4^{\cdot-}$ (Fig. 4B).

Finally, we examined the activation of even stronger 3° C–H bonds in protonated aliphatic amines (BDE ~ 96 kcal mol^{-1}). The oxygenation of analogous substrates was carried out previously by Sanford and co-workers *via* thermal

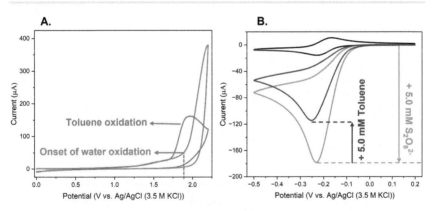

Fig. 5 (A) Cyclic voltammetry in the absence (green) and presence of 5.0 mM toluene (magenta). (B) CVs of a solution containing: (black) 0.50 mM $Ru(NH_3)_6^{3+}$; (red) 0.50 mM $Ru(NH_3)_6^{3+}$ and 5.0 mM $S_2O_8^{2-}$; and (blue) 0.50 mM $Ru(NH_3)_6^{3+}$, 5.0 mM $S_2O_8^{2-}$, and 5.0 mM toluene. All CVs were recorded using a 0.07 cm^2 GC working electrode, a Pt mesh auxiliary electrode, and a Ag/AgCl (3.5 M KCl) reference electrode. The measurements were made at $\nu = 100$ mV s^{-1} in an O_2-free H_2O/MeCN solution (80% H_2O v/v) containing 0.1 M Na_2SO_4 (pH = 6.5).

activation of $S_2O_8^{2-}$ to functionalize 3° $C(sp^3)$–H bonds in a dilute H_2SO_4 solution.[23] This homogeneous chemical process was shown to have a high yield and to be applicable to a large scope of substrates.

Electrochemical hydroxylation was carried out at constant potential of −0.5 V vs. Ag/AgCl in a phosphate buffer solution (pH = 3.1) containing 1.0 mM $Ru(NH_3)_6^{3+}$, 30 mM $S_2O_8^{2-}$, and 5.0 mM of an aliphatic amine. As shown in Fig. 4C, three aliphatic amines were hydroxylated at the 3° site, demonstrating the utility of mediated reductive oxidation for this class of substrates.

Analogous to the use of $S_2O_8^{2-}$ reduction for carrying out oxidations, $C_2O_4^{2-}$ oxidation can be used for performing reduction of organic species. The direct oxidation of $C_2O_4^{2-}$ ($E^{0'}$ ($C_2O_4^{\cdot-}/C_2O_4^{2-}$) = 1.2 V vs. Ag/AgCl (3.5 M KCl)) results in the formation of the $CO_2^{\cdot-}$ very close to the electrode surface (<1 μm), which leads to the preferential oxidation of the $CO_2^{\cdot-}$ at the electrode rather than in solution with a reaction partner.[11] To circumvent this problem, a redox mediator can be used to generate $CO_2^{\cdot-}$ further away from the electrode surface. Potassium hexachloroiridate(II) ([$IrCl_6$]$^{4-}$), tris(2,2′-bipyridine)iron(II) tetrafluoroborate ($Fe(bpy)_3^{2+}$), and tris(2,2′-bipyridine)ruthenium(II) chloride ($Ru(bpy)_3^{2+}$) were screened for the ability of their oxidized forms to oxidize $C_2O_4^{2-}$ (Fig. 6). In general, as $E^{0'}$ of the metal-based mediator becomes more positive, the rate of mediation increases, in agreement with prior results reported by Bard and co-workers.[24] As shown in Fig. 6A, a minimal increase in the anodic peak current for [$IrCl_6$]$^{4-}$ oxidation occurs upon addition of $C_2O_4^{2-}$, indicating that transfer of an electron from $C_2O_4^{2-}$ to [$IrCl_6$]$^{3-}$ is negligible. Upon addition of $C_2O_4^{2-}$ to a solution containing $Fe(bpy)_3^{2+}$, a sigmoidal CV shape representative of sluggish homogeneous electron transfer is observed, as previously reported.[24] However, the CV of $Ru(bpy)_3^{2+}$ in the presence of $C_2O_4^{2-}$ shows that electrogenerated $Ru(bpy)_3^{3+}$ can rapidly accept an electron from $C_2O_4^{2-}$, in agreement with previous reports.[24]

The $Ru(bpy)_3^{2+}/C_2O_4^{2-}$ system was used to generate the $CO_2^{\cdot-}$ for the reduction of benzyl bromide and Zn^{2+} ions. Fig. 7A shows a decrease in the anodic peak current for $Ru(bpy)_3^{2+}$ oxidation upon addition of 3.0 mM benzyl bromide to a solution of 0.5 mM $Ru(bpy)_3^{2+}$ and 2.0 mM $C_2O_4^{2-}$. Based on the reported ECL mechanism for the $Ru(bpy)_3^{2+}/C_2O_4^{2-}$ system along with analysis of the CV of

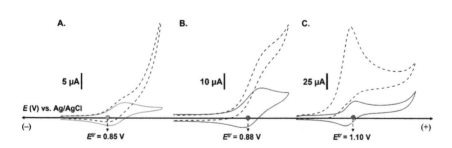

Fig. 6 Cyclic voltammetry in an O_2-free H_2O/MeCN solution (90% H_2O v/v) containing 0.1 M PBS (pH = 7.3) with (A; red) 1.0 mM $IrCl_6^{4-}$; (B; violet) $Fe(bpy)_3^{2+}$; and (C; blue) $Ru(bpy)_3^{2+}$, in the absence and presence of 2.0 mM $C_2O_4^{2-}$ (dashed black traces). All CVs were recorded using a 0.07 cm² GC working electrode, a Pt mesh auxiliary electrode, and a Ag/AgCl (3.5 M KCl) reference electrode. The measurements were made at v = 10 mV s^{-1}.

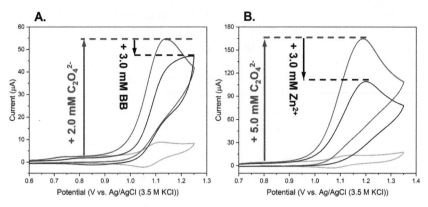

Fig. 7 (A) Cyclic voltammetry in solutions containing: (red) 0.50 mM $Ru(bpy)_3^{2+}$; (blue) 0.50 mM $Ru(bpy)_3^{2+}$ and 2.0 mM $C_2O_4^{2-}$; and (black) 0.50 mM $Ru(bpy)_3^{2+}$, 2.0 mM $C_2O_4^{2-}$, and 3.0 mM benzyl bromide. (B) Cyclic voltammetry in solutions containing: (red) 0.50 mM $Ru(bpy)_3^{2+}$; (blue) 0.50 mM $Ru(bpy)_3^{2+}$, and 5.0 mM $C_2O_4^{2-}$; and (black) 0.50 mM $Ru(bpy)_3^{2+}$, 5.0 mM $C_2O_4^{2-}$, and 3.0 mM $ZnCl_2$. Voltammogram A was recorded in a H_2O/ MeCN solution (90% H_2O v/v) containing 0.1 M PBS (pH = 7.3), and B was recorded in an aqueous solution containing 0.1 M PBS (pH = 7.3). All CVs were carried out with a 0.07 cm^2 GC working electrode, a Pt mesh auxiliary electrode, and a Ag/AgCl (3.5 M KCl) reference electrode at $\nu = 100$ mV s^{-1} in O_2-free solutions.

$Ru(bpy)_3^{2+}$ in the presence of $C_2O_4^{2-}$, a 7-step mechanism for the oxidation of $C_2O_4^{2-}$ by electrogenerated $Ru(bpy)_3^{3+}$ (eqn (11)–(17)) is proposed to occur.[24,25] Briefly, $Ru(bpy)_3^{3+}$ is produced from the 1e oxidation of $Ru(bpy)_3^{2+}$ (eqn (11)). Then, $Ru(bpy)_3^{3+}$ homogeneously oxidizes $C_2O_4^{2-}$ to yield $C_2O_4^{\cdot-}$ (eqn (12)), which rapidly decomposes (i.e., $\tau_{1/2} \approx 1$ μs) into CO_2 and $CO_2^{\cdot-}$ (eqn (13)). The $CO_2^{\cdot-}$ in solution can reduce $Ru(bpy)_3^{3+}$ to $Ru(bpy)_3^{2+}$ (eqn (14)), reduce $Ru(bpy)_3^{2+}$ to $Ru(bpy)_3^+$ (eqn (15)), or be oxidized to CO_2 at the electrode (eqn (16)). Any $Ru(bpy)_3^+$ that is formed will react with $Ru(bpy)_3^{3+}$ to generate two equivalents of $Ru(bpy)_3^{2+}$ (eqn (17)).[27]

$$Ru(bpy)_3^{2+} \rightleftharpoons Ru(bpy)_3^{3+} + e^- \tag{11}$$

$$Ru(bpy)_3^{3+} + C_2O_4^{2-} \rightleftharpoons Ru(bpy)_3^{2+} + C_2O_4^{\cdot-} \tag{12}$$

$$C_2O_4^{\cdot-} \rightleftharpoons CO_2 + CO_2^{\cdot-} \tag{13}$$

$$Ru(bpy)_3^{3+} + CO_2^{\cdot-} \rightleftharpoons Ru(bpy)_3^{2+} + CO_2 \tag{14}$$

$$Ru(bpy)_3^{2+} + CO_2^{\cdot-} \rightleftharpoons Ru(bpy)_3^+ + CO_2 \tag{15}$$

$$CO_2^{\cdot-} \rightleftharpoons CO_2 + e^- \tag{16}$$

$$Ru(bpy)_3^{3+} + Ru(bpy)_3^+ \rightleftharpoons 2Ru(bpy)_3^{2+} \tag{17}$$

An alternative reaction pathway for the $CO_2^{\cdot-}$ becomes available upon the addition of benzyl bromide to the $Ru(bpy)_3^{2+}/C_2O_4^{2-}$ system (eqn (18)) wherein the C–Br bond contained in benzyl bromide is reductively cleaved *via* a homogeneous one-electron transfer from $CO_2^{\bullet-}$. In the absence of benzyl bromide, the $CO_2^{\bullet-}$ acts to regenerate the primary current-contributing species, $Ru(bpy)_3^{2+}$ (eqn (14), (15), and (17)). In the presence of benzyl bromide, however, the $CO_2^{\bullet-}$ is intercepted (*i.e.*, eqn (18)), thereby avoiding the occurrence of eqn (14)–(17), which leads to the decrease in anodic peak current (i_{pa}) observed in Fig. 7A. In very preliminary studies, the electrolysis of the $Ru(bpy)_3^{2+}/C_2O_4^{2-}$ system in the presence of benzyl bromide predominately gives benzaldehyde as the major product. Although this result demonstrates the capability of $CO_2^{\bullet-}$ to reductively cleave the benzylic C–Br moiety, a net oxidation occurs. At the potential required to generate $CO_2^{\bullet-}$, the benzylic radical shown in eqn (18) and Br^- can both undergo heterogeneous oxidation to generate a benzylic cation and Br_2.[26] The oxygen contained in the product likely originates from water, which can act as a nucleophile to add to the electrophilic benzylic cation. This yields benzyl alcohol that is either oxidized at the electrode (or by Br_2) to give the product, benzaldehyde. Oxidative reduction reactivity can also be applied to the reduction of zinc ions (eqn (19)) ($E^{0\prime}$ (Zn^{2+}/Zn) $= -0.95$ V *vs.* Ag/AgCl (3.5 M KCl)), where two equivalents of electrogenerated $CO_2^{\bullet-}$ reduce Zn^{2+} to neutral zinc metal (Fig. 7B). Although it is hypothesized that Zn^0 is formed in this reaction, isolation and analysis of product following bulk electrolysis is still required.

$$\text{benzyl bromide} + CO_2^{\bullet-} \longrightarrow \text{benzyl radical} + Br^- + CO_2 \tag{18}$$

$$Zn^{2+} + 2CO_2^{\bullet-} \rightarrow Zn^0 + 2CO_2 \tag{19}$$

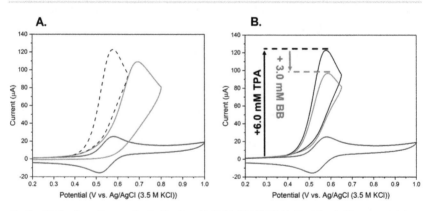

Fig. 8 (A) Cyclic voltammetry in O_2-free H_2O/MeCN (90% H_2O v/v) solutions containing: (blue) 1.0 mM TEMPO; (green) 6.0 mM TPA; and (black dashed line) 1.0 mM TEMPO and 6.0 mM TPA. (B) Cyclic voltammetry in O_2-free H_2O/MeCN (90% H_2O v/v) solutions containing: (blue) 1.0 mM TEMPO; (black) 1.0 mM TEMPO and 6.0 mM TPA; and (green) 1.0 mM TEMPO, 6.0 mM TPA, and 3.0 mM benzyl bromide. All voltammograms were recorded at $v = 100$ mV s^{-1} using a 0.07 cm^2 GC working electrode, a Pt mesh auxiliary electrode, and a Ag/AgCl (3.5 M KCl) reference electrode. All solutions contained 0.1 M PBS (pH $= 10.5$).

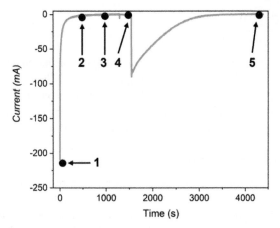

Fig. 9 The sequence of events in mediated electrolysis including: (1) applying -0.5 V vs. Ag/AgCl, (2) injecting benzyl alcohol, (3) introduction of $S_2O_8^{2-}$, (4) injection of $Ru(NH_3)_6^{3+}$, and (5) end of the electrolysis.

Tri-*n*-propylamine (TPA) also presents an opportunity to conduct oxidative reduction as its oxidation gives rise to a strong reducing agent. The electrochemical behavior of TPA has been studied previously.[26] When TPA is oxidized, the $TPA^{\bullet+}$ ($\tau_{1/2} \approx 1$ ms) is rapidly deprotonated to yield $TPA^{\bullet}$ ($E^{0\prime}$ ($TPA^{\bullet}/TPA^{+}$) $= -1.7$ V vs. Ag/AgCl (3.5 M KCl)), which is then capable of homogeneously reducing organic compounds. Based on the half-life for the $TPA^{\bullet+}$, the $TPA^{\bullet}$ is predicted to be generated approximately 1 μm from the electrode surface where it is susceptible to oxidation. Voltammetric analysis of TPA in the presence of 2,2,6,6-tetramethyl-1-piperidinyloxy (TEMPO) demonstrates that the oxidized form of TEMPO acts as an electrocatalyst for TPA oxidation (Fig. 8A). Thus, TEMPO was used to mediate TPA oxidation to form $TPA^{\bullet}$ further from the electrode. Then, upon introduction of benzyl bromide to the TEMPO/TPA system, a decrease in peak current was observed, characteristic of oxidative reduction chemistry (Fig. 8B).

Conclusion

In this work, we showcased the utility of the electrochemically generated intermediates $SO_4^{\bullet-}$ and $CO_2^{\bullet-}$ to carry out electroorganic reactions close to the water potential window. In both reductive oxidation and oxidative reduction, a mediator was required to produce $CO_2^{\bullet-}$ and $SO_4^{\bullet-}$ away from the working electrode to allow these radical ions to react with the organic substrate. Our analysis shows that $Ru(NH_3)_6^{3+}$ is the optimal electrocatalyst for the homogeneous reduction of $S_2O_8^{2-}$. The mediated reductive oxidation using $Ru(NH_3)_6^{3+}$ and $S_2O_8^{2-}$ was employed to oxidize various benzylic and aliphatic alcohols, toluene, and aliphatic amines at -0.5 V vs. Ag/AgCl in H_2O/MeCN solutions with high yield. Analogous to previous findings, oxalate oxidation can be mediated by $Ru(bpy)_3^{3+}$ to generate $CO_2^{\bullet-}$ in water. The utility of the $CO_2^{\bullet-}$ for electrosynthesis was examined by reductive cleavage of the C–Br moiety in benzyl bromide as well as in the reduction of Zn ions.

Experimental

Cyclic voltammetry studies

Cyclic voltammetry studies were carried out using a Biologic Dual Channel SP300 Potentiostat and a three-electrode configuration including a 0.07 cm^2 glassy carbon (GC) working electrode, a Pt mesh auxiliary electrode, and a Ag/AgCl (3.5 M KCl) reference electrode in an undivided cell. The GC electrode was polished with a slurry of 1 μm alumina in H_2O and then dried with stream of N_2.

Controlled-potential electrolysis

Controlled-potential electrolysis was performed at room temperature using a Biologic Dual Channel SP300 Potentiostat. A detailed description for the preparation of the electrolysis cell and the electrolysis procedure is outlined elsewhere.[28] An example i–t trace for the mediated electrolysis of benzyl alcohol and the sequence of addition of reagents during electrolysis is shown in Fig. 9. First, 20 mL of a deoxygenated H_2O–MeCN (80% H_2O v/v) was placed in the cathode compartment of the electrolysis cell. The cathode compartment was maintained under Ar for the duration of the electrolysis. Next, a constant potential of −0.5 V $vs.$ Ag/AgCl (3.5 M KCl) was applied to the cell which is marked by a current spike (1, Fig. 9). Once the charging current reached a steady baseline, benzyl alcohol was dissolved in 2 mL of MeCN and injected into the cathode compartment (2, Fig. 9). Next, 1 mL of the deoxygenated catholyte was removed from the electrolysis cell and was used to dissolve $Na_2S_2O_8$. The solution of $Na_2S_2O_8$ was then injected into the cathode compartment (3, Fig. 9). The same procedure was carried out to introduce $Ru(NH_3)_6^{3+}$ into the electrolysis cell. A current spike was observed upon injection of $Ru(NH_3)_6^{3+}$ to the cathode compartment of the electrolysis cell (4, Fig. 9). The electrolysis was concluded once the current returned to the baseline level and remained steady for ~15 min (5, Fig. 9).

A similar procedure was carried out for mediated reductive oxidation of benzylic and aliphatic alcohols, toluene, aliphatic amines, and benzyl bromide. Electrolyses of aliphatic amines were carried out in an aqueous phosphate buffer (pH = 3.10). Electrolysis of benzyl bromide was carried out at E_{app} = 1.2 V $vs.$ Ag/AgCl (3.5 M KCl) using $Ru(bpy)_3^{3+}$ as the mediator for oxalate oxidation.

Product isolation

A similar procedure was carried out for the product purification and yield determination upon reductive oxidation of benzylic and aliphatic alcohols, toluene, and benzyl bromide.[12] In summary, the catholyte was mixed with 20 mL of brine solution and 30 mL of ethyl acetate in a separatory funnel, and the mixture was shaken vigorously. The mixture in the separatory funnel was left to separate into immiscible organic and aqueous layers. Next, the organic phase was removed from the separatory funnel, and the aqueous phase was extracted with ethyl acetate two more times. The three ethyl acetate portions were mixed and dried over anhydrous Na_2SO_4. After one-hour, the dried organic phase was separated from Na_2SO_4 and was condensed under reduced pressure to give a 1 mL organic solution. Next, the components of the organic solution were separated via normal phase chromatography.

Post-electrolysis workup and separation of hydroxylated aliphatic amines followed the procedure reported by Sanford and co-workers.[23] In summary, once electrolysis was complete, the pH of the catholyte was lowered to 1.98 using dropwise additions of dilute H_3PO_4. Next, the catholyte solution was mixed with 20 mL toluene, and the mixture was completely evaporated at 60 °C under reduced pressure. The remaining solid in the round bottom flask was mixed with 50 mL MeCN, 12 mL of tri-*n*-ethylamine, and 30 µL of benzoyl chloride. The mixture was stirred at room temperature overnight. Next, MeCN was evaporated under reduced pressure and the solid materials were extracted with 20 mL CH_2Cl_2. Finally, the organic phase was concentrated under reduced pressure to 1 mL, and components of the organic phase were separated using normal phase chromatography.

Conflicts of interest

The authors declare no competing financial interest.

Acknowledgements

This work was supported by National Science Foundation Center for Synthetic Organic Electrochemistry (CHE-2002158).

References

1 A. J. Bard, R. Parsons and J. Jordan, *Standard Potentials in Aqueous Solution*, Marcel Dekker, New York, 1985.

2 A. J. Bard, L. R. Faulkner and H. S. White, *Electrochemical Methods: Fundamentals and Applications*. John Wiley & Sons, New York, 2022.

3 O. Hammerich and B. Speiser, *Organic Electrochemistry*, CRC Press, Boca Raton, 2016.

4 A. J. Bard, *Electrogenerated Chemiluminescence*. CRC Press, Boca Raton, 2004.

5 A. D. Armstrong, R. E. Huie, W. H. Koppenol, S. V. Lymar, G. Merényi, P. Neta, B. Ruscic, B. D. M. Stanbury, S. Steenken and P. Wardman, *Pure Appl. Chem.*, 2015, **87**, 1139.

6 K. C. F. Araújo, E. V. dos Santos, P. V. Nidheesh and C. A. Martínez-Huitle, *Curr. Opin. Chem. Eng.*, 2022, **38**, 100870.

7 R. Memming, *J. Electrochem. Soc.*, 1969, **116**, 785.

8 W.-D. Oh, Z. Dong and T.-T. Lim, *Appl. Catal., B*, 2016, **194**, 169.

9 M.-M. Chang, T. Saji and A. J. Bard, *J. Am. Chem. Soc.*, 1977, **99**, 5399.

10 F. Kanoufi, C. Cannes, Y. Zu and A. J. Bard, *J. Phys. Chem. B*, 2001, **105**, 8951.

11 T. Kai, M. Zhou, S. Johnson, S. H. Ahn and A. J. Bard, *J. Am. Chem. Soc.*, 2018, **140**, 16178.

12 S. Hosseini, J. N. Janusz, M. Tanwar, A. D. Pendergast, M. Neurock and H. S. White, *J. Am. Chem. Soc.*, 2022, **144**, 21103.

13 H.-Y. Thu, W.-Y. Yu and C.-M. Che, *J. Am. Chem. Soc.*, 2006, **128**, 9048.

14 C. Huang, J.-H. Wang, J. Qiao, X.-W. Fan, B. Chen, C.-H. Tung and L.-Z. Wu, *J. Org. Chem.*, 2019, **84**, 12904.

15 J. Jin and D. W. MacMillan, *Angew. Chem., Int. Ed.*, 2015, **54**, 1565.

16 A. F. Chmiel, O. P. Williams, C. P. Chernowsky, C. S. Yeung and Z. K. Wickens, *J. Am. Chem. Soc.*, 2021, **143**, 10882.

17 C. M. Hendy, G. C. Smith, Z. Xu, T. Zian and N. T. Jui, *J. Am. Chem. Soc.*, 2021, **143**, 8987.

18 S. N. Alektiar and Z. K. Wickens, *J. Am. Chem. Soc.*, 2021, **143**, 13022.

19 P. Brandi, C. Galli and P. Gentili, *J. Org. Chem.*, 2005, **70**, 9521.

20 X.-S. Xue, P. Ji, B. Zhou and J.-P. Cheng, *Chem. Rev.*, 2017, **117**, 8622.

21 D. Mazzarella, G. E. Crisenza and P. Melchiorre, *J. Am. Chem. Soc.*, 2018, **140**, 8439.

22 F. Li, D. Tian, Y. Fan, R. Lee, G. Lu, Y. Yin, B. Qiao, X. Zhao, Z. Xiao and Z. Jiang, *Nat. Commun.*, 2019, **10**, 1774.

23 M. Lee and M. S. Sanford, *Org. Lett.*, 2017, **19**, 572.

24 F. Kanoufi and A. J. Bard, *J. Phys. Chem. B*, 1999, **103**, 10469.

25 I. Rubinstein and A. J. Bard, *J. Am. Chem. Soc.*, 1981, **103**, 512.

26 D. D. Wayner, D. McPhee and D. Griller, *J. Am. Chem. Soc.*, 1988, **110**, 132.

27 W. Miao, J.-P. Choi and A. J. Bard, *J. Am. Chem. Soc.*, 2002, **124**, 14478.

28 E. C. R. McKenzie, S. Hosseini, A. G. C. Petro, K. K. Rudman, B. H. R. Gerroll, M. S. Mubarak, A. L. Baker and R. D. Little, *Chem. Rev.*, 2022, **122**, 3292.

New strategies in organic electrosynthesis: general discussion

Mickaël Avanthay, [ID] Joshua A. Beeler, [ID] Belen Batanero, [ID]
Dylan G. Boucher, Richard C. D. Brown, [ID] Victoria Flexer, [ID]
Robert Francke, [ID] Bernardo Antonio Frontana-Uribe, [ID]
Seyyedamirhossein Hosseini, [ID] Long Luo, [ID] Shelley D. Minteer, [ID]
Robert Price, [ID] Naoki Shida, [ID] José Manuel Ramos-Villaseñor [ID]
and Thomas Wirth [ID]

DOI: 10.1039/d3fd90039h

Robert Francke opened a general discussion of the paper by Bernardo Antonio Frontana-Uribe: The plot of the ionic conductivity *vs.* the supporting electrolyte concentration in your paper (https://doi.org/10.1039/d3fd00064h), appears to follow a discontinuous function in cyrene–EtOH (1 : 1). How can this unusual behavior be explained?

Bernardo Antonio Frontana-Uribe answered: A possible explanation is that the intermolecular aggregations of the cyrene–EtOH solvent mixture allow the dissociation of a greater amount of supporting electrolyte, increasing the conductivity. But further studies are required to clarify this hypothesis.

Robert Francke suggested: The behavior of the cyrene–EtOH-based electrolyte solutions in the conductivity measurements is very interesting. Molecular dynamics simulations could be taken into consideration for a better understanding of the unusual properties.

Bernardo Antonio Frontana-Uribe replied: Thanks for the suggestion. Molecular dynamic simulations will be taken into consideration for understanding the possible intermolecular aggregations of cyrene–EtOH solvent mixtures interacting with the supporting electrolyte.

Mickaël Avanthay said: The conductivity of the solvent is very low, so it seems it is not suitable to be used on its own. You detail the use of ethanol co-solvent – what is the minimum amount of co-solvent possible, and are other co-solvents appropriate?

Bernardo Antonio Frontana-Uribe responded: We think that the minimum quantity of co-solvent should be 30% ethanol in a cyrene/ethanol mixture, to avoid conductivity issues. The viscosity value of this solvent mixture is 8.4 cP, which we consider as the limit that does not cause conductivity issues. Using less than 30% of ethanol as co-solvent could lead to conductivities issues. Other co-solvents are also possible such as methanol and propylene carbonate.

Mickaël Avanthay addressed José Manuel Ramos-Villaseñor and Bernardo Antonio Frontana-Uribe: Given the solvent is naturally enantiopure, can you use it as a chiral solvent to perform enantioselective electrochemistry from achiral/racemic reagents?

José Manuel Ramos-Villaseñor and **Bernardo Antonio Frontana-Uribe** answered: It can be used for these purposes. In fact, the cathodic reduction of benzaldehyde using the cyrene–EtOH (1 : 1) solvent mixture afforded the diasteromeric DL pinacol product in greater proportion than the *meso* product. But further studies are required to know the potential of cyrene in asymmetric reactions.

Dylan G. Boucher addressed José Manuel Ramos-Villaseñor and Bernardo Antonio Frontana-Uribe: What is the mechanism you propose for the differing selectivity in different electrolytes?

José Manuel Ramos-Villaseñor and **Bernardo Antonio Frontana-Uribe** replied: The corresponding anion radical formed at the electrode surface is stabilized by the Bu_4N^+ cation present at the electric double layer when cyrene is present at high concentration, this process facilitates the second electron transfer near the electrode surface, yielding the diphenylmethanol derivative. On the other hand, the use of $LiClO_4$ favors the stabilization of the radical anion of benzophenone through its coordination with the lithium cation, promoting the C–C bond formation and obtaining the pinacol product.

Victoria Flexer commented: The solvent you're presenting shows a certain similarity in its chemical structure to DOL (dioxalane), which is used in lithium–sulfur batteries. In the assembly of these batteries, DOL is most often used in combination with 1,2-dimethoxyethane (DME). Have you considered the possibility of mixing cyrene with DME? I was thinking that the mixture could potentially widen the electrochemical window of the cyrene.

Bernardo Antonio Frontana-Uribe replied: Not for the moment, but we have been exploring other solvent mixtures to form aprotic solvent mixtures with cyrene.

Victoria Flexer said: I'm surprised by the high market price of this solvent, particularly given the raw materials from where it is synthesized. Do you foresee any price decrease in the medium to long term?

Bernardo Antonio Frontana-Uribe responded: Yes, I have heard that there are plans to build a new plant for European consumption. This should decrease this solvent price.[1]

1 https://cen.acs.org/environment/green-chemistry/Cyrene-solvent-plant-set-France/98/i8.

Shelley D. Minteer highlighted: Your work focuses on undivided cells. Can you use this solvent in a divided cell? Do you know if the common separators (Nafion, alkaline exchange membranes, *etc.*) are stable in this new solvent?

Bernardo Antonio Frontana-Uribe replied: Yes, the solvent mixture can be used in a divided cell. We think that the Nafion membrane could be stable in this solvent mixture. We have not used alkaline membranes, but we keep in mind that cyrene has a carbonyl functionality and can undergoes an aldolic reaction itself with inorganic bases. Therefore, alkaline exchange membranes used in alkaline media could promote the aldolic reaction of cyrene.

Thomas Wirth added: The current cost of cyrene is very high (approx. £200/ litre). This is due to the production capacity as there is not yet a large scale manufacturing plant available. It was reported that these are currently being built.[1]

The solvent cyrene has a carbonyl and a ketal functionality. How are these functionalities surviving the electrochemical reaction conditions? In combination with EtOH as co-solvent, different reaction might be possible which could degrade the solvent.

1 https://renewable-carbon.eu/news/resolute-project-update-after-15-years-of-work/.

Bernardo Antonio Frontana-Uribe answered: Under electrochemical conditions the ketal functionality survives due to DABCO being used as a sacrificial reagent to prevent the degradation of cyrene. In combination with EtOH, it could react with the carbonyl functionality of cyrene affording the corresponding ketal that improves the stability of cyrene. However, after passing 4 F mol^{-1}, the cathodic reduction of EtOH produces the corresponding alkoxide that can react with cyrene, providing the enolate of cyrene and promoting the aldol reaction of cyrene itself.

Robert Price said: You mentioned that the viscosity of cyrene poses issues for conductivity of the solution when used as a single solvent, hence the requirement for addition of ethanol as a co-solvent. Have you performed any rheological analyses on the mixtures of cyrene : ethanol? In particular, for the 1 : 1 cyrene : ethanol solution, do you see any evidence of dilatant (shear-thickening), pseu-doplastic (shear-thinning), rheopectic (time-dependent shear-thickening) or thixotropic (time-dependent shear-thinning) behaviour that could impact these solutions in terms of mass transport and conductivity?

Bernardo Antonio Frontana-Uribe answered: We have not performed rheo-logical studies indicating that the solvent mixture behaves as a non-Newtonian fluid; however, to answer completely this question further studies are required. We will take them into consideration in future experiments.

Robert Price said: How does the viscosity of the co-solvent mixture vary as a function of cyrene : ethanol ratio? Is a 1 : 1 ratio optimal for this electrochem-istry and how does it behave, rheologically, over time?

Bernardo Antonio Frontana-Uribe answered: The viscosity of the solvent mixture decreases with an increase in the quantity of ethanol. The cyrene–EtOH $1:1$ mixture has the viscosity value of 7 cP, representing half the value of pure cyrene. The viscosity value of the solvent mixture did not exhibit any change after one hour.

Belen Batanero opened a general discussion of the papers by Bernardo Antonio Frontana-Uribe: I note the surprising effect of this novelty solvent depending on the electrolyte used. In one case (with R_4NBF_4 as electrolyte) the alcohol is the main product, however the dimer (pinacol) is obtained when benzophenone is reduced with $LiClO_4$ as the electrolyte. What explanation do you have to explain these results? I wonder about the stability of this acetal (solvent) that should be hydrolyzed under acidic conditions to provide a 1,2-dicarbonyl compound. Did you observe any change in solvent appearance when exposed to oxygen or light? Did you previously deoxygenate the cathodic solution in order to avoid superoxide anion formation (which would probably react with the solvent)?

Bernardo Antonio Frontana-Uribe answered: The anion radical formed at the electrode surface is stabilized by the Bu_4N^+ cation present at the electric double layer when cyrene is present in an equal or major proportion, this process facilitates the second electron transfer at the electrode surface, yielding the diphenylmethanol derivative. Otherwise, the coordination of the lithium cation with the radical anion of benzophenone favors $1e^-$ transfer for C–C bond formation.

Long Luo opened a general discussion of the paper by Seyyedamirhossein Hosseini:Reductive oxidation and oxidative reduction are very interesting reaction schemes. The different redox environments on the electrode surface *versus* in the solution near the electrode surface, can be utilized to achieve chemical transformations that cannot be accomplished otherwise. What are the unique reactions you think can be potentially achieved using your reductive oxidation or oxidative reduction (https://doi.org/10.1039/d3fd00067b)?

Seyyedamirhossein Hosseini and **Joshua A. Beeler** responded: Opportunities in this system lie in the follow-up reactions that would be impossible to achieve using direct electrolysis. For example, the halogenation of toluene would not be possible using direct toluene oxidation since halogen oxidation to the halide would occur simultaneously. However, reductive oxidation can be used to oxidize toluene at an applied potential of -0.2 V, meaning that direct halogen oxidation can be avoided.

Long Luo queried: Sulfate radicals can be conveniently generated by mild heating, UV irradiation, or sonication of a $Na_2S_2O_8$ solution. What are the advantages of using electrochemistry?

Richard C. D. Brown commented: I really like the concept described in the paper, where strong oxidants (or reductants) are generated by electrochemical reduction (or oxidation) under relatively mild conditions using mediators. This could be a very powerful and useful approach in electrosynthesis.

As a comment in response to Long Luo's question on the difference between the electrochemical approach to generating sulfate radical anions from peroxydisulfate and the thermal alternative (heating peroxydisulfate): assuming even temperature distribution in the thermal reaction, the sulfate radical anion will be generated throughout the solution, wherever peroxydisulfate is present. On the other hand, the electrochemical method will give different concentration profiles of sulfate radical anions depending on the flux of the mediator (reduction) from the electrode and the relative rates of the follow-up electron-transfer and chemical steps. Thus, some additional opportunities for control may be possible in the electrochemical method described in your presented paper.

Joshua A. Beeler responded: This comment from Richard C. D. Brown answers Long Luo's question. The non-electrochemical methods used previously generate an even distribution of sulfate radical anions throughout the solution. However, using the electrochemical method, sulfate radical anion is generated at a high concentration away from the electrode surface *via* the reaction with the mediator. This allows one to control the reaction kinetics and provides opportunities for follow-up chemical steps.

Naoki Shida asked: For the reductive oxidation system, is the reactivity of active species closer to a single-electron oxidant *via* outer-sphere electron transfer? Or does it show the hydrogen atom abstraction type of reactivity?

Joshua A. Beeler answered: The reductive oxidation system shows hydrogen atom abstraction reactivity. A detailed explanation can be found in our previous publication.[1]

1 S. Hosseini, J. N. Janusz, M. Tanwar, A. D. Pendergast, M. Neurock and H. S. White, *J. Am. Chem. Soc.*, 2022, **144**, 21103–21115.

Naoki Shida addressed Joshua A. Beeler and Seyyedamirhossein Hosseini: In the manuscript, you demonstrate the oxidation of tertiary C–H bonds by avoiding the oxidation of primary amines. Primary amines are generally easily oxidized, but this reductive oxidation system avoids such reaction by generating reactive species under very mild anodic potential. Is there any oxidatively weak functional group that is compatible with this reaction, such as carboxylic acid (which easily undergoes Kolbe-type reaction) and olefins?

Joshua A. Beeler responded: We protected the amine groups *via* protonation. We were able to carry out C–H activation in the presence of a number of functional groups such as halides, sulfonyl, carboxyl, and amine groups using reductive oxidation.

Richard C. D. Brown asked: For the mediated electrosyntheses described in your work, did you investigate different mediators that are more difficult to reduce or oxidise relative to the substrate in the preparative experiments? For example, did you try to perform the mediated oxidation of oxalate in the presence of a mediator possessing a more positive oxidation potential?

Joshua A. Beeler answered: All mediators that were screened for persulfate reduction and oxalate oxidation are discussed in our paper. For the reductive oxidation system (*i.e.*, mediated persulfate reduction), we screened four different mediators, see Fig. 2 in our paper. For the oxidative reduction system (*i.e.*, mediated oxalate oxidation), three mediators were screened, as shown in Fig. 6 in our paper.

Richard C. D. Brown addressed Seyyedamirhossein Hosseini and Joshua A. Beeler: Following on from my previous question, I emphasise that the question related to the preparative experiment rather than the cyclic voltammetry.

Joshua A. Beeler replied: We did not try other mediators in the preparative experiment.

Robert Francke addressed Joshua A. Beeler and Seyyedamirhossein Hosseini: In your work, one electron reduction/oxidation of radical intermediates derived from peroxydisulfate and oxalate, respectively, are described as unproductive pathways. The radicals seem to be rapidly converted at the electrode surface, but react slowly with the activated form of the mediator. Do you have an explanation for the difference between the rates of these competing processes?

Joshua A. Beeler responded: The reaction rate for the $1e^-$ quenching of SO_4^- and CO_2^- by the active form of the electrocatalyst is very rapid and occurs at approximately the same rate for both species.

Robert Francke addressed Joshua A. Beeler and Seyyedamirhossein Hosseini: Since the generation of radical intermediates has to proceed with a certain spatial distance to the electrode, one seems to face a dilemma: on the one hand, mediators have to be employed, which react relatively slowly (otherwise, radical formation occurs close to the electrode); on the other hand, high overall reaction rates are desired. Is there a way to address this issue or do you think that this compromise is acceptable?

Joshua A. Beeler answered: Indeed, we would like to employ an electrocatalyst that exhibits a large rate constant for the homogeneous reduction of peroxydisulfate. As facile electrocatalysis enables the generation of a higher concentration of SO_4^-, which in return increases the efficiency of the oxidation reaction.

Robert Francke addressed Joshua A. Beeler and Seyyedamirhossein Hosseini: How many equivalents peroxydisulfate/oxalate were typically employed? How much was consumed during electrolysis?

Joshua A. Beeler responded: The optimal value of peroxydisulfate equivalents depends on the oxidation reaction. For example, 10 and 20 equivalents were used for alcohol oxidation to the aldehyde and carboxylic acid, respectively. For specifics concerning reaction yield *vs.* equivalents of persulfate, see Table S2–S4 in ref. 1. All persulfate initially present is consumed by the end of the electrolysis.

1 A. Hosseini, J. N. Janusz, M. Tanwar, A. D. Pendergast, M. Neurock and H. S. White, *J. Am. Chem. Soc.*, 2022, **144**, 21103–21115.

Belen Batanero addressed Seyyedamirhossein Hosseini and Joshua A. Beeler: How difficult is it to solve organic compounds under your experimental aqueous conditions; a process subjected to oxidative reactions, water, which limit organic synthesis. Have you tried to apply it to some specific organic compound? How were the results?

Joshua A. Beeler answered: As outlined in our paper, various organic compounds are dissolved through the addition of 10 to 20% acetonitrile by volume, to the aqueous solutions. For examples of specific organic compounds, see Fig. 4 in our paper.

Conflicts of interest

There are no conflicts to declare.

Faraday Discussions

PAPER

Core–shell nanostructured Cu-based bi-metallic electrocatalysts for co-production of ethylene and acetate

Jeannie Z. Y. Tan, [ID] * Ashween Kaur Virdee, [ID] John M. Andresen [ID] and M. Mercedes Maroto-Valer [ID]

Received 3rd March 2023, Accepted 19th June 2023

DOI: 10.1039/d3fd00058c

Direct electrocatalytic CCU routes to produce a myriad of valuable chemicals (*e.g.*, methanol, acetic acid, ethylene, propanol, among others) will allow the chemical industry to shift away from the conventional fossil-based production. Electrofuels need to go beyond the current electroreduction of CO_2 to CO, and we will here demonstrate the continuous flow electroreduction of syngas (*i.e.*, CO and H_2), which are the products from CO_2-to-CO, with enhanced product selectivity (~90% towards ethylene). To overcome current drawbacks, including bicarbonate formation that resulted in low CO_2 utilisation and low C_{2+} product selectivity, the development of nanostructured core–shell bi-metallic electrocatalysts for direct electrochemical reduction of syngas to C_{2+} is proposed. Electrosynthesis of ethylene is performed in a state-of-the-art continuous flow three-compartment cell to produce ethylene (cathodic gas phase product) and acetate (cathodic liquid phase product), simultaneously.

1. Introduction

Meeting the Paris Agreement[1] will require a wide range of mitigation and removal strategies, among which carbon capture and utilization (CCU) is particularly appealing as it can curb emissions while creating economic value. Notably, according to estimates, the large-scale deployment of CCU could help to decarbonise industrial activities by saving up to 3.5 $GtCO_{2eq}$ per year in 2030, that is, an 83% reduction compared with conventional fossil-based technologies.[2] Chemical industry is one of the top three emitting industries and is among the most difficult to decarbonise. The most effective way to decarbonise the chemical industry is to reduce the reliance on fossil fuels and gas demand as the feedstocks to the sector as the largest-volume chemicals to produce commodity chemicals to satisfy global markets. In addition, fossil fuels are also used to provide heat and pressure to drive those chemical reactions.

Research Centre for Carbon Solutions (RCCS), Heriot-Watt University, Edinburgh, EH14 4AS, UK. E-mail: j. tan@hw.ac.uk

Amongst the commodity chemicals, acetic acid, ethylene and ethylene glycol (6.5, 164 and 30.2 Mt per year, respectively) are the most important chemicals for the chemical and pharmaceutical sectors. Ethylene as the major platform chemical has a global production increased from 185 Mt in 2018 to 214 Mt in 2021 and the average price of ethylene worldwide has increased to 1235 USD per ton.[3] Unfortunately, it is produced *via* steam cracking of naphtha, currently the predominant production route followed by the thermal cracking of ethane,[4] that emits 1.51 $kgCO_{2eq}$ kg^{-1} of ethylene (cradle-to-gate),[5] resulting in 0.26 $GtCO_{2eq}$ (in order to satisfy the 2021 production volume) and accounting for 30% of the total energy needs of the chemical industry.[6]

CO_2 electrochemical reduction (CO2RR) into value added products using renewable energy offers a promising approach to reduce anthropogenic CO_2 emissions and development of sustainable energy systems. Coupling the intermittent electricity generated from solar and wind energy for CO2RR has been explored extensively for storing renewable electricity in chemicals, such as ethylene. CO2RR can produce carbon monoxide (CO), methane (CH_4), methanol (CH_3OH), formic acid (HCOOH), ethylene (C_2H_4) ethanol (CH_3CH_2OH) and chemicals with longer chain hydrocarbons. However, implementations of this technology for large-scale applications are challenging as the molecule is thermodynamically stable. Some other bottlenecks, including carbonate formation, poor product selectivity, low carbon utilisation rate, competition of H_2 evolution reaction at low potentials, and poor stability, need to be overcome to make low temperature CO2RR competitive with other CO_2 conversion technologies or chemical and energy production processes.[7]

Cu is the only electrocatalyst with acceptable activity and selectivity for C_{2+} products due to the unique property of binding *CO and *H.[8] However, CO2RR requires a high bias (approx. −1.0 V *vs.* reversible hydrogen electrode, RHE). Furthermore, under high biases (−0.9 to −1.1 V) up to 12 different C2 or C3 products can be identified at the Cu polycrystalline cathode.[9] To improve the efficiency and selectivity of C_{2+} products, the alkaline flow electrolyser has been proposed because it is one of the best electrolysers with high faradaic efficiency (90%) as well as superior single pass conversion efficiency (40%).[10] However, the electrolyser consumes a lot of the electrolyte due to reductive disproportionation into CO and CO_3^{2-}. As a result, the CO_2 utilisation rate is significantly limited, and large voltage is required.[7] To date, the best performing electrolyser for CO_2 to ethylene displayed a 15% energy efficiency, 2% carbon efficiency, 60 h of steady state operation at 3.5 V and 500 mA cm^{-2}.[11]

Addressing the carbonate formation is vital for this technology to make it a viable option for renewable chemical and fuel production. To date, the best reported CO production performance could be achieved at 3 V with approx. 4000 h at steady state operation under 200 mA cm^{-2} with 98% product selectivity.[12] The carbon efficiency reported is 43%. With this outstanding performance of CO production, the production of C_{2+} products in the sequential step needs to be developed to produce more valuable products, such as ethylene and acetic acid, which are ~$1200 and ~$800 per Mt in 2021. Furthermore, the electro-conversion of these C_{2+} products are among the potential commodity chemicals with high revenue.[10] Hence, the optimisation of the co-production of these commodity chemicals is the aim of the first part of this study.

The use of bimetallic electrocatalysts has been proposed previously to enhance the product selectivity with low overpotential required. For instant, Wang *et al.* reported that bimetallic Cu–Ag nanoflowers showed enhanced product selectivity towards acetaldehyde (~70%) from CO at −0.536 V *vs.* RHE.[13] Kuhn *et al.* also pointed out that a Cu–Ag bimetallic electrocatalyst had shown enhanced product selectivity towards ethylene (43% at −0.75 V *vs.* RHE) due to increased conductivity, stability and concentration of adsorbed CO when compared to Ag and Cu single metallic electrocatalysts.[14] Furthermore, Gao *et al.* demonstrated that the Cu–Co core–shell bimetallic electrocatalysts with Co-rich samples were more selective towards hydrocarbons, whereas Cu-rich samples were prone to produce oxygenated compounds from syngas.[15] Herein, three types of core–shell bimetallic electrocatalysts were synthesized. The syngas conversion performances of these electrocatalysts were analysed as the second part of this study.

2. Experimental

2.1 Chemicals

Oleylamine (OAm, R&D grade); copper nanoparticles (Cu$_{Ref.}$, 60–80 nm, ≥99.5%); copper(II) acetylacetonate (Cu[acac]$_2$, ≥99.9%); palladium(II) trifluoroacetate (Pd[OCOCF$_3$]$_2$, 97%); silver trifluoroacetate (AgOCOCF$_3$, 98%); tetrahydrofuran (THF, anhydrous ≥99.9%); isopropanol (IPA, ACS reagent ≥99.5%); Nafion™ (20 wt%) and dimethyl sulfoxide (DMSO, ≥99.9%) were purchased from Sigma-Aldrich. Potassium hydroxide (KOH, 85%) was purchased from Acros. Methanol (HPLC grade ≥99.8%); hexane (HPLC grade ≥95%) and deuterated water (D$_2$O) were purchased from Fisher Chemical. Diphenyl ether (DPE) and cobalt(II) acetylacetonate (Co[acac]$_2$, ≥99.0%) were purchased from Merck. All chemicals were used without further purification. Carbon gas diffusion layer (GDL) was Teflonated Toray TGP-60 from Alfa Aesar GmbH & Co; nickel foam (Nanoshel, 99.9%). Milli Q water (18 ΩM) was used throughout the study.

2.2 Methods

2.2.1 Synthesis of Cu nanoparticles (Cu$_{core}$).
The synthesis of Cu$_{core}$ followed a previous published report.[14] Briefly, the growth of the Cu$_{core}$ nanoparticles was achieved in a single pot under a CO gas environment. 16 mg of Cu(acac)$_2$ was dissolved in OAm (9 mL) in a three-neck flask. Then, DPE (120 μM) was added into the resulting turquoise-blue solution, which was placed in an oil bath at 25 °C with continuous stirring throughout the following steps. A Schlenk line was connected with N$_2$, CO and a vacuum pump (KNF, Model: N86KT.18). Before switching to N$_2$, the line was held under vacuum for 5 min to remove the air from the system. This process was repeated for 4 times. CO was then introduced into the system at a flow rate of 5 mL min^{-1} throughout the following steps. Caution: CO is toxic and should be handled with extra care in a well-vented chemical hood equipped with a CO detector. The temperature was maintained at 30 °C for 30 min, followed by heating to 220 °C with a ramp rate of 1.6 °C min^{-1} using the oil bath. The colour of the solution changed from transparent blue green to yellow to reddish-brown during heating. The temperature of the solution was kept at 220 °C for 2 h before letting it cool down to 25 °C naturally.

2.2.2 Synthesis of core–shell bi-metallic nanoparticles. The synthesis of bi-metallic core–shell nanoparticles followed a previous published report.[14] Briefly, an $AgOCOCF_3$, $Co[acac]_2$ or $Pd[OCOCF_3]_2$ solution was prepared by adding pre-determined amount of OAm (2 mg mL^{-1}) in a 10 mL vial. The vial was sonicated for 30 min to ensure the salt was completely dissolved. A syringe pump (Kent Scientific, Model: Genie Touch) was used to inject 5 mL of the resulted salt solution into the three-neck flask that contained the suspension of Cu$_{core}$ at 30 °C while CO was flowing. The infusion rate of the metal (*i.e.*, Ag, Co or Pd) precursor was set at 0.5 mL h^{-1} to ensure uniform growth of the shell layer. After 5 mL of the metal precursor solution was added, the mixture was kept under stirring for another 12 h while maintaining the temperature at 30 °C and a flowrate of CO gas at 5 mL min^{-1}. The formed NPs were collected by centrifuge (Eppendorf 5810) at 11 000 rpm for 3 min, followed by washing with hexane as the dispersing agent (3 mL) and methanol as the antisolvent (17 mL). The samples, labelled as Cu@Ag, Cu@Co or Cu@Pd, were dispersed in hexane for further use.

2.2.3 Preparation of electrodes. The cathode catalysts were deposited *via* hand-painting onto a carbon GDL. 2 mg of Cu$_{Ref.}$ sample was mixed with THF (200 μL), IPA (200 μL), and Nafion (5.2 μL). The solutions were sonicated for 1 h to obtain a homogeneous ink. The ink was then hand-painted using an airbrush on the GDL at a loading of 1.0 mg cm^{-2}. This procedure was repeated for the Cu$_{core}$, Cu@Ag, Cu@Co and Cu@Pd electrocatalysts.

2.3 Electrocatalytic testing

The electroreduction of syngas (CO/H_2) was tested using an alkaline flow cell setup (Fig. 1). Gaseous CO/H_2 (varied with different ratio) was continuously supplied to the back of the cathode at various rates controlled by a mass flow control (Bronkhorst EL-Flow Select). A syringe pump continuously fed KOH (refer to Table 1) as the catholyte and anolyte through two silicone rubber (SiloCell White from Polymax, 9 mm thickness) electrolyte chambers separated by an anion exchange membrane (AEM, AHA Membrane from Eurodia Ltd). The cell

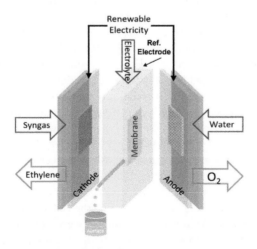

Fig. 1 Schematic diagram of the alkaline flow cell setup.

Table 1 Parameters manipulated in DOE using commercially manufactured Cu nanoparticles

Run no.	$CO:H_2$	Concentration KOH (M)	Flowrate (mL min^{-1})
1	1:0	0.1	0.54
2	1:1	0.1	0.41
3	1:0	0.1	1.07
4	1:1	0.1	1.07
5	1:0	2.0	0.54
6	1:1	2.0	0.41
7	1:0	2.0	1.07
8	1:1	2.0	1.07
9	1:1	1.0	0.78

potential was varied *via* a potentiostat (Autolab PGSTAT302N) and the corresponding current density was continuously measured. Multimeters hooked up to the flow cell were used to measure cathode and anode potentials against an Ag/AgCl reference electrode (Ossila, C2015B1). For a typical procedure, the cell was first hooked up to the CO and/or H_2 feed, electrolyte pump, potentiostat, multimeters, and gas chromatograph (GC, Agilent Technologies 7890B equipped with TCD-FID detectors). Each experiment was begun with two successive linear sweeps between −1.0 and −1.8 V *versus* Ag/AgCl to ensure proper functioning of the cell. The linear sweeps were followed by a potentiostatic preconditioning step at −1.0 V *versus* Ag/AgCl for 60 s. The cell was then stepped galvanostatically with one potentiostat and the cathode potentials were measured *versus* the Ag/AgCl reference electrode. Each galvanostatic step was 9 min in duration. Electrochemical impedance spectroscopy (EIS) measurements were conducted with the frequency range from 100 kHz to 0.1 Hz with an amplitude of 0.1 mA. Ten points were recorded per frequency decade.

2.4 Characterisation

The morphology of the synthesized products was examined using a field emission scanning electron microscope (FE-SEM, Quanta 200 F FEI), a high-resolution transmission electron microscope (HRTEM, FEI Titan Themis 200) equipped with an energy-dispersive X-ray spectroscopy (EDX) detector operated at 200 kV. To investigate the interior structures of the nanospheres, samples were embedded in TAAB 812 resin and sliced into ∼90 nm thick sections. The sliced sections were mounted on the TEM Ni grid. Crystallinity and phase identification of the synthesized products were conducted using powder X-ray diffraction XRD (Malvern Panalytical Empyrean Diffractometer) equipped with Cu Kα radiation (λ = 1.5418 Å) and compared with the ICDD-JCPDS powder diffraction file database. Nuclear magnetic resonance spectroscopy (NMR, Bruker AVIII-400) was used to analyse the liquid products. Typically, 100 μL of the catholyte sample was mixed with 100 μL of DMSO standard solution and 400 μL of D_2O in an NMR tube. The DMSO standard solution was made by diluting the DMSO with D_2O to obtain the final concentration (1.262 mM). The samples were analyzed using ^1H NMR equipped with solvent suppression. The spectra were integrated and compared against a DMSO standard to quantify the concentrations of the liquid products.

3.   Results and discussion

The electroreduction of syngas parameters were optimised through a series of experiments using design of experiment (DOE). A blank GDL was loaded onto the Cu flow plate and the experiment was run following the procedure above. No observable product was detected based on the GC and NMR results.

The parameters, including the concentration of electrolyte (*i.e.*, 0.1, 1 and 2 M), ratio of $CO:H_2$ and flow rate of syngas (Table 1), were manipulated to optimise the parameters for the electrocatalytic process. Using the stepped galvanostatic reduction method, the current applied was manipulated ranging from 1.0 to 4.5 mA. Each step (0.5 mA) was held for 9 min to allow sufficient time for sampling into GC. The optimised current applied was determined with the highest ethylene yielded from syngas. The experimental results were tabulated, and a Pareto chart was used to visualise the most significant factor in the DOE (Fig. 2). The Pareto chart revealed that the concentration of the electrolyte was the most significant factor affecting ethylene production efficiency. Meanwhile, −2.5 mA was the optimised applied current that resulted in the highest single pass production of ethylene, which was ∼2.5 mM cm^{-2}.

Electronic impedance analysis was performed to justify the role of H_2 in the electroreduction of CO to ethylene (Fig. 3a). The Nyquist plot was obtained for commercial Cu being exposed to different gases, namely N_2, $CO-N_2$, $CO-H_2$ and CO. In particular, the high frequency process was found to be not dependent on electrode material and applied current; on the contrary the low frequency one was different for the measurements under different gases and highly dependent on the current applied. The former was attributed to the ionic migration toward the reaction site, whereas the latter reflected the charge transfer due to the reduction reaction.[16] The charge transfer resistance (R_{ct}) was found to be decreasing in the order of N_2 > CO > $CO-N_2$ > $CO-H_2$. When CO and $CO-N_2$ were introduced into the electrolyser, the reduction of R_{ct} was not significant. However, the exposure under $CO-H_2$ largely decreased R_{ct}. This result was attributed to the strong reducing

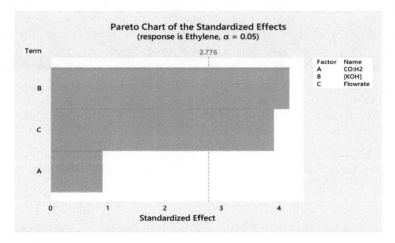

Fig. 2   Pareto chart of the standardized effects obtained for the factorial design optimization of the variables (A) $CO:H_2$, (B) concentration of KOH, and (C) flowrate of syngas.

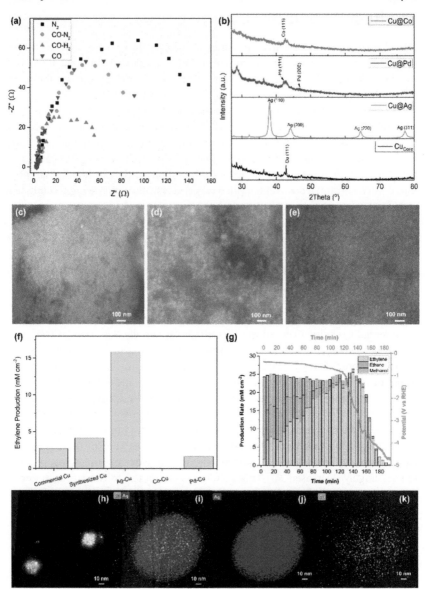

Fig. 3   (a) Impedance (Nyquist plot) of the commercial Cu measured in the presence of N_2 (black ■), CO : $N_2 = 1:1$ (red ●), CO : $H_2 = 1:1$ (magenta ▲), and CO only (blue ▼) under -2.5 mA of applied current and a total flow rate of 0.78 mL min^{-1}. (b) XRD pattern of synthesized Cu_{core} (black), Cu@Ag (red), Cu@Pd (blue) and Cu@Co (magenta). SEM images of (c) Ag@Cu, (d) Pd@Cu and (e) Co@Cu. HRTEM of (h) synthesized Cu_{core} and (i) Ag@Cu with elemental mapping of (j) Ag and (k) Cu. (f) Single pass ethylene production rate using commercial Cu, synthesized Cu_{core}, Ag@Cu, Co@Cu and Pd@Cu. (g) Stability test of Ag@Cu under 2.5 mA for 200 min.

power of CO and H_2. When CO and/or H_2 were introduced into the electrolyser, the oxidised Cu electrocatalyst was reduced to Cu. Being a stronger reducing agent, H_2 tended to reduce more O_2 that adsorbed on the electrocatalyst compared to CO, providing more electrochemically active area, thus, enhancing the conductance.[17] Furthermore, a recent study concluded that the adsorption of H_2 would reconstruct the surface Cu atoms in the electroreduction conditions based on the simulation, thus, reducing the potential barrier.[18]

The optimised experimental conditions, which were $1 : 1$ CO $:$ H_2, 1.0 M of KOH and 0.78 mL min^{-1} of syngas, were applied for the following experiments using different bi-metallic electrocatalysts. The flow of electrolyte was maintained at 0.15 mL min^{-1}.

Three different types of bi-metallic electrocatalysts were synthesized using the sol–gel method as detailed in the experimental procedure section 2.2.2. The crystallinity of the synthesized electrocatalysts was evaluated using XRD (Fig. 3b). The synthesized Cu_{core} with metallic dark brownish colour showed a very weak XRD signal probably due to the small nano-crystallite size. When Ag was loaded as the shell layer, Ag metal produced a strong XRD pattern at 38.1, 44.2, 64.4 and 77.4°, which corresponded to (110), (200), (220) and (311) phases, respectively. Pd and Co were also synthesized as the shell materials of Cu_{core} using the same experimental conditions. Unfortunately, the XRD patterns of these metals were very broad and poor compared to Ag although the Pd (111), Pd (200) and Co (111), which were located at 42.3, 47.0 and 42.6°, respectively, were identified. This was due to very small nanocrystallite size deposited on Cu_{core}.

The morphology of the nanoparticles was characterised using SEM. The synthesized samples appeared as very small nanoparticles. The synthesized Cu_{core} was hardly seen under SEM, thus, the TEM image of Cu_{core} was provided (Fig. 3h). The highly crystallized Ag@Cu revealed the largest particle size ($\sim$25 nm, Fig. 3c) among the bi-metallic electrocatalysts (*i.e.*, $\sim$15 and $\sim$12 nm for Pd@Cu and Co@Cu (Fig. 3d and e), respectively).

The synthesized electrocatalysts were used as the cathodic material in the tailor-made alkaline flow cell using the optimised conditions. Each sample was tested using the galvanostatic method as detailed in DOE experiments. The single conversion efficiency of Ag@Cu was the highest among the samples, which was $\sim$15.0 mM cm^{-2}, when 2.5 mA was applied. Meanwhile, Cu_{core} produced $\sim$4.9 mM cm^{-2}, followed by commercial Cu ($\sim$2.5 mM cm^{-2}) and Pd@Cu ($\sim$2 mM cm^{-2}, Fig. 3f). Unfortunately, Co@Cu showed trace amount of ethylene production. The underperforming of Co@Cu and Pd@Cu was due to the strong electronegativity of Ag compared to Co and Pd. In addition, the low crystallinity and coverage of the Co and Pd compared to Ag@Cu led to unsatisfactory performance.

To further understand the morphology of the synthesized Cu_{core} and Ag@Cu, HRTEM was utilised to examine these electrocatalysts. The synthesized Cu_{core} exhibited a self-assembled nanosphere with diameter $\sim$10 nm that was formed from a group of nanoparticles (Fig. 3h). The elemental analysis on the electrocatalysts was conducted using EDX-HRTEM. The shell layer of Ag (Fig. 3i and j) exhibited a much stronger signal than Cu as the core material (Fig. 3k). This suggested that Ag had formed a thick, solid layer over the Cu.

The best performing sample Ag@Cu was used to perform a stability test. The current of -2.5 mA was supplied continuously to the cell, while the potential was measured simultaneously with the gas sampling through the connected GC. The production rate of ethylene was analysed and plotted (Fig. 3g, yellow bars). The production rate increased immediately after the first 5 min from 1 to $\sim$7.5 mM cm^{-2} (10–30 min). The production rate was doubled after 40 min and stabilised at

~ 22 mM cm^{-2}. The maximum production rate achieved was 25 mM cm^{-2} at the 140th min before the measured potential started to descend. The potential dropped quickly and approached -4.8 V *vs.* RHE after 140 min. This had also resulted in the decrease of ethylene production concurrently and eventually approached 0. The overall ethylene selectivity was 75% with $\sim 15\%$ of ethane (Fig. 3g, green bars) and 10% of methanol (Fig. 3g, purple bars) based on the gaseous product analysis using GC. The total production of ethylene in the single pass conversion was calculated to be 0.5 M after 180 min, which was $\sim 40\%$ efficiency.

To elucidate the liquid products from the cathode compartment, the catholyte, which was flowing through the middle compartment throughout the experiment, was analysed using NMR (Fig. 4). The catholyte obtained was analysed without purification. The NMR spectrum revealed that acetate was the only product in the catholyte. No other by-product was detected. The concentration of acetate was estimated to be 2.65 g L^{-1}.

The impedance of Ag@Cu before and after the stability test was recorded as shown in the Nyquist plots (Fig. 5). The R_{ct} of the post-run sample increased 5–6 times when compared to the fresh sample. In addition, the post-run electrocatalyst revealed some white deposition on the GDE (inset of Fig. 5), which was due to the accumulation of acetate. Hence, the increase in R_{ct} was postulated to be due to the accumulation of acetate that increased the potential barrier and hindered the electrocatalytic sites. This explained the phenomenon of rapid reduction in products obtained from the electroreduction process after 140 min. The accumulation of acetate on the GDE over the time had also reduced and constrained the adsorption of CO on the GDE that probably shifted the product selectivity from methanol to ethylene as suggested by a recent study.[19]

The co-production of high selective ethylene from syngas in the gas compartment and potassium acetate in the middle compartment exhibited the advantage of the three-compartment electrolyser, in which the products in gas and liquid phases were prohibited to cross-over, which prevented the oxidation of reduced products. As a result, the product selectivity was promoted. Further enhancement of the single pass conversion efficiency with high product selectivity (*i.e.*, avoid product purification/separation processes) would significantly enhance the economic feasibility of the electrolyser for CCU applications.

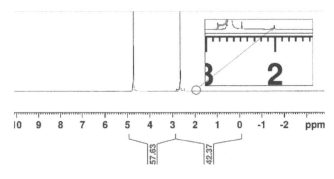

Fig. 4 NMR spectrum of the solution obtained after a single pass conversion from the middle compartment of the electrolyser.

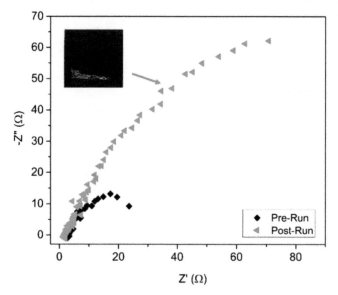

Fig. 5   Impedance (Nyquist plot) of Ag@Cu measured before (◆) and after (◀) the stability test.

4.   Conclusions

A three-compartment alkaline flow-through electrolyser was successfully developed and the single pass conversion of syngas to ethylene achieved was 2.5 mM cm^{-2} in the optimised conditions (*i.e.*, $1:1$ of $CO:H_2$; 1.0 M of KOH and 0.78 mL min^{-1} of total gas flow under -2.5 mA cm^{-2}). The single pass conversion of syngas to ethylene was dramatically enhanced when the fabricated core–shell Cu@Ag sample (15 mM cm^{-2}) was employed as the electrocatalyst, the co-production of ethylene (~40%) and potassium acetate (2.65 g L^{-1}) was obtained as the gas and liquid products, respectively, with high purity (>70%). The impedance measurements suggested that the shifting of product selectivity from methanol to ethylene over the time and deterioration of ethylene production after 2 h were due to the accumulated acetate that hindered the electrocatalytic sites. Further mechanistic study and optimisation are required to fully understand the reaction pathways. Although the core–shell Cu–Ag bimetallic electrocatalyst showed promising conversion efficiency, the stability of ethylene production would need to be explored further.

Conflicts of interest

There are no conflicts to declare.

Acknowledgements

The authors/we would like to acknowledge that this work was supported by the UKRI ISCF Industrial Challenge within the UK Industrial Decarbonisation Research and Innovation Centre (IDRIC) award number: EP/V027050/1.

References

1 United Nations Framework Convention on Climate Change (UNFCCC), *Adoption of the Paris Agreement*, 2015.

2 A. Kätelhön, R. Meys, S. Deutz, S. Suh and A. Bardow, *Proc. Natl. Acad. Sci. U. S. A.*, 2019, **116**, 11187–11194.

3 https://www.statista.com/statistics/1170573/price-ethylene-forecast-globally/.

4 M. Ghanta, D. Fahey and B. Subramaniam, *Appl. Petrochem. Res.*, 2014, **4**, 167–179.

5 G. Wernet, C. Bauer, B. Steubing, J. Reinhard, E. Moreno-Ruiz and B. Weidema, *Int. J. Life Cycle Assess.*, 2016, **21**, 1218–1230.

6 E. Worrell, D. Phylipsen, D. Einstein and N. Martin, *Energy Use and Energy Intensity of the U.S. Chemical Industry*, 2000.

7 J. A. Rabinowitz and M. W. Kanan, *Nat. Commun.*, 2020, **11**, 5231.

8 A. Bagger, W. Ju, A. S. Varela, P. Strasser and J. Rossmeisl, *ACS Catal.*, 2019, **9**, 7894–7899.

9 A. R. Woldu, A. H. Shah, H. Hu, D. Cahen, X. Zhang and T. He, *Int. J. Energy Res.*, 2020, **44**, 548–559.

10 J. Sisler, S. Khan, A. H. Ip, M. W. Schreiber, S. A. Jaffer, E. R. Bobicki, C.-T. Dinh and E. H. Sargent, *ACS Energy Lett.*, 2021, **6**, 997–1002.

11 F. P. García de Arquer, C.-T. Dinh, A. Ozden, J. Wicks, C. McCallum, A. R. Kirmani, D.-H. Nam, C. Gabardo, A. Seifitokaldani, X. Wang, Y. C. Li, F. Li, J. Edwards, L. J. Richter, S. J. Thorpe, D. Sinton and E. H. Sargent, *Science*, 2020, **367**, 661–666.

12 Z. Liu, H. Yang, R. Kutz and R. I. Masel, *J. Electrochem. Soc.*, 2018, **165**, J3371.

13 L. Wang, D. C. Higgins, Y. Ji, C. G. Morales-Guio, K. Chan, C. Hahn and T. F. Jaramillo, *Proc. Natl. Acad. Sci. U. S. A.*, 2020, **117**, 12572–12575.

14 A. N. Kuhn, H. Zhao, U. O. Nwabara, X. Lu, M. Liu, Y.-T. Pan, W. Zhu, P. J. A. Kenis and H. Yang, *Adv. Funct. Mater.*, 2021, **31**, 2101668.

15 W. Gao, Y. Zhao, H. Chen, H. Chen, Y. Li, S. He, Y. Zhang, M. Wei, D. G. Evans and X. Duan, *Green Chem.*, 2015, **17**, 1525–1534.

16 A. Sacco, *J. CO2 Util.*, 2018, **27**, 22–31.

17 D. P. Nagmani, A. Tyagi, T. C. Jagadale, W. Prellier and D. K. Aswal, *Appl. Surf. Sci.*, 2021, **549**, 149281.

18 Z. Zhang, Z. Wei, P. Sautet and A. N. Alexandrova, *J. Am. Chem. Soc.*, 2022, **144**, 19284–19293.

19 J. Li, Z. Wang, C. McCallum, Y. Xu, F. Li, Y. Wang, C. M. Gabardo, C.-T. Dinh, T.-T. Zhuang, L. Wang, J. Y. Howe, Y. Ren, E. H. Sargent and D. Sinton, *Nat. Catal.*, 2019, **2**, 1124–1131.

Faraday Discussions

PAPER

Self-assembly of random networks of zirconium-doped manganese oxide nanoribbons and poly(3,4-ethylenedioxythiophene) flakes at the water/chloroform interface†

Subin Kaladi Chondath and Mini Mol Menamparambath (iD) *

Received 13th April 2023, Accepted 2nd May 2023

DOI: 10.1039/d3fd00077j

Owing to their magnificent chemical and physical properties, transition metal-based heterostructures are potential materials for applications ranging from point-of-care diagnostics to sustainable energy technologies. The cryptomelane-type octahedral molecular sieves (K-OMS-2) are extensively studied porous materials with a hollandite (2 × 2 tunnel of dimensions 4.6 × 4.6 $Å^2$) structure susceptible to the isovalent substitution of metal cations at the framework of MnO_6 octahedral chains. Here we report a facile *in situ* synthesis of framework-level zirconium (Zr)-doped K-OMS-2 nanoribbons in poly(3,4-ethylenedioxythiophene) (PEDOT) nanoflakes at a water/ chloroform interface at ambient conditions. An oxidant system of $KMnO_4$ and $ZrOCl_2·8H_2O$ initiated the polymerisation at temperatures ranging from 5° to 50 °C. The lattice distortions arising from the framework-level substitution of Mn^{4+} by Zr^{4+} in the K-OMS-2 structure were evidenced with powder X-ray diffraction, Raman spectroscopy, X-ray photoelectron spectroscopy, and N_2 adsorption–desorption studies. Transmission electron microscopic and mapping images confirmed that PEDOT/Zr-K-OMS-2 comprises a highly crystalline random network of two-dimensional PEDOT flakes and Zr-doped K-OMS-2 nanoribbons. In this regard, the proposed interfacial strategy affirms an *in situ* method for the morphological tuning of heterostructures on polymer supports at low temperatures.

1 Introduction

Over the past decades, phenomenal research has been focused on manganese oxide (MnO_x)-based materials because of their exceptional electroactive functionality for lithium-ion batteries,[1,2] supercapacitors,[3,4] redox catalysts,[5,6] bio-

Department of Chemistry, National Institute of Technology Calicut, Calicut-673601, Kerala, India. E-mail: minimol@nitc.ac.in

† Electronic supplementary information (ESI) available. See DOI: https://doi.org/10.1039/d3fd00077j

sensors,[7,8] and more.[9,10] Expeditious efforts have been made to develop MnO_x nanostructures, which offer strong redox couples of Mn^{2+}/Mn^{3+} and Mn^{3+}/Mn^{4+} associated with the multivalent Mn^{x+} species on the MnO_x surface.[11-13] Among the wide varieties of MnO_x catalysts, manganese dioxide (MnO_2) is identified as the most promising due to its easily reversible multiple oxidation states,[14] low cost,[15] environmentally-benign nature,[16] controllable morphology,[9] and high-level of polymorphism.[11]

All allotropes emerge through the assembly of octahedral $[MnO_6]$ building blocks interconnected through oxygen atoms to form distinct layered or tunneled-type structures,[11,14] known as manganese octahedral molecular sieves (OMS).[17] The most extensively studied OMS material is K-OMS-2 because of its rich porous and tunnel structure.[11] K-OMS-2 has a hollandite 2×2 tunnel structure of dimensions 4.6×4.6 Å^2 formed by corner and edge sharing $[MnO_6]$ units, acquiring a tetragonal $I4/m$ crystal structure.[6,11,14] The framework manganese centers are multivalent (4+, 3+, and minor 2+), where the charge neutrality is retained by K^+ ions occupying the 2×2 tunnels.[11]

The hollandite tunnel structure of K-OMS-2, designated as $K_xMn_8O_{16}$, is susceptible to doping at varying levels.[11,15,17,18] Novel physicochemical properties are introduced by doping potent metal cations and successive fine-tuning of the crystal structure, morphology, and lattice parameters.[11] The framework-level doping in K-OMS-2 involves the isomorphic substitution of framework Mn^{4+} by transition metal cations such as Cu^{2+}, Zn^{2+}, Al^{3+}, Mg^{2+}, Fe^{3+}, Ce^{4+}, Zr^{4+}, and V^{5+} (but is not limited to these).[18,19] For example, the V^{5+}-substituted MnO_2 nano-flakes prepared by the hydrothermal method exhibited significant lithium-storage capacity.[4] Defect-induced non-stoichiometry in MnO_2 generated by varying amounts of Ni^{2+} resulted in $MnNi_x$ nanoflakes which were subsequently used for propane deep oxidation.[20] Selective incorporation of Zr^{4+} in α-MnO_2 was utilized as an effective oxygen reduction reaction (ORR) catalyst.[21] Similarly, Zn, Fe, Zr, and V-doped K-OMS-2 materials were synthesized using a reflux method for nitrogen oxide abatement.[17]

Various reports have endorsed the profound role of synthesis methods in controlling the morphology, crystallographic structure, and physicochemical properties of doped OMS-2 materials.[11] The hydrothermal method is the widely accepted synthesis method, where the manganese precursors and the dopant reside in the same reaction mixture.[22] However, hydrothermal routes consume high amounts of energy since they require longer reaction times (90 min to 4 days) and elevated temperatures (>100 °C).[22] Hitherto, simple preparation methods that need only facile experimental setups are still in their infancy and yet to be explored. The relatively less explored liquid/liquid (L/L) interface-assisted synthesis method proved to be an ideal strategy for effective morphological tuning of nano/micro-scale two-dimensional (2-D) materials. The L/L interface-assisted synthesis method uses two (or more) immiscible liquids separated by a quasi-2-D interface, where the desired reaction occurs.[23] Additionally, it ensures a simple, mild, and time-saving experimental setup using a conventional approach to synthesize 2-D materials.[23]

It has been reported that the L/L interface was successful in synthesizing a few nanometer-thick 2-D sheets of transition metal chalcogenides/oxides, conducting polymers (CPs), molecular organic frameworks (MOFs), and more.[24-26] All immiscible liquids have a high energy L/L interface due to the surface tension compared to bulk liquids.[27] The resultant interfacial tension decreases by the preferential adsorption of nanoparticle (NP)-attached oligomers or short-chain polymers towards the interface.[28–30] Furthermore, the polymerisation of CPs at the L/L interface can be utilized to self-assemble metal and metal oxide NPs at the L/L interface.[28–30] The L/L interfacial methods were highly advantageous in fabricating 2-D architectures by directing the self-assembly of nanostructures (NS) in a confined 2-D space.[30–32]

Here, we report a facile one-pot synthesis of framework-level Zr-doped K-OMS-2 in PEDOT at a solvent/non-solvent interface of water and chloroform. The choice of monomer (3,4-ethylenedioxythiophene (EDOT)) and oxidants (potassium permanganate ($KMnO_4$) and zirconium oxychloride octahydrate ($ZrOCl_2\cdot 8H_2O$)) was based on the standard reduction potentials favouring a spontaneous redox reaction between the reactants. The $ZrOCl_2\cdot 8H_2O$ catalyzes the thermodynamically favourable redox reaction between EDOT and $KMnO_4$ by lowering the pH and standard Gibbs free energy. The Zr-doped PEDOT/K-OMS-2 was synthesized at the water/chloroform interface by an oxidant mixture of a $1:0.2$ mole ratio of $KMnO_4$ and $ZrOCl_2\cdot 8H_2O$ at temperatures ranging from 5° to 50 °C. The powder X-ray diffraction, Raman, and X-ray photoelectron spectroscopy studies of Zr-doped PEDOT/K-OMS-2 confirmed the framework-level doping of Zr^{4+} by replacing Mn^{4+} in MnO_6 octahedral chains. The optical and field emission scanning electron microscopic images demonstrated the morphology transition from agglomerates to microscale films upon Zr doping in PEDOT/K-OMS-2 samples. The atomic force microscopic images validated the ~21 nm thickness for the microscale thin films. The high-resolution transmission electron microscopic images and selected area diffraction pattern confirmed the assembly of highly crystalline random networks of 2-D PEDOT flakes and Zr-doped K-OMS-2 nanoribbons in microscale films.

2 Experimental section

2.1 Chemicals and reagents

3,4-Ethylene dioxythiophene (Alfa Aesar, 56 533, 97%), potassium permanganate (Merck, 640 079, 99%), zirconium oxychloride octahydrate (Loba Chemie, 3260, 98%) and chloroform (Thermo Fisher, 67 663, 99.5%) were used without any further purification. All the reagents were of analytical grade. Extra pure deionized water (ELGA Purelab Quest UV, 18.2 MΩ) was adopted throughout the experiments.

2.2 Synthesis of PEDOT/Zr-K-OMS-2 at the water/chloroform interface

The polymerisation of the EDOT monomer was carried out at a solvent/non-solvent interface of water and chloroform utilizing a mixture of oxidants $KMnO_4$ and $ZrOCl_2\cdot 8H_2O$. The experiment was carried out by fixing the mole ratio of EDOT, $KMnO_4$, and $ZrOCl_2\cdot 8H_2O$ as $1:1:0.2$. Initially, a 0.06 M EDOT monomer solution was prepared in 15 mL chloroform. A mixture of 0.12 M $KMnO_4$ (0.142 g in 7.5 mL water) and 0.024 M $ZrOCl_2\cdot 8H_2O$ solution (0.058 g in 7.5 mL water) was added to the monomer solution. The reaction mixture was kept undisturbed at room temperature (RT) for 12 hours to complete the polymerisation. Finally, the products confined at the interface of the bisolvent system were washed several times with an ethanol–water mixture and vacuum filtered. The filtered products were vacuum dried at room temperature resulting in finely powdered composites. In addition, control experiments were performed at varying temperatures, 5 °C and 50 °C for 15 hours and 2 hours, respectively. The PEDOT/Zr-K-OMS-2 (PZrK) samples were prepared at different temperatures, 5 °C, RT, and 50 °C, from now on referred to as PZrK 5, PZrK RT, and PZrK 50, respectively.

2.3 Synthesis of PEDOT/K-OMS-2 at the water/chloroform interface

The undoped PEDOT/K-OMS-2 was prepared by the polymerisation of EDOT in the presence of $KMnO_4$ in the water/chloroform bisolvent system. The mole ratio of EDOT monomer and $KMnO_4$ was fixed as $1:1$. A 0.06 M solution of EDOT monomer was prepared in 15 mL chloroform; to this, 0.06 M $KMnO_4$ solution (0.142 g $KMnO_4$ in 15 mL water) was added and kept undisturbed until the completion of polymerisation. The polymerisation reaction was not experimentally feasible at 5 °C and RT; however, the reaction was completed within 4 hours at 50 °C. After 4 hours, the synthesized product was vacuum filtered using an ethanol–water mixture, followed by room temperature vacuum drying. The PEDOT/K-OMS-2 sample (denoted as PK 50 throughout the work) prepared at the water/chloroform interface was employed for comparative studies with PZrK samples. Interestingly, visible product formation was not observed in the presence of pure $ZrOCl_2 \cdot 8H_2O$ as the oxidant.

2.4 Characterization techniques

The morphology and composition of the as-synthesized samples were characterized using an inverted optical microscope (OLYMPUS DP80 IX73), field emission scanning electron microscope (FESEM, Gemini SEM 300, and SU6600 HITACHI), and a high-resolution transmission electron microscope (HR-TEM) with energy dispersive spectroscopy (EDS) and elemental mapping (Thermo Fisher Talos F200 S). The surface morphology was analyzed using an atomic force microscope (AFM, Flex-Axiom) in a tapping mode. For all the above-mentioned characterizations, the products confined at the interface were transferred to the substrates after mild washing in water. The crystalline nature of the powdered samples was analyzed using a powder X-ray diffraction (XRD) technique using a Rigaku Miniflex-600 diffractometer with Cu Kα radiation. The Raman spectroscopy of powdered samples was carried out using an FT-Raman spectrometer (Bruker RFS27, a laser source with an excitation wavelength of 785 nm). The surface chemical characterization and chemical state mapping were performed using an X-ray photoelectron spectroscope (XPS, Omicron Nanotechnology Ltd, Germany) equipped with a monochromatic Mg Kα X-ray source ($h\nu = 1253.6$ eV). The survey scans were recorded with a pass energy of 50 eV, whereas the high-resolution spectra of the major elements were recorded with a pass energy of 20 eV. The binding energies of all elements were referenced to the C 1s peak at 284.8 eV. The Brunauer–Emmett–Teller (BET) surface area of the samples was analyzed in the P/P_0 range of 0–1 using Belsorp-max. Cyclic voltammograms (CVs) were recorded in an Origalys OGF500 instrument using a three-electrode setup, consisting of a glassy carbon electrode as the working electrode, Ag/AgCl (1 M KCl) as the reference electrode, and platinum wire as the counter electrode.

3 Results and discussion

Conducting polymer (CP) films were primarily synthesized by conventional electrochemical polymerisation.[23] The experimental setup of a typical electrochemical polymerisation is an electrochemical cell consisting of a three-electrode setup (working electrode, reference electrode, and counter electrode) dipped in a solution of monomer and dopant (Fig. 1A).[23] An applied voltage attributed to the oxidation potential of the monomer must be supplied to initiate the

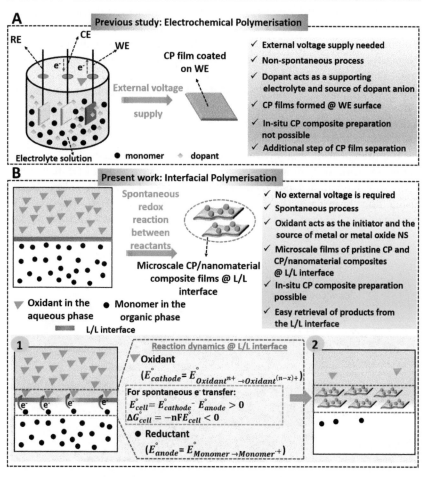

Fig. 1 Schematic representation of (A) electrochemical polymerisation comprising a three-electrode setup (RE: reference electrode, WE: working electrode, and CE: counter electrode); polymer film gets electrodeposited on the WE surface. (B) Interfacial polymerisation involving a 2-D interface formed between two immiscible liquids, where the microscale polymer/nanomaterial composites are confined at the 2-D L/L interface. In the initial step of interfacial polymerisation (B.1), the reactants, monomer, and oxidant meet at the interface, where a spontaneous flow of electrons from monomer to oxidant molecules occurs. This electron flow is feasible only if the standard Gibbs free energy $(\Delta G^{\circ}_{cell} = -nFE^{\circ}_{cell})$ of the redox reaction is negative. The subsequent steps (B.2) result in polymerisation and furnish 2-D microscale films of CP/nanomaterial composites at the L/L interface.

polymerisation, and the CP films are deposited on the working electrode surface (Fig. 1A).[33–35] However, electrochemical methods are limited to synthesizing pristine CP films, and polymer/nanomaterial composites are achieved only by *ex situ* incorporation of nanomaterials into the polymeric matrix.[33–35] To overcome these limitations, facile *in situ* polymerisation methods for synthesizing CP-nanomaterial composites are required, and the L/L interfacial polymerisation method represents one such efficient method. In a L/L interfacial polymerisation method, the monomer is dispersed in an organic solvent, and the oxidant is dissolved in water (Fig. 1B).[23] The monomer and oxidant meet at the L/L interface, where the spontaneous redox initiation reaction occurs (Fig. 1B).[23] The spontaneous electron flow from monomer to oxidant facilitates the thermodynamically favorable redox initiation polymerisation, forming monomer radical cations and eventually forming microscale CP/nanomaterial composites at the L/L interface (Fig. 1B). The as-prepared microscale CP/nanomaterial composites are characterized by a uniform distribution of nanomaterials (metal and metal oxide NS) in microscale CP sheets or films (Fig. 1B).[28,30] The uniform distribution of NS in CP films is ideal for preparing stable processable CP/nanomaterial dispersions in suitable solvents.[29] The distinct features of the two polymerization methods are also listed in Fig. 1.[23,33–35]

The present work employed a water/chloroform interface-assisted polymerisation to synthesize composites of Zr-doped K-OMS-2 and PEDOT flakes. The schematic illustration depicting the experimental feasibility of the polymerisation using different oxidants is shown in Fig. 2. The EDOT polymerisation was initiated by three different oxidant systems (pure $KMnO_4$, pure $ZrOCl_2 \cdot 8H_2O$, and

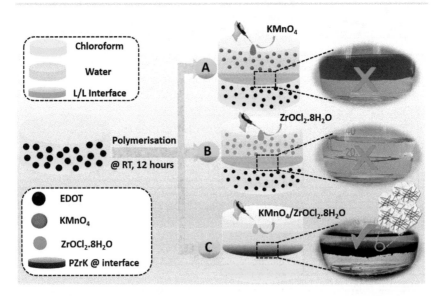

Fig. 2 Schematic illustration of the water/chloroform interface-assisted method for the synthesis of Zr-doped K-OMS-2 in PEDOT at RT. The polymerisation was performed for 12 hours in three different pathways, (A) the oxidant chosen was $KMnO_4$, (B) the oxidant was $ZrOCl_2 \cdot 8H_2O$ and (C) the oxidant was a mixture of 1 : 0.2 mole ratio of $KMnO_4$ and $ZrOCl_2 \cdot 8H_2O$. The photographic images in path A and path B confirmed no visible product formation, however path C showed clear interfacial confinement of microscale PZrK films.

a mixture of $(1:0.2$ mole ratio) $KMnO_4$ and $ZrOCl_2 \cdot 8H_2O)$ at RT for 12 hours. As observed in the photographic images (Fig. 2A and B), no polymerisation was achieved for individual oxidants. However, the introduction of a 0.2 mole ratio of $ZrOCl_2 \cdot 8H_2O$ into $KMnO_4$ emanated polymerisation, yielding the confinement of PZrK RT at the water/chloroform interface (Fig. 2C). The mechanism of the product confinement at the L/L interface can be briefly summarised as follows.

(1) Initially, the reactants, EDOT monomer, and oxidants meet at the L/L interface, driving the redox reaction between the monomer and oxidant.

(2) The subsequent steps lead to the formation of Zr-doped K-OMS-2 attached PEDOT oligomeric chains.

(3) The L/L interface acts as an active template where the competition between solvent polarity-controlled diffusions and interfacial tension-driven NS adsorption occurs.[28,29]

(4) Due to the inherent nature of the interface to minimize its high interfacial tension and interfacial free energy, the NS-attached oligomers or short-chain polymers get adsorbed at the solvent/non-solvent interface.[30]

(5) Subsequently, the L/L interface promotes the confined 2-D growth of NS-attached oligomers or short-chain polymers,[30] forming a random network of 2-D PEDOT flakes and Zr-doped K-OMS-2 nanostructures.

The thermodynamic feasibility of the redox-type polymerisation can be corroborated with the thermodynamic Gibbs free energy of the redox initiation reaction between the monomer and oxidant.[28] The redox reaction between the EDOT monomer and oxidant initiates cationic-type polymerisation, where the EDOT acts as an anode, and the oxidant acts as the cathode. Hence, the EDOT undergoes one-electron oxidation to EDOT radical cations, having a standard reduction potential (E_P°) of -1.197 V as shown in Table 1.[36,37] Table S1† shows the sign conventions of the electrode potentials of EDOT $vs.$ Ag/AgCl and the standard reduction potential of EDOT $vs.$ SHE.[36,37] Concurrently, the oxidant reduces to a lower oxidation state, which acts as a nuclei for the self-assembly of nanostructures.

The standard reduction potential of the primary oxidant, $KMnO_4$ to MnO_2, is 0.595 V, which may spontaneously initiate the redox reaction at the interface. The oxidant conversion reaction,[38,39] standard reduction potential value of the oxidant (E_M°),[38,39] pH of the reaction medium, and the standard Gibbs free energy (ΔG°) of the redox initiation are tabulated in Table 1. The standard Gibbs free energies corresponding to the pure oxidants $KMnO_4$ and $ZrOCl_2 \cdot 8H_2O$ are -124 kcal mol^{-1} and -31 kcal mol^{-1}, respectively (Table 1). Despite the calculated negative ΔG° of pure oxidants, the polymerisation was not feasible at standard reaction conditions, as shown in the photographic images in Fig. 2. However, the oxidant system comprising a $1:0.2$ mole ratio of $KMnO_4$ and $ZrOCl_2 \cdot 8H_2O$ induced polymerisation (Fig. 2C and Table 1). The mechanism of polymerisation using Zr precursor as a catalyst along with $KMnO_4$ is proposed as follows.

(1) Dissolution of $ZrOCl_2 \cdot 8H_2O$ furnished excess H^+ ions in the reaction medium.

(2) The presence of excess H^+ ions altered the pH of the aqueous $KMnO_4$ medium (pH = 7.6) to acidic (pH = 2.08) (Table 1).

(3) At high acidic conditions, the standard reduction potential for the conversion of $KMnO_4$ to MnO_2 increases to 1.679 V (Table 1).[38]

Table 1 Table showing the oxidant conversion reaction,[38] the standard reduction potential of the metal ions,[38,39] the number of electrons involved in the reaction, the pH of the reaction medium, the electrochemical cell potential, and the standard Gibbs free energy of the redox reaction between EDOT and oxidant. Pure oxidant systems failed to initiate polymerisation, whereas a mixture of oxidants led to polymerisation under the given experimental conditions

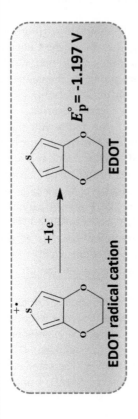

EDOT radical cation $\xrightarrow{+1e^-}$ EDOT

$E_p^\circ = -1.197\ V$

Oxidant system	Oxidant conversion reaction	Standard reduction potential of the oxidant, E_M° (V)	Number of electrons involved, n	pH of the reaction medium	E_{cell}° (V) $= E_M^\circ - E_P^\circ$	Standard Gibbs free energy, ΔG° (kcal mol^{-1})
KMnO$_4$/ ZrOCl$_2$·8H$_2$O	MnO$_4^-$ + 2H$_2$O + 3e$^-$ ⇌ MnO$_2$ + 4OH$^-$	0.595 (ref. 38)	3	7.6	1.792	−124
	Zr^{2+} + 2e$^-$ ⇌ Zr^0	−0.525 (ref. 39)	2	1.96	0.672	−31
KMnO$_4$/ ZrOCl$_2$·8H$_2$O	MnO$_4^-$ + 4H$^+$ + 3e$^-$ ⇌ MnO$_2$ + 2H$_2$O	1.679 (ref. 38)	3	2.08	2.876	−200

(4) Correspondingly, the $\Delta G°$ for the initiation of polymerisation decreases from -124 kcal mol^{-1} to -200 kcal mol^{-1} (Table 1).

(5) The free Zr^{4+} ions in the reaction medium are doped to the MnO_2 structures to yield Zr-doped K-OMS-2 nanostructures attached small chain PEDOT at the interface.

As reported previously, these nanostructures adsorb at a high-energy L/L interface and may confine the growth of PZrK structures to 2-D morphology.[30]

The polymerisation reaction was also performed at higher (50 °C) and lower (5 °C) temperatures, demonstrating the competence of the oxidant mixture for polymerisation at other than standard temperature–pressure conditions (Fig. 2 and Table 1). For comparison, PEDOT polymerisation was performed at an elevated temperature, 50 °C using $KMnO_4$ as oxidant, yielding PK 50. The optical and photographic images of PK 50 and PZrK samples are shown in Fig. S1.† The transition of composite morphology from agglomerates of PK 50 to micrometer-sized films of PZrK is evident from optical microscopic and FESEM images (Fig. S1†).

The physicochemical structural evolution upon Zr doping in K-OMS-2 nano-structures embedded in PEDOT flakes was elucidated by powder XRD, Raman, and BET studies (Fig. 3 and 4). The powder XRD patterns in Fig. 3A depict the crystallinity, phase purity, and doped structure of the PZrK samples analogous to PK 50. The XRD spectrum of PK 50 matched well with the standard JCPDS patterns of cryptomelane, $K_{1.33}Mn_8O_{16}$ (JCPDS no. 00-077-1796).[40,41] Fig. 3B shows the hollandite crystal structure of cryptomelane with 2×2 tunnels.[40,41] The major distinct peaks at 12.63°, 46.9°, 52.4°, 59.8°, 67.2°, 74.61°, and 89.42° can be assigned to intense reflections caused by (110), (150), (440), (251), (550), (701), and (532) planes of $K_{1.33}Mn_8O_{16}$ with a pure tetragonal phase [space group: $I4/m$ (87)] (Fig. 3A).[40,41] The PK 50 showed another reflection at 18.3° attributed to the tetragonal phase of α-MnO_2.[15] The peaks at 31.2° and 45.2°, and at 23.9° and 38.22° correspond to potassium manganese oxide (K_3MnO_4, 00-031-1049)[42] and Mn_3O_4 (00-016-0350),[43] respectively. The formation of impurity peaks at 16.7° and 81.75° revealed the presence of $K_{0.51}Mn_{0.93}O_2$, attributed to the conversion of cryptomelane phases at 50 °C (Fig. 3A).[44] The presence of three various manganese oxides in notable amounts could be due to the inhomogeneous and random assembly of NS-attached polymer chains at elevated temperatures.

The XRD spectra corresponding to PEDOT/Zr-K-OMS-2 samples prepared at varying temperatures showed major peaks assigned to cryptomelane, $K_{1.33}Mn_8O_{16}$ (JCPDS no. 00-077-1796) (Fig. 3A and B).[40,41] In all PZrK samples, the emergence of dissimilar peaks and disappearance of few peaks compared to PK 50 suggests lattice distortions caused by the possible doping of the Zr^{4+} ions in the tetragonal crystal structure of K-OMS-2 (Fig. 3A).[15] Furthermore, the cryptomelane peaks in PZrK samples significantly shifted towards lower 2θ angles compared to the standard JCPDS of $K_{1.33}Mn_8O_{16}$.[3,6,15,45,46] This could be attributed to the lattice expansions resulting from the replacement of smaller framework Mn^{4+} (53 pm) with larger Zr^{4+} ions (80 pm).[3,47,48] The PZrK samples prepared at a lower temperature, PZrK 5, exhibited a peak at 18.4° (200) corresponding to the crystal structure of α-MnO_2, which may act as a site for Zr doping. The peak at 49.8° (411) for KMn_8O_{16} was shifted to lower 2θ (JCPDS no. 00-029-1020), characteristic of the lattice expansions resulting from Zr doping.[19] The controlled assembly of NS-attached oligomers and concomitant Zr substitution at lower temperatures (5 °

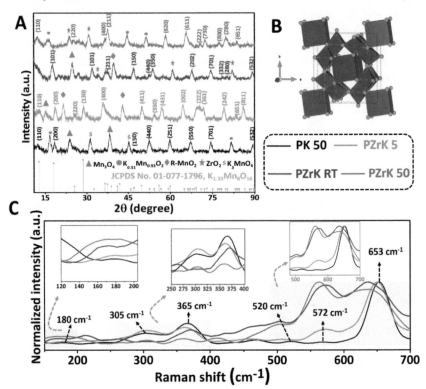

Fig. 3 (A) Powder XRD patterns of PK 50, and PZrK samples. The standard JCPDS pattern of cryptomelane (01-077-1796) is also shown. (B) Tetragonal 2 × 2 tunnel structure of cryptomelane, $K_{1.33}Mn_8O_{16}$. The K^+ ions reside in the 2 × 2 tunnels formed by MnO_6 octahedral chains. (C) Raman spectra of PK 50 and PZrK samples synthesized at water/chloroform interface. The insets show the enlarged spectra of the selected regions represented in the form of colored rectangular boxes.

C) reflected highly crystalline XRD patterns with minimal impurity phases in the PZrK 5 samples. The peaks at 21.54°, 42.73°, and 15° are assigned to the impurity phase of R-MnO$_2$ (JCPDS no. 01-073-1539)[48] and Mn$_3$O$_4$ (01-075-1560),[49] respectively (Fig. 3A). On the other hand, PZrK RT and PZrK 50 showed notable impurity phases of Mn$_3$O$_4$,[49] R-MnO$_2$,[48] K$_{0.51}$Mn$_{0.93}$O$_2$,[44] and ZrO$_2$,[50] which reveals the uncontrolled self-assembly of Zr-doped K-OMS-2 nanostructures-attached oligomers or short-chain polymers at the L/L interface (Fig. 3A). The absence of any crystalline impurity phases of ZrO$_2$ in PZrK 5 further emphasizes the efficient Zr doping in the K-OMS-2 crystallites at lower temperatures.[5,15] The PK 50 and PZrK samples exhibited a broad peak at 26.8° corresponding to the interplanar spacing of PEDOT chains assigned to the (020) plane (Fig. 3A).[51]

Raman spectroscopy is a powerful technique for probing different structural and crystalline disorders or defects related to MnO$_x$-based materials.[6,52] The Raman spectra of PK 50 and PZrK samples showed distinct bands related to various vibrational modes of the well-developed tetragonally-structured K-OMS-2 in the Raman shift range 100–700 cm^{-1} (Fig. 3C). For clarity for readers, zoomed portions are shown in the insets. The bands at 572 cm^{-1} and 653 cm^{-1} are characteristic of symmetrical Mn–O lattice vibrations and are assigned to A_g symmetry.[52] The band at 572 cm^{-1} originated from vibrations of O–Mn–O–Mn along the lengths of MnO_6 chains. In contrast, vibrations of O–Mn–O perpendicular to MnO_6 octahedral chains resulted in a strong band at 653 cm^{-1}.[5,6,20] The consistent downshift for this band, from 653 cm^{-1} to 634 cm^{-1} indicates the structural distortion upon Zr doping in PZrK samples (inset of Fig. 3C).[13] A distinct peak broadening is also observed for PZrK RT and PZrK 50, resulting from the presence of impurity phases. A considerable increase in the intensity of the Raman shift around 572 cm^{-1} and 520 cm^{-1} accounts for increased distortions associated with the oxygen vacancies created by Zr doping (inset of Fig. 3C).[52] The strong band at 365 cm^{-1} is ascribed to Mn–O bending vibrations.[20] A gradual decrease in the intensity of the strong band at 365 cm^{-1} also affirms the highly distorted tunnel structure of K-OMS-2 in PZrK samples due to Zr substitution

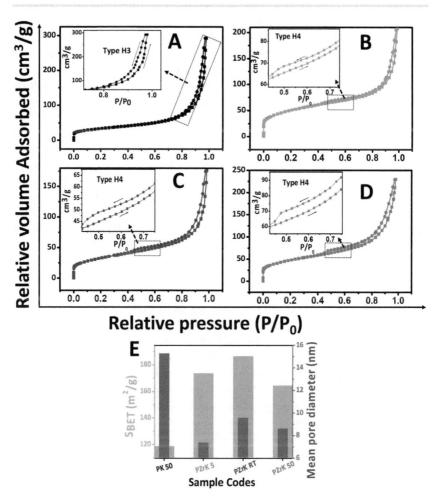

Fig. 4 (A–D) N$_2$ adsorption–desorption isotherms of PK 50 and PZrK samples. The inset plots of PK 50 and PZrK samples indicate a type H3 and type H4 hysteresis loop, respectively. (E) The S_{BET} and mean pore diameter were calculated from BET equations and plotted against sample codes.

(inset of Fig. 3C).[20] A strong band at 305 cm^{-1} becomes broadened and weakened in PZrK RT and PZrK 50 due to the highly distorted K-OMS-2 spatial structure. The low-frequency band at 180 cm^{-1} (E_g symmetry) represents external vibration derived from the translational motion of the MnO_6 octahedral structure.[52] Considerable increase in its intensity is accounted for by increased distortions associated with the oxygen vacancies created with Zr doping. The consistent variations in intensity and band positions of Raman shifts could be due to the confinement of phonons by crystal defects and local lattice distortions arising from the doping of Zr into the tetragonal structure of K-OMS-2.[52]

The doping of transition metal ions in K-OMS-2 frameworks alters the mean pore size and specific surface area (S_{BET}) of the PK 50 agglomerates. The variations in the mean pore size and S_{BET} of mesoporous PZrK samples were quantitatively estimated from N_2 adsorption–desorption isotherms using the BET equation (Fig. 4). A type II adsorption isotherm with a distinct hysteresis loop of type H3 between $P/P_0 > 0.6$ is obtained for undoped PK 50,[16,19] implying an aggregated morphology forming slit-like pores.[16,19,53] In contrast with PK 50, PZrK samples showed a drastic change in adsorption isotherm (type IV) and hysteresis loops (type H4), indicative of narrow slit-like pores.[16,53] Zoomed portions of isotherms are shown in the insets of Fig. 4A–D. Compared to PK 50, the relatively higher slopes at low P/P_0 for PZrK samples indicate increased surface area upon cation substitution (the insets of Fig. 4B–D).[16] The changes in BET surface area and average pore diameter upon Zr substitution are analyzed and plotted in Fig. 4E. The pore size distribution curves of all samples are provided in Fig. S2.† The PZrK samples showed an increase in S_{BET} compared to PK 50, which proves the beneficial role of Zr cation substitution in expanding the surface area (Fig. 4E).[5] The substantial increase in S_{BET} upon Zr substitution in PZrK samples is consistent with the evident morphological transition (agglomerates to microscale films) observed in the optical and FESEM images (Fig. S1†). A decrease in the mean pore size of PZrK samples is due to the formation of oxygen vacancy defects (OVDs) created by Zr substitution in the MnO_6 octahedral framework (Fig. 4E and S2†).[17]

Despite the product confinement at the L/L interface, a clear-cut transition of morphology from agglomerates (PK 50) to microscale films (PZrK 5) was evident from optical microscopic and FESEM images (Fig. S1†). Besides, the HR-TEM techniques were employed to investigate the distribution of Zr-K-OMS-2 nano-structures in the PEDOT matrix, as shown in Fig. 5, 6, and S3.† The HR-TEM images evidenced that random assembly of Zr-K-OMS-2 nanoribbons (NRs) and PEDOT nanoflakes led to the microscale films (Fig. S1†). The nanoribbons in the random network of PEDOT flakes possess an average length of ~32 nm with a narrow length distribution of 10–69 nm and an average diameter of ~4 nm (Fig. 5B and S3†). The atomic composition determined with HR-TEM-EDS (Fig. 5C and S3†) confirmed the presence of the characteristic elements of PZrK 5 (carbon, sulphur, oxygen, potassium, manganese, and zirconium). The at% of the respective elements confirms <1% Zr doping into K-OMS-2 NRs, uniformly distributed in PEDOT nanoflakes. The highly crystalline nature of Zr-K-OMS-2 NRs is demonstrated by diffraction fringes in Fig. 5D1–3. The lattice fringes with a d-spacing of 2.2 Å, 2.3 Å, and 2.43 Å are assigned to (240), (330), and (400) planes of K-OMS-2 (cryptomelane, $K_{1.33}Mn_8O_{16}$, JCPDS no. 00-077-1796).[40,41] The interplanar d-spacing of 2.31 Å attributed to the reflections caused by the (020) plane of π–π stacked PEDOT chains further confirmed the crystalline nature of 2-

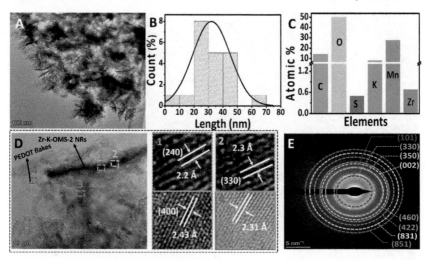

Fig. 5 (A) The HR-TEM image of PZrK 5 shows a random network of PEDOT flakes and Zr-K-OMS-2 nanoribbons. (B) Histogram profile showing the nanoribbon length distribution of PZrK 5. (C) The at% was obtained from the EDS spectrum of PZrK 5. (D) HR-TEM image of PZrK 5. The PEDOT flakes and Zr-K-OMS-2 nanoribbons are separately marked. Grey-colored portions indicated PEDOT flakes, whereas dark-colored portions in the shape of nanoribbons correspond to Zr-K-OMS-2. The zoomed portions marked as D1–D3 show distinct lattice fringes of Zr-K-OMS-2 NRs corresponding to d-spacing of 2.2 Å (240), 2.3 Å (330), and 2.43 Å (400). The PEDOT flakes also show crystalline lattice fringes of (020) plane with a d-spacing of 2.31 Å (D4). (E) The SAED pattern of PZrK 5 marked with concentric circles attributed to crystal planes of Zr-K-OMS-2 NRs.

D PEDOT flakes in PZrK 5 (Fig. 5D4).[54] The selected area electron diffraction (SAED) pattern in Fig. 5E shows the polycrystalline nature of PZrK 5 with characteristic diffraction Bragg spots with rings.[30] The reflections caused by the planes of K-OMS-2 are represented as different concentric circles in Fig. 5E. The Bragg reflections caused by (101), (330), (350), (002), and (460) planes are assigned to the tetragonal structure of cryptomelane, $K_{1.33}Mn_8O_{16}$ (JCPDS no. 00-077-1796) (Fig. 5E).[40] Moreover, the planes (422), (831), and (851) are represented by Bragg reflections corresponding to the tetragonal structure of cryptomelane, KMn_8O_{16} (JCPDS no. 00-029-1020).[12] Overall, the HR-TEM analysis emphasizes the role of Zr-K-OMS-2 as a one-dimensional crystalline template adsorbed at the L/L interface, confining the growth of the crystalline PEDOT matrix in a two-dimensional fashion.

The high-angle annular dark-field scanning transmission electron microscopy (HAADF-STEM) image and elemental mapping images of various elements in PZrK 5 are also provided in Fig. 6A–F. The elemental mapping images in Fig. 6B–F (carbon, manganese, oxygen, and zirconium) and Fig. S3† (sulphur and potassium) further support the model of the formation of Zr-K-OMS-2 in the PEDOT matrix with uniform distribution of various elements. Additionally, the AFM images (Fig. 6G, H and S4†) confirm the microscale film-like morphology of PZrK 5. The detectable nanoscale-long lateral protuberances in the 2-D and three-dimensional (3-D) AFM images support the model of Zr-K-OMS-2 templated

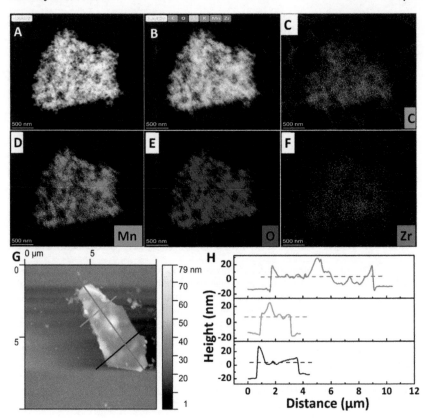

Fig. 6 (A) HAADF-STEM image of PZrK 5 showing a random network of PEDOT nanoflakes and Zr-K-OMS-2 nanoribbons. (B) The elemental mapping image of the mix with elements such as carbon (cyan), manganese (green), oxygen (blue), and zirconium (magenta). Elemental mapping images of individual elements, (C) carbon, (D) manganese, (E) oxygen, and (F) zirconium. 2-D AFM image (G) and height profiles (H) of the microscale PZrK 5 film with an average thickness of ~21 nm.

growth of the PEDOT matrix. In addition, the height profiles of the 2-D images show an average thickness of ~21 nm for the PZrK 5 films (Fig. 6H).

The surface chemical composition, valence states of various elements, and Zr doping in PZrK 5 films were determined with XPS (Fig. 7 and S5†). The survey scan in Fig. 7A shows characteristic elements; manganese, carbon, sulphur, potassium, oxygen, and zirconium in PZrK 5 films. The high-resolution spectra with deconvoluted peaks was used to further investigate the oxidation states of respective elements and the doping of Zr^{4+} in the K-OMS-2 structure (Fig. 7B–E and S5†). The high-resolution deconvoluted spectrum of the C 1s in Fig. 7B shows the contribution from two non-equivalent carbon atoms at 284.7 eV and 285.5 eV, corresponding to C–S and C=C–O of PEDOT chains.[55] The S 2p spectrum attributed to the thiophene ring of PEDOT shows a weak signal at 164.6 eV (Fig. S5†).[55] The high-resolution XPS spectrum of Mn 2p shown in Fig. 7C reveals two peaks centered at binding energies 641.1 eV and 652.8 eV, which correspond to Mn $2p_{3/2}$ and Mn $2p_{1/2}$, respectively.[52,56] A peak separation of 11.7 eV between

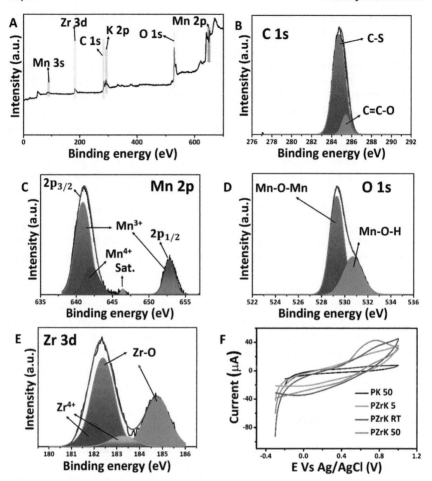

Fig. 7 (A) XPS survey scan spectrum, (B) high-resolution C 1s spectrum, (C) high-resolution Mn 2p spectrum, (D) high-resolution O 1s spectrum, and (E) high-resolution Zr 3d spectrum. (F) Cyclic voltammograms of the PK 50 and PZrK modified GCE in the presence of 5 mM $K_4[Fe(CN)_6]$ in 0.1 M KCl.

these two peaks is a fingerprint of K-OMS-2.[4] Furthermore, the deconvolution of Mn peaks confirmed the co-existence of different valence states of manganese (Fig. 7C).[52] The significant peaks at 641.0 eV ($2p_{3/2}$) and 652.8 eV ($2p_{1/2}$) are attributed to Mn^{3+}.[52] The peaks at 642.12 eV ($2p_{3/2}$)[52] and 646.45 eV [56] represent Mn^{4+}, and the satellite of Mn^{4+}. Prominence in intensity corresponding to Mn^{3+} compared to Mn^{4+} also affirms the Zr^{4+} doping in K-OMS-2. It was reported that oxygen vacancies (created upon Zr doping) might contribute to Mn^{3+} formation by converting Mn^{4+} to Mn^{3+},[52] as observed in the Raman spectrum of PZrK samples. Furthermore, the tunnel cations (K^+ ions) residing in the 2×2 tunnel of K-OMS-2 [57] may also execute Mn^{4+}/Mn^{3+} transformation to maintain charge neutrality.[52] The evolution of mixed valence for Mn in PZrK 5 can be correlated with the average oxidation state (AOS) of Mn.[5,20] The AOS of Mn is calculated by the equation: $8.956 - 1.126\Delta E_s$, where ΔE_s is the binding energy difference of multiplet of the Mn 3s spectrum.[5,20] The AOS of Mn in PZrK 5 is 3.03 eV (Fig. S5†), lower than the AOS reported for pristine K-OMS-2, which indicates a possible structural defect caused by Zr^{4+} doping that leads to multiple oxidation states of Mn.[15] The deconvoluted spectrum of O 1s (Fig. 7D) shows two major peaks at 529.31 eV and 530.7 eV (ref. 52) representing lattice oxygen (Mn–O–Mn) and surface-adsorbed oxygen (Mn–O–H), respectively.[52] The relatively intense peak for surface adsorbed oxygen at 530.7 eV confirms oxygen vacancy defects formed by Zr^{4+} doping (Fig. 7D). The Zr 3d spectrum of PZrK 5 shows a doublet peak at binding energies, 182.3 eV and 184.75 eV (Fig. 7E).[21] The peak at 182.3 eV is resolved to three prominent peaks at 181.9 eV, 183.24 eV, and 182.38 eV (Fig. 7E).[21] The peaks at 182.38 and 184.75 eV correspond to Zr–O bonds (Fig. 7E).[21] The other two peaks are attributed to lattice Zr^{4+}, occupying the positions of elemental manganese in the octahedral framework, resulting in oxygen vacancies.[21] The effect of Zr^{4+} on the electrochemical performance of PZrK samples was assessed from the cyclic voltammograms in 0.1 M KCl containing 5 mM $K_4[Fe(CN)_6]$, as shown in Fig. 7F. The CVs show an improved electrochemical activity of PZrK samples in the form of distinct redox peaks, in comparison to PK 50. The PK 50 showed a broad oxidation peak at 0.72 V with an oxidation current (I_{pa}) of 7 μA. Upon Zr doping, a precise observance of redox peaks with increased I_{pa} (23 μA: PZrK 5, 32.8 μA: PZrK RT, and 40 μA: PZrK 50) was observed in PZrK samples. This could be due to the microscale film nature of PZrK samples that exposes more electroactive sites, Mn^{3+}/Mn^{4+} resulting from the Zr doping in K-OMS-2. Furthermore, the phase impurities formed at elevated temperatures may contribute to the electroactive surfaces for redox reactions.

4 Conclusions

In summary, a relatively less explored interfacial polymerisation technique was employed to synthesize composites of Zr-doped K-OMS-2 and PEDOT at a water/chloroform interface at room temperature. The polymerisation of EDOT is carried out in the presence of mixed oxidants of $KMnO_4$ and $ZrOCl_2 \cdot 8H_2O$, where $ZrOCl_2 \cdot 8H_2O$ promotes polymerisation by reducing pH. The addition of $ZrOCl_2 \cdot 8H_2O$ into $KMnO_4$ lowers the standard Gibbs free energy of the system by 76 kcal mol^{-1}, thereby enabling the polymerisation even at lower temperatures. The powder XRD, Raman, and XPS studies demonstrated that $ZrOCl_2 \cdot 8H_2O$ acts as a source for Zr^{4+} ions for the framework-level substitution of Mn^{4+} ions in K-OMS-2. The significant variations in the peak positions of XRD patterns and Raman shift establish the Zr substitution-induced distortions of the tetragonal K-OMS-2 structure. The determination of active surface area from N_2 adsorption-desorption isotherms proved there was an increased surface area upon Zr doping in PEDOT/Zr-K-OMS-2 composites. The substantial increase in S_{BET} and a decrease in average pore size upon Zr substitution in PZrK samples were confirmed from BET isotherms. The optical and FESEM images ensured a transition in the morphology from agglomerates to microscale films when Zr is doped in PEDOT/K-OMS-2. The HR-TEM images illustrated the distribution of crystalline nanoribbons (length ~32 nm and diameter ~4 nm) in the random network of PEDOT flakes. The AFM images evidenced an average thickness of ~21 nm for the microscale PEDOT/Zr-K-OMS-2 films. Thus, the water/chloroform interface acts as a potential template for the self-assembly of Zr-K-OMS-2 nanostructures attached to PEDOT oligomers or short-chain polymers. The proposed interfacial synthesis strategy opens an attractive avenue for synthesizing the hetero-structures for potential applications.

Author contributions

Subin Kaladi Chondath – conceptualization; data curation; formal analysis; investigation; methodology; writing – original draft, review & editing, Mini Mol Menamparambath – conceptualization; project administration; supervision; validation; visualization; funding acquisition; writing – original draft, review & editing.

Conflicts of interest

The authors declare no conflicts of interest.

Acknowledgements

MMM greatly acknowledges the funding from the Science and Engineering Research Board (EEQ/2019/000606) by the Department of Science and Technology (DST), India.

Notes and references

1 H. N. Yoo, D. H. Park and S. J. Hwang, *J. Power Sources*, 2008, **185**, 1374–1379.
2 J. Huang, X. Hu, A. B. Brady, L. Wu, Y. Zhu, E. S. Takeuchi, A. C. Marschilok and K. J. Takeuchi, *Chem. Mater.*, 2018, **30**, 366–375.
3 C. L. Tang, X. Wei, Y. M. Jiang, X. Y. Wu, L. N. Han, K. X. Wang and J. S. Chen, *J. Phys. Chem. C*, 2015, **119**, 8465–8471.
4 Z. Hu, X. Xiao, L. Huang, C. Chen, T. Li, T. Su, X. Cheng, L. Miao, Y. Zhang and J. Zhou, *Nanoscale*, 2015, **7**, 16094–16099.
5 G. Chen, D. Hong, H. Xia, W. Sun, S. Shao, B. Gong, S. Wang, J. Wu, X. Wang and Q. Dai, *Chem. Eng. J.*, 2022, **428**, 131067.
6 C. H. Chen, E. C. Njagi, S. Y. Chen, D. T. Horvath, L. Xu, A. Morey, C. Mackin, R. Joesten and S. L. Suib, *Inorg. Chem.*, 2015, **54**, 10163–10171.
7 Y. Ye, X. Sun, Y. Zhang, X. Han and X. Sun, *Biosens. Bioelectron.*, 2022, **202**, 113990.
8 M. Q. Khan, R. A. Khan, A. Alsalme, K. Ahmad and H. Kim, *Biosensors*, 2022, **12**, 849.
9 M. F. Warsi, K. Chaudhary, S. Zulfiqar, A. Rahman, I. A. Al Safari, H. M. Zeeshan, P. O. Agboola, M. Shahid and M. Suleman, *Ceram. Int.*, 2022, **48**, 4930–4939.
10 M. Ikram, R. Asghar, M. Imran, M. Naz, A. Haider, A. Ul-Hamid, J. Haider, A. Shahzadi, W. Nabgan, S. Goumri-Said, M. B. Kanoun and A. Rafiq Butt, *ACS Omega*, 2022, **7**, 14045–14056.
11 F. Sabaté and M. J. Sabater, *Catalysts*, 2021, **11**, 1147.
12 H. C. Genuino, Y. Meng, D. T. Horvath, C. H. Kuo, M. S. Seraji, A. M. Morey, R. L. Joesten and S. L. Suib, *ChemCatChem*, 2013, **5**, 2306–2317.

13 D. M. Lutz, M. R. Dunkin, K. R. Tallman, L. Wang, L. M. Housel, S. Yang, B. Zhang, P. Liu, D. C. Bock, Y. Zhu, A. C. Marschilok, E. S. Takeuchi and K. J. Takeuchi, *Inorg. Chem.*, 2021, **60**, 10398–10414.

14 L. M. Housel, L. Wang, A. Abraham, J. Huang, G. D. Renderos, C. D. Quilty, A. B. Brady, A. C. Marschilok, K. J. Takeuchi and E. S. Takeuchi, *Acc. Chem. Res.*, 2018, **51**, 575–582.

15 D. Jampaiah, V. K. Velisoju, P. Venkataswamy, V. E. Coyle, A. Nafady, B. M. Reddy and S. K. Bhargava, *ACS Appl. Mater. Interfaces*, 2017, **9**, 32652–32666.

16 J. Ma, C. Wang and H. He, *Appl. Catal., B*, 2017, **201**, 503–510.

17 X. Wu, X. Yu, Z. Chen, Z. Huang and G. Jing, *Catal. Sci. Technol.*, 2019, **9**, 4108–4117.

18 P. F. Smith, L. Wang, D. C. Bock, A. B. Brady, D. M. Lutz, S. Yang, X. Hu, L. Wu, Y. Zhu, A. C. Marschilok, E. S. Takeuchi and K. J. Takeuchi, *Inorg. Chem.*, 2020, **59**, 3783–3793.

19 M. Sun, L. Yu, F. Ye, G. Diao, Q. Yu, Z. Hao, Y. Zheng and L. Yuan, *Chem. Eng. J.*, 2013, **220**, 320–327.

20 L. Chen, J. Jia, R. Ran and X. Song, *Chem. Eng. J.*, 2019, **369**, 1129–1137.

21 Y. Wang, Y. Li, Z. Lu and W. Wang, *RSC Adv.*, 2018, **8**, 2963–2970.

22 D. H. Park, S. H. Lee, T. W. Kim, S. T. Lim, S. J. Hwang, Y. S. Yoon, Y. H. Lee and J. H. Choy, *Adv. Funct. Mater.*, 2007, **17**, 2949–2956.

23 S. K. Chondath and M. M. Menamparambath, *Nanoscale Adv.*, 2021, **3**, 918–941.

24 L. Wang, H. Sahabudeen, T. Zhang and R. Dong, *npj 2D Mater. Appl.*, 2018, **26**, 1–7.

25 R. Dong, T. Zhang and X. Feng, *Chem. Rev.*, 2018, **118**, 6189–6325.

26 J. Neilson, M. P. Avery and B. Derby, *ACS Appl. Mater. Interfaces*, 2020, **12**, 25125–25134.

27 U. I. Premadasa, Y. Z. Ma, R. L. Sacci, V. Bocharova, N. A. Thiele and B. Doughty, *J. Colloid Interface Sci.*, 2022, **609**, 807–814.

28 S. K. Chondath, J. S. Gopinath, R. R. Poolakkandy, P. Parameswaran and M. M. Menamparambath, *Macromol. Mater. Eng.*, 2022, **307**, 2100705.

29 S. K. Chondath, R. R. Poolakkandy, R. Kottayintavida, A. Thekkangil, N. K. Gopalan, S. T. Vasu, S. Athiyanathil and M. M. Menamparambath, *ACS Appl. Mater. Interfaces*, 2019, **11**, 1723–1731.

30 S. K. Chondath, A. P. K. Sreekala, C. Farzeena, S. N. Varanakkottu and M. M. Menamparambath, *Nanoscale*, 2022, **14**, 11197–11209.

31 S. Shi and T. P. Russell, *Adv. Mater.*, 2018, **30**, 1800714.

32 Y. Chai, J. Hasnain, K. Bahl, M. Wong, D. Li, P. Geissler, P. Y. Kim, Y. Jiang, P. Gu, S. Li, D. Lei, B. A. Helms, T. P. Russell and P. D. Ashby, *Sci. Adv.*, 2020, **6**, eabb8675.

33 E. Mazzotta, A. Caroli, E. Primiceri, A. G. Monteduro, G. Maruccio and C. Malitesta, *J. Solid State Electrochem.*, 2017, **21**, 3495–3504.

34 J. Kawakita, J. M. Boter, N. Shova, H. Fujihira and T. Chikyow, *Electrochim. Acta*, 2015, **183**, 15–19.

35 A. R. Zanganeh and M. K. Amini, *Electrochim. Acta*, 2007, **52**, 3822–3830.

36 V. S. Vasantha, R. Thangamuthu and S. M. Chen, *Electroanalysis*, 2008, **20**, 1754–1759.

37 L. R. F. Allen and J. Bard, *Electrochemical Methods: Fundamentals and Applicationsed*, ed. Allen J. Bard, Larry R. Faulkner and Henry S. White, 2000.

38 P. Vanysek, *Electrochemical Series, CRC Handbook of Chemistry and Physics*, CRC Press, Boca Raton, 2002.

39 R. Ahluwalia, T. Q. Hua and H. K. Geyer, *Nucl. Technol.*, 1999, **126**, 289–301.

40 J. Huang, A. S. Poyraz, K. J. Takeuchi, E. S. Takeuchi and A. C. Marschilok, *Chem. Commun.*, 2016, **52**, 4088–4091.

41 J. R. Li, W. P. Zhang, C. Li and C. He, *J. Colloid Interface Sci.*, 2021, **591**, 396–408.

42 I. Staničić, M. Hanning, R. Deniz, T. Mattisson, R. Backman and H. Leion, *Fuel Process. Technol.*, 2020, **200**, 106313.

43 C. E. Simion, O. G. Florea, M. Florea, F. Neatu, S. Neatu, M. M. Trandafir and A. Stanoiu, *Mater*, 2020, **13**, 2196.

44 O. Ghodbane, J. L. Pascal, B. Fraisse and F. Favier, *ACS Appl. Mater. Interfaces*, 2010, **2**, 3493–3505.

45 L. R. Pahalagedara, S. Dharmarathna, C. K. Kingondu, M. N. Pahalagedara, Y. T. Meng, C. H. Kuo and S. L. Suib, *J. Phys. Chem. C*, 2014, **118**, 20363–20373.

46 J. Gao, C. Jia, L. Zhang, H. Wang, Y. Yang, S. F. Hung, Y. Y. Hsu and B. Liu, *J. Catal.*, 2016, **341**, 82–90.

47 J. M. Lee and S. J. Hwang, *J. Solid State Chem.*, 2019, **269**, 354–360.

48 S. K. Godara, V. Kaur, K. Chuchra, S. B. Narang, G. Singh, M. singh, A. chawla, S. Verma, G. R. Bhadu, J. C. Chaudhari, P. D. Babu and A. K. Sood, *Results Phys.*, 2021, **22**, 103892.

49 L. Ben Said, A. Inoubli, B. Bouricha and M. Amlouk, *Spectrochim. Acta, Part A*, 2017, **171**, 487–498.

50 T. Blanquart, J. Niinistö, N. Aslam, M. Banerjee, Y. Tomczak, M. Gavagnin, V. Longo, E. Puukilainen, H. D. Wanzenboeck, W. M. M. Kessels, A. Devi, S. Hoffmann-Eifert, M. Ritala and M. Leskelä, *Chem. Mater.*, 2013, **25**, 3088–3095.

51 B. Anothumakkool, R. Soni, S. N. Bhange and S. Kurungot, *Energy Environ. Sci.*, 2015, **8**, 1339–1347.

52 E. Hastuti, A. Subhan, P. Amonpattaratkit, M. Zainuri and S. Suasmoro, *RSC Adv.*, 2021, **11**, 7808–7823.

53 S. Yurdakal, C. Garlisi, L. Özcan, M. Bellardita and G. Palmisano, *(Photo) catalyst characterization techniques: Adsorption isotherms and BET, SEM, FTIR, UV-Vis, photoluminescence, and electrochemical characterizations*, Elsevier, 2019, pp. 87–152.

54 K. Su, N. Nuraje, L. Zhang, I. W. Chu, R. M. Peetz, H. Matsui and N. L. Yang, *Adv. Mater.*, 2007, **19**, 669–672.

55 N. Puthiyottil, S. Kanakkayil, N. P. Pillai, A. Rajan, S. K. Parambath, R. G. Krishnamurthy, R. Chatanathodi and M. M. Menamparambath, *J. Mater. Chem. B*, 2023, **11**, 1144–1158.

56 J. Hao, Y. Liu, H. Shen, W. Li, J. Li, Y. Li and Q. Chen, *J. Mater. Sci.: Mater. Electron.*, 2016, **27**, 6598–6605.

57 P. Hao, T. Zhu, Q. Su, J. Lin, R. Cui, X. Cao, Y. Wang and A. Pan, *Front. Chem.*, 2018, **6**, 195.

Faraday Discussions

DISCUSSIONS

Materials for electrosynthesis: general discussion

Mickaël Avanthay, Belen Batanero, Pim Broersen,
Anthony Choi, Robert Francke, Mini Mol Menamparambath,
Shelley D. Minteer, Eniola Sokalu and Jeannie Z. Y. Tan

DOI: 10.1039/d3fd90040a

Mickaël Avanthay opened the discussion of the paper by Jeannie Z. Y. Tan: The reaction produces acetylene. How is this separated from the syngas?

Jeannie Z. Y. Tan answered: This is not the focus of this study. However, purifying ethylene from syngas could be achieved using pressure-swing-based absorption techniques.

Eniola Sokalu commented: Silver/copper alloys are widely reported as an optimal material for the CO_2 reduction reaction. In the paper (https://doi.org/10.1039/d3fd00058c), you study various other alloys (cobalt/copper and palladium/copper). What was your rationale for choosing and testing these other alloys for ethylene production?

Jeannie Z. Y. Tan answered: There are studies that have shown that Co and Pd can be used as catalysts for ethylene production. To justify the performance of Co and Pd in ethylene production and perform a techno-economic analysis of the system, we synthesized these 3 composites in this study.

Pim Broersen asked: Did you do any tests to see where the hydrogen atoms in the reduction products come from? Such as deuterium incorporation tests? It seems that the hydrogen could also come from the water in the catholyte, as is quite common in electroreduction reactions.

Jeannie Z. Y. Tan answered: No, I have not done this. This would be the future work to justify the source of H_2 in ethylene.

Pim Broersen queried: Did you perform any control experiments to see whether the addition of hydrogen gas after the reduction of the catalyst is necessary? Or if the reaction can proceed without the use of electricity? Since syngas formally does not need to be reduced to form hydrocarbon products, because hydrogen gas can act as the reductant.

Jeannie Z. Y. Tan responded: Yes, control experiments, including an unbiased reaction, have been performed. However, no product was detected. External stimuli, such as temperature or electricity, are always required to convert syngas to hydrocarbons. Natural reduction of syngas would require a much longer time and product selectivity is not tunable.

Robert Francke remarked: It was shown that feeding a syngas mixture $(1:1)$ to the cathode instead of pure CO leads to a decrease of the charge transfer resistance (and thereby to a decreased cell voltage), likely due to removal of Cu oxides on the electrode surface. What if this syngas mixture was sequentially fed to the electrodes, first to the anode and then to the cathode (or *vice versa*)? Thereby, the OER would be replaced by the HER, further lowering the cell voltage and reducing the hydrogen content in the product mixture.

Jeannie Z. Y. Tan answered: It would be interesting to see if this approach would significantly reduce the HER on the cathode side.

Robert Francke remarked: It is quite remarkable that despite negative polarization of the cathode and abundance of protons, a chemical reducing agent (H_2) is needed to reduce CuO particles.

In the present work, Cu/Ag core shell nanoparticles have been used for the CO_2RR. Copper is known for the CO_2-to-ethylene conversion process, while silver is known to make C1 products. In the course of the electrolysis, the product distribution shifted from C1 to C2 products. What would happen in the reverse approach, *i.e.*, a catalyst made of a Ag core and a Cu shell?

Jeannie Z. Y. Tan answered: This is an interesting suggestion. It would be great to see what the product distribution would be when the core and shell materials are swapped.

Shelley D. Minteer commented: The scale of your potential system would be larger than what most of us are considering in the organic electrosynthesis community. Is it possible to scale to this large of a system? What are the biggest issues with scale-up that you are considering?

Jeannie Z. Y. Tan answered: The system demonstrated in the paper can feasibly be applied to organic electrosynthesis processes. The scaling-up issues, including the optimized reacting area of the electrodes (*i.e.*, 10 cm^2 might not give higher conversion efficiency than 5 cm^2), the best configuration of stacked cells and the stability of the electrodes, would need to be addressed at this stage.

Robert Francke remarked: In industry, there are long-established and highly important electrochemical processes, such as metal deposition and chloralkali electrolysis. The industrial developments in hydrogen production from water electrolysis look promising, but still seem to be in their infancy despite decades of extensive research. In this context, what potential do you see for industrial-scale electro-reduction of CO_2? Which of the possible products are the most desirable? Which technology is most promising?

Jeannie Z. Y. Tan answered: Extensive research has been conducted to develop these sustainable processes to generate green fuels. However, exploration of the downstream market has not been done much previously. Furthermore, the policy and market competitiveness do not show high preferences towards renewable H_2 and hydrocarbons. Hence, the technology uptake in this field has been retarded. Global market demand would be the key to determine the priority in technology development and uptake. Both H_2 and CO_2/CO electro-reduction technology will play an important role in mitigating climate change and reduce the dependency on fossil-based products.

Robert Francke said: Following up on my previous question, I would be interested to know: assuming that your presented system goes to an industrial scale, where would you get the syngas from in the future?

Jeannie Z. Y. Tan replied: Renewable approaches, including steam reforming, solid-oxide electrolysis, and bio-conversion, are able to produce syngas. However, their ratio of $CO:H_2$ would be slightly different. This might impact the efficiency of ethylene production from syngas.

Anthony Choi opened the discussion of the paper by Mini Mol Menamparambath: It's good to see that these zirconium films can be produced using the methods you described in your presentation/paper (https://doi.org/10.1039/d3fd00077j). I'm curious about what you are planning on doing next with these films? Are you planning on coating any specific materials or can you provide more explanation of where you would like to utilise these films?

Mini Mol Menamparambath answered: The uniform distribution of multifunctional nanostructures in the polymeric matrix is one of the major challenges faced during composite fabrication. For example, consider the transparent electrode technology, where it is desirable to disperse conductive nanostructures in conductive polymers devoid of insulative coverings on conductive fillers. However, in conventional composite synthesis methods, the uniform distribution of small-sized nanostructures is complex due to their aggregation tendencies. Due to the solvent–polymer interaction, the composites reported in the paper exhibit highly stable dispersions in appropriate solvents and are also printable on various surfaces. Hence, these composites can be employed for flexible/wearable devices. We have recently developed PEDOT/Au–Ag alloys capable of detecting neurotransmitters from real samples aiming to develop health monitoring devices for point-of-care diagnostics. The authors focused on exploring the self-assembly at liquid/liquid aiming to disperse multi-functional nanostructures in the polymer matrix for flexible electronics.

Robert Francke commented: Since polymerization and composite formation do not work in the absence of $ZrOCl_2$, the latter seems to do more than being a doping agent. What role does $ZrOCl_2$ play during materials synthesis?

Mini Mol Menamparambath responded: The introduction of $ZrOCl_2$ decreases the pH of the reaction medium, and this would improve the oxidative nature of $KMnO_4$. It should also be noted that the polymerization is feasible with a meager elevation in reaction temperatures (50 °C), indicating that the redox reaction is feasible at elevated temperatures without $ZrOCl_2$. From these observations, it can be concluded that under standard conditions, the increase in the acidic nature of the reaction medium could generate an energetic pathway for the electron transfer between monomer and oxidant. However, the exact mechanistic pathway for the polymerization needs to be investigated in detail.

Robert Francke said: Although addition of $ZrOCl_2$ is desired for doping of the material, it may be interesting to explore its role in the mechanism of the synthetic process by performing a control experiment, in which a comparable pH is adjusted using a standard Brønsted acid.

Mini Mol Menamparambath responded: The authors have already reported $KMnO_4$ as an oxidant for the polymerization of pyrrole under standard temperature and pressure conditions at a liquid/liquid interface,[1] where uniform distribution of MnO_2 nanoparticles in the PPy matrix was very evident. However, $KMnO_4$ is not efficient in inducing the polymerization of EDOT under standard conditions. Surprisingly, adding $ZrOCl_2$ could lower the pH of the reaction medium due to the ionization; furthermore, the pH was lowered from 7.6 to 2.08. It is very well-known that acidic $KMnO_4$ is a better oxidizing agent than neutral or basic $KMnO_4$. We have not tried control experiments in the presence of Brønsted acids; this could help in understanding the polymerization mechanism at comparable pH. The authors will explore this point appropriately to devise the mechanism of product formation at liquid/liquid interfaces.

1 S. K. Chondath, A. P. K. Sreekala, C. Farzeena, S. N. Varanakkottu and M. M. Menamparambath, *Nanoscale*, 2022, **14**, 11197–11209.

Belen Batanero asked: Have you applied these self-assembly surfaces to any electrosynthesis processes? For any particular purpose? How useful is the surface? Is it better than an unmodified surface?

Mini Mol Menamparambath answered: As discussed in the work, the authors explored the spontaneous electron transfer between the monomer and oxidant (metal ions), which initiates polymerization at the interface of water and chloroform. The authors assume that all the supplied monomers undergo polymerization at the given monomer concentrations, and the reaction is complete. However, there is a considerable possibility of hampering the polymerization at higher concentrations of monomers during the reaction due to the mass transfer limitation at the interface. The products accumulated at the interface could limit the diffusion of monomers to the interface at higher concentrations. As pointed out by Prof. Batanero, the reaction can be conducted by applying external potentials to different phases to ensure the complete polymerization of the monomers. However, applying potentials at different phases may decrease the efficiency in dispersing the nanoparticles in the polymer matrix, which is one of the advantages of the current *in situ* polymerization technique. It would be fascinating to look at the efficiency of polymerization with and without external electric fields in a comparative manner. The reported polymer nanocomposites are important in several applications where the uniform dispersion of a few nanometer-sized nanostructures in the polymeric matrix is essential. For example, the dispersion of smaller-sized conductive nanostructures is essential for transparent electrodes/flexible electronics applications. Furthermore, the microstructure tuning of composites to 2D morphology is another highlight of the reported work. The authors also reported the possibility of uniform distribution of various metal nanoparticles, metal alloys, and metal oxides into conductive polymeric matrices and reported potential electrochemical applications. From the physicochemical characterisations, it was obvious that the unique properties achieved could be advantageous over unmodified conductive films formed.

Robert Francke remarked: The presented Zr-doped PEDOT–Mn oxide composites seem to be very promising for application in supercapacitors. First, upon positive electrode polarization, the polymer becomes electrically conductive. Second, the Mn oxide part is highly porous and reveals a large surface area. Third, the degree of Zr doping can be used to adjust the pore size to the ion of the supporting electrolyte. Fourth, the $Mn(III)/Mn(IV)$ couple could introduce pseudocapacitance and thereby increase the energy density.

Mini Mol Menamparambath answered: Thank you for the valuable suggestion. The authors have not focussed on the supercapacitance performance of the reported composites so far. The current research article (https://doi.org/10.1039/d3fd00077j) reported the synthesis of Zr-doped MnO_2 under standard temperature and pressure conditions to alleviate the challenges associated with conventional synthetic strategies. In order to explore the advantages of the 2D sheet-like morphology, extremely thin nanoribbon characteristics, and the Zr doping in MnO_2, the authors will be looking in detail into the supercapacitance/pseudocapacitance properties.

Robert Francke commented: The synthetic process that was presented is not an electrosynthesis in the classical sense; at least, no external circuits and power sources are used to drive the reaction. Could you discuss the similarities and differences between your method and electrosynthesis, and outline the benefits of your approach?

Mini Mol Menamparambath responded: The authors do agree that the *in situ* polymerization is not an electrosynthesis in the classical sense; however, a standard reduction potential controlled redox reaction is reported in the present work. A comparative evaluation can be summarised as follows.
Similarities:
(1) A redox reaction is involved during the polymerization, where monomers and metal ions serve as the anode and cathode, respectively.
(2) The standard reduction potentials of the cathode and anode govern the spontaneity of the reaction.
(3) Polymerization is achieved successfully in both methods.
Differences:
(1) No external circuits and power sources are used in the present work, while an external power supply is required to initiate polymerization.
(2) Anodes and cathodes are placed in two completely immiscible phases separated by a distinct interface (bi-phase reaction) in the present work, while the

anodes and cathodes are placed in the electrochemical cell containing a common electrolyte solution (single-phase reaction).

(3) Incorporation of metal/alloy/metal oxide nanoparticles into polymeric matrices is achieved in an *in situ* manner in the present work, while nano-structures are deposited in the polymer matrix by a two-step process.

The current approach is beneficial in polymerizing monomers in the presence of metal ions, resulting in the *in situ* incorporation of nanostructures in the desired polymers. This would enhance the interaction between multi-functional fillers and the matrix. Morphology tuning is purely attributed to the interaction between polymer and solvents (aqueous or organic phases chosen). Therefore, compared to conventional electrosynthesis, the appropriate choice of solvents and oxidants could control the morphology and physicochemical properties of polymer nanocomposites.

Conflicts of interest

There are no conflicts to declare.

Faraday Discussions

PAPER

Electrochemical decarboxylation of acetic acid on boron-doped diamond and platinum-functionalised electrodes for pyrolysis-oil treatment†

Talal Ashraf, [ID] [a] Ainoa Paradelo Rodriguez,[a] Bastian Timo Mei [ID] *[ab] and Guido Mul [ID] [a]

Received 13th March 2023, Accepted 4th April 2023

DOI: 10.1039/d3fd00066d

Electrochemical decarboxylation of acetic acid on boron-doped-diamond (BDD) electrodes was studied as a possible means to decrease the acidity of pyrolysis oil. It is shown that decarboxylation occurs without the competitive oxygen evolution reaction (OER) on BDD electrodes to form methanol and methyl acetate by consecutive reaction of hydroxyl radicals with acetic acid. The performance is little affected by the applied current density (and associated potential), concentration, and the pH of the solution. At current densities above 50 mA cm^{-2}, faradaic efficiencies (FEs) of 90% towards the decarboxylation products are obtained, confirmed by in situ electrochemical mass spectrometry (ECMS) investigation showing only small amounts of oxygen formed by water oxidation. Using platinum-modified BDD electrodes, it is shown that selectivity to ethane, the Kolbe product, strongly depends on the shape and geometry of the platinum particles. Using nano-thorn-like Pt particles, a faradaic efficiency of approx. 40% towards ethane can be obtained, whereas 3D porous platinum nanoparticles showed high selectivity towards the OER. Using thin platinum layers, a high FE of >70% towards ethane was obtained, which is thickness-independent at layer thicknesses above 20 nm. Comparison with other substrates revealed that BDD is an ideal support for Pt functionalisation, giving advantages of stability and high-value-product formation (ethane and methanol). In short, this work provides guidelines for electrode fabrication in the context of the electrochemical upgrading of biomass feedstocks by acid decarboxylation.

Introduction

For a sustainable and circular bioeconomy, biomass has been used as a feedstock for various applications, including production of chemicals and fuels. Pyrolysis

[a]PhotoCatalytic Synthesis Group (PCS-TNW), University of Twente, Drienerlolaan 5, 7522 NB Enschede, The Netherlands. E-mail: b.t.mei@utwente.nl

[b]Industrial Chemistry, Ruhr-Universität Bochum, Universitätsstraße 150, 44801 Bochum, Germany

† Electronic supplementary information (ESI) available. See DOI: https://doi.org/10.1039/d3fd00066d

oil obtained from the thermochemical conversion of biomass is a low-value biogenic liquid, consisting of carboxylic acids, sugars, phenolics and lignin. Carboxylic acids are a significant fraction of pyrolysis oil with a low energy content (44 MJ kg^{-1}), and their catalytic conversion requires harsh conditions ($\geq$225 °C and 60 bar).[1-3] Electrochemical treatment of pyrolysis oil enables the conversion of carboxylic acids into hydrocarbons by (non-)Kolbe electrolysis. Kolbe chemistry was first reported by Michael Faraday in 1834 and recently gained attention as a potential enabling technology. Thus, product distributions and reaction mechanisms have been analysed.

Electrochemical decarboxylation occurs by the discharge of the deprotonated acid (influenced by the electrolyte pH) at potentials >2.1 V$_{NHE}$, which is critical for forming alkyl radicals.[4-6] Moreover, mechanistic studies suggest that the coverage of the carboxylate on the electrode surface inhibits the oxygen evolution reaction.[7,8] The reaction proceeds via acetate adsorption on the electrode surface followed by a proton-coupled electron transfer (PCET)[7] reaction that results in the formation of carboxyl species (Fig. S1†). The one-electron transfer decarboxylation process produces carbon dioxide as the by-product along with an alkyl radical. Subsequent alkyl-radical conversion depends on the decisive conditions (radical concentration, electrolyte and pH (ref. 9)). Dimerisation results in alkane formation, whereas by disproportionation alkanes and alkenes are formed. The overoxidation of alkyl radicals leads to the formation of carbocations, which deprotonate to form alkenes or react with hydroxyl ions to form alcohols via the so-called Hofer–Moest reaction. Moreover, the carbocation can react with the deprotonated acetate to form the respective ester. Several parameters influence the fate and selectivity of (non-)Kolbe electrolysis, such as pH, applied current density, solvent, and electrolyte composition.[5,10]

The electrode material is also crucial, and platinum electrodes have been reported to be ideal for Kolbe electrolysis at negligible water discharge rates. For Pt electrodes, the formation of a barrier layer consisting of carboxylates is suggested to be essential to prevent water oxidation.[9] Electrode materials such as nickel and gold are considered inactive for (non-)Kolbe electrolysis due to the absence of the aforementioned barrier layer. Electrochemical impedance spectroscopy (EIS) has revealed that the pseudo-adsorption capacitance of CH_3COO^- is absent for gold electrodes.[11] Due to an immediate formation of ~30 Å-thick Au_2O_3 (at 0.93 V$_{SCE}$)[11] upon water discharge, the C–C coupling reaction is inhibited. The acetate discharges (~10 Å-thick layer) on Pt–O monolayers (~10 Å)[11] at around 2 V, nearby to the OER region;[12] therefore, Kolbe electrolysis dominates at potentials higher than that of the OER. Carbon-based electrodes show higher activity towards alcohol products, including esters.[5] Thus, the product selectivity of (non-)Kolbe electrolysis is highly dependent on the anode material. Various strategies have been proposed to minimize the use of platinum; among others, platinised electrodes, platinum nanoparticles (NPs) on carbon substrates, nanostructured platinum electrodes, and carbon-based electrodes have been used.[13-16] For Pt-nanoparticle-based electrodes, it has been suggested that the selectivity of the decarboxylation reaction depends on the shape and geometry of the Pt nanoparticles.[15] Yet the activity of platinum-nanoparticle electrodes for (non-)Kolbe electrolysis has to be confirmed[14,17] and the impact of NP loading or the influence of the substrate material has not been fully disclosed. Despite the high activity of platinum electrodes for (non-)Kolbe electrolysis, Pt is prone to dissolving under (non-)Kolbe conditions, as has been shown by Ranninger *et al.*[18] In non-aqueous electrolyte, dissolution of approx. 13.3% of a 10 μm-thick Pt electrode was estimated (over a span of one year of electrolysis with a 1 m^2 electrode), whereas a slightly lower material loss of 8.1% would occur in water-containing electrolytes (Kolbe electrolysis was performed in 1 M acetic acid in ethanol-based solvent at 3 V *vs.* Fc/Fc$^+$).[18]

Boron-doped diamond (BDD) has been used as a substrate material for deposition of metal nanoparticles used in sensing and degradation applications.[19–21] In particular, BDD is widely used due to its inertness towards poisoning within a wide potential window and high stability at high current densities.[22–25] For example, excessive polarisation of BDD for 250 hours in 1 M H$_2$SO$_4$ at 1 A cm^{-2} resulted in no sign of etching,[26] confirming the stability of BDD. Generally, BDD electrodes show higher corrosion resistance than Pt electrodes under Kolbe electrolysis conditions and thus extended operation can be achieved.[5,27] Still, only a few reports are available related to acetic acid oxidation on BDD,[28] and usually electrooxidation is performed in methanol-, sulphate- or perchlorate-containing electrolytes, while the product distribution is not well described. The decarboxylation of acetic acid on BDD is considered to occur *via* the formation of weakly adsorbed hydroxyl radicals or other radicals (such as peroxysulphates),[10] which oxidise the acetate, omitting a direct PCET. A direct electron-transfer reaction is only observed for compounds with thermodynamic potentials well below those of water oxidation and formation of OH radicals. For example, formic acid is reported to be fully oxidised to CO$_2$ *via* direct electron transfer (DET).[29] Interestingly, BDD is considered unstable at high current densities >50 mA cm^{-2} in the presence of alkyl radicals,[30] which produce dangling bonds by abstraction of OH groups from C–OH functional groups. Kashiwada *et al.*[30] reported that corrosion can be effectively prevented at high pH or moderate current densities. Despite the details known about the reaction mechanism of acetic acid on BDD, the impact of the BDD surface morphology has not yet been considered.[31–33]

In this work, we aim to address the reaction mechanism of (non-)Kolbe/indirect oxidation of acetic acid in the absence of foreign anion species and explore the stability of BDD during electrolysis in a flow cell. We show that an indirect oxidation occurs in the presence of BDD and moreover we reveal that functionalisation of the BDD surface with thin platinum films or electrodeposited Pt nanoparticles allows tuning of the reaction mechanism towards Kolbe product formation. In comparison to other substrates (graphite, nickel foam and fluorine-doped tin oxide (FTO)), BDD is shown to be an ideal material for functionalisation. In addition, we discuss how the product distribution is affected by current density and pH in a flow cell.

Materials and methods

Acetic acid (glacial, ReagentPlus®, ≥99%, Sigma Aldrich), sulphuric acid (reagent grade, 95–98%, Sigma Aldrich), perchloric acid (reagent grade, 70%, Sigma Aldrich), sodium acetate (ACS reagent, ≥99.0%, Sigma Aldrich), acetone (technical grade, BOOM B.V.), isopropanol (technical grade, BOOM B.V.) and ethanol (technical grade, BOOM B.V.) were used. Hexachloroplatinic acid, potassium tetrachloroplatinate(II) and potassium hexachloroplatinate(IV) were acquired from

Sigma Aldrich. Acetic acid was chosen as a model compound due to its abundancy in pyrolysis oil.[34]

Boron-doped-diamond electrodes (DIACHEM®), with a coating thickness of 15 nm BDD on 2 mm-thick tantalum substrates and a doping level of 2000–5000 ppm of boron incorporated into the diamond lattice, were acquired from CONDIAS Gmbh. The procedure for manufacturing the electrodes is reported in detail elsewhere.[35] Surface cleaning prior to electrochemical testing was performed by washing with Milli-Q water (18.2 MΩ cm) and isopropanol, prior to ultrasonication in Milli-Q water for 10 minutes, anodic polarisation in 1 M $HClO_4$ for 30 minutes and subsequent rinsing with Milli-Q water, following a protocol reported in the literature.[36] For comparison, graphite (LFYGY Industry Materials Professional Supplier, China), FTO (redox.me, Sweden) and nickel foam (IKA, Germany) electrodes were used as both electrodes and substrates. Prior to use, the graphite was polished with carbide paper and sonicated for 10 minutes in Milli-Q water. Impurities were removed electrochemically in 1 M H_2SO_4 by potential cycling in between 0.2–1 V_{RHE} at 50 mV s^{-1} for 100 cycles. The FTO electrodes were cleaned in an ultrasonic bath in acetone, isopropanol, ethanol and Milli-Q water for 15 minutes each. The nickel foam was cleaned by ultrasonication in 2 M HCl solution for 10 minutes prior to ultrasonication in ethanol and Milli-Q water for five minutes each.

Thin platinum layers (5, 20, 50 and 100 nm) were deposited on cleaned BDD substrates using a sputtering system (AJA International, USA). Electrodeposition of Pt nanoparticles was carried out with different platinum salt solutions. Porous nanoflower-like platinum nanoparticles were electrodeposited from a solution of 8 mM H_2PtCl_6 in 0.5 M H_2SO_4 by applying a constant potential of -0.24 $V_{Ag/AgCl}$ for 15 minutes, and are denoted as ED-A hereafter. Dispersed platinum nanoparticles were obtained from a solution of 1 mM K_2PtCl_4 in 0.5 M H_2SO_4 during a constant potential treatment (-1 $V_{Ag/AgCl}$ for 200 seconds), and labelled as ED-B. Thorn-like platinum nanoparticles (active-crystal-plane particles grown on 2D substrates[37,38]) were produced in 2.5 mM H_2PtCl_6 in 0.5 M H_2SO_4 at -0.2 $V_{Ag/AgCl}$ for 15 minutes, and labelled as ED-C. Platinum nanocrystals were produced by applying -0.22 $V_{Ag/AgCl}$ for 15 minutes using a solution of 4.0 mM K_2PtCl_6 in 0.5 M H_2SO_4, and labelled as ED-D.

The electrochemical decarboxylation reactions were performed in 1 M acetate solution at pH 5 (acetic acid/sodium acetate). BDD, graphite, FTO, NiF (bare), Pt-sputtered BDD, and Pt nanoparticles on BDD were used as working electrodes. Platinised titanium (Magneto Special Anodes B.V.) and Ag/AgCl (3 M NaCl, ProSense) were used as counter and reference electrodes, respectively. Electrolysis was performed in a custom-made glass cell at a constant stirring rate of 600 rpm with a helium (99.99%) purge of 30 mL min^{-1}. Prior to batch electrolysis, the electrochemical cell and electrode holders were thoroughly rinsed with ethanol, washed five times with Milli-Q water and subsequently boiled in Milli-Q water. Gas analysis during the reaction was performed by online gas chromatography (GC). A flame ionisation detector (FID) and an RTX-1 column (15 m × 0.32 mm I.D., 5 μm fused silica film thickness at 45 °C) were used for detection of hydrocarbons. CO_2, O_2 and H_2 were detected with a thermal conductivity detector (TCD) coupled with a Carboxen® 1010 column (15 m × 0.32 mm I.D. at 70 °C). External calibration was conducted for CH_4, C_2H_4, C_2H_6, H_2, O_2, and CO_2 with r^2 (coefficient of determination) > 0.995. Hofer–Moest and other liquid products (methanol and methyl acetate) were detected by liquid analysis. Aliquots were collected from the electrolyte, diluted 10 times with water, neutralised and detected using a headspace GC-FID (Agilent) using a Zebron 7HG-G013-11 column (0.25 μm I.D., 250 μm × 40 m) for product separation.

Cyclic voltammetry (CV) in the potential range of 0.05–1.2 V_{RHE} was performed in 1 M H_2SO_4 solution to determine the electrochemically active surface area (ECSA) of the platinised electrodes. CV was also performed in 1 M acetic acid/ sodium acetate solution (0–3 V_{RHE}) before and after electrolysis, followed by current interruption (CI) for iR (ohmic drop) correction. Chronopotentiometry was performed at 25 mA cm^{-2} at room temperature and pressure, and the measured potentials were converted to RHE scale. The calculations for faradaic efficiency (FE) are explained in the ESI.† Flow experiments were conducted using the commercial Condias Synthesis cell kit (see Fig. S2†).

Electrochemical mass spectrometry (ECMS, instrument from Spectro Inlets, Copenhagen, Denmark) was performed for instantaneous detection of volatile products. In particular, signals related to methane, ethane, ethylene, methanol, methyl acetate, oxygen and CO_2 were recorded during cyclic voltammetry in 1 M acetic acid/sodium acetate electrolyte. A platinum mesh and an Ag/AgCl electrode (sat. KCl, CH Instruments) were used as the counter and reference electrodes, respectively. The morphology and surface structure of the platinum nanoparticles and BDD electrodes were visualised using scanning electron microscopy (SEM) performed with a ZEISS MERLIN HR-SEM or a JEOL JSM 7610F, respectively. The stability of the BDD electrodes was analysed by SEM and using a custom-made Raman spectroscopy setup equipped with a 650 nm laser, before and after electrolysis. Due to the surface morphology and uneven BDD crystal height, three measurements were taken at different depth levels from the surface to inside of the sample. The scanned area is 10 × 10 μm and measurements were performed at 35 mW, for 150 ms at 50 kHz. The Raman spectrometer is equipped with an optical microscope (BX41 Olympus, objective Olympus 40× NA 0.95).

Results and discussion

(Non-)Kolbe electrolysis at various electrodes

Kolbe electrolysis of acetic acid/sodium acetate at pH 5 was performed at 25 mA cm^{-2} using BDD, graphite, FTO and nickel foam as working electrodes. Fig. 1a shows the cyclic voltammograms obtained in acetate solution. Clearly, the lowest onset potential and the highest oxidative currents (39 mA cm^{-2} at 3 V_{RHE}) are achieved with nickel foam (NiF), which is primarily due to its characteristic high surface area. The onset potential of 1.8 V_{RHE} is related to the oxygen evolution reaction (OER) as also confirmed by the high faradaic efficiency for oxygen obtained during electrolysis (Fig. 1b). For BDD, the CV resembles previously reported data in the literature,[39] and only above 2.5 V_{RHE} is an exponential increase in currents observed. The CV obtained with FTO electrodes closely resembles that obtained with BDD, but the increase in current is less steep (attaining 5.7 mA cm^{-2} at 3 V_{RHE}). Similarly, for graphite electrodes, the oxidation current density at high potential is low (5.7 mA cm^{-2} at 3 V_{RHE}), contrasting with its early current onset. The product selectivities, as shown by the faradaic efficiencies in Fig. 1b, are clearly electrode-material specific. Kolbe electrolysis products are not observed for NiF electrodes. With a very high FE towards oxygen of up to 95%,

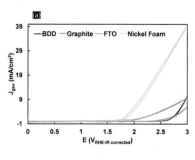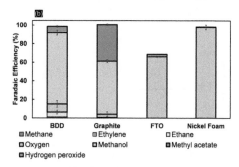

Fig. 1 Effect of different electrodes on (non-)Kolbe electrolysis: (a) CV of BDD, graphite, FTO and nickel foam conducted in 1 M acetic acid/sodium acetate at pH 5; (b) average FE towards (non-)Kolbe/Hofer–Moest/indirect oxidation products during electrolysis of 1 M acetic acid/sodium acetate for 50 minutes in a batch cell at 25 mA cm^{-2} and pH 5.

formation of oxides on the nickel surface favours the OER[40] (Fig. S3†). When using FTO, the OER is also dominant and only small quantities of H$_2$O$_2$ were detected, besides oxygen, resulting in a faradaic efficiency of 2%. Considering that acetic acid oxidation on FTO and nickel foam is inhibited, both materials are inadequate electrode materials for (non-)Kolbe electrolysis. Formation of methanol and methyl acetate in equivalent FEs is achieved at graphite electrodes. Thus, overoxidation of the methyl radical to the carbocation leads to Hofer–Moest product formation.[5] During chronopotentiometry (Fig. S4†), the rise in potential within the electrolysis time frame (50 minutes) observed for graphite and FTO electrodes suggests that both materials exhibit instability under (non-)Kolbe electrolysis conditions. The instability of FTO was reflected in the working electrode potential reaching ∼7 V$_{RHE}$. This contrasts with the high stability observed for BDD, where the electrode potential was stable at 3.6 V$_{RHE}$ (Fig. S4†). Furthermore, when employing BDD electrodes, the electrooxidation of acetic acid is more selective to methanol as compared to graphite electrodes. Considering that the separation of azeotropic methanol and methyl acetate mixtures is challenging, the oxidation on BDD can be considered more efficient.[41]

The stability of BDD electrodes during Kolbe electrolysis was analysed by Raman spectroscopy (Fig. S5†) and SEM (Fig. S6†). The BDD coatings were investigated before and after Kolbe electrolysis as shown in Fig. 2. The Raman spectra are dominated by bands at 488, 1137, 1216, 1323, and 1147 and 1550 cm^{-1}, in agreement with literature reports.[42,43] The spectra were taken across random areas on the BDD surface and averaged for analysis (Fig. S5†). The Raman spectrum of the as-received BDD electrode confirms a high boron doping, as in general with increasing boron concentration in the diamond lattice the first-order diamond phonon line[44] shifts from 1332 cm^{-1} to lower wavenumber. Comparison of the BDD surface before and after electrolysis reveals a slight change in the signal of the diamond phonon mode, which shifts from 1312 cm^{-1} to 1323 cm^{-1} after electrolysis. In addition, the scattering intensity at 488 cm^{-1} and 1216 cm^{-1} decreases, corresponding to the weak band due to metallic and non-metallic impurities. Considering that the signal at 488 cm^{-1} is due to the boron doping level and vibration modes of boron clusters and pairs,[19,42] Raman analysis suggests that the concentration of boron in the diamond lattice decreases with

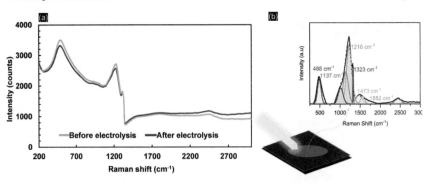

Fig. 2 (a) Raman spectra of BDD before and after 2 h electrolysis in 1 M acetic acid/sodium acetate pH 5 at 25 mA cm^{-2}. The BDD electrodes used have a concentration of 2000–5000 ppm boron atoms in the lattice. (b) Deconvoluted Raman spectra of BDD reveal the positions of the bands, in accordance with the existing literature.

electrolysis. Degradation of BDD can occur by either boron leaching or the graphitization of carbon at the grain boundaries. The G-band peak observed at 1552 cm^{-1} is due to the bond stretching of sp^2 carbon atoms in rings and chains and is in agreement with the literature.[45] The band present at 1473 cm^{-1} corresponds to the sp^2 carbon of *trans*-polyacetylene lying in the grain boundaries. Interestingly, due to the high quality of the BDD coating, polymeric carbon species[46] are not prominent before electrolysis as highlighted by the minor signal intensities at 1470 cm^{-1}. Still, it is worth noting that the post-treatment (chronopotentiometery at 25 mA cm^{-2} for 30 minutes in 1 M HClO$_4$) (see Materials and methods) removed graphitized carbon effectively (Fig. S6†). Overall, Raman analysis confirmed that the electrolytic corrosion on BDD is negligible during acetic acid electrolysis at low current densities. Therefore, we conducted all subsequent experiments below 100 mA cm^{-2}. The amount of boron content (within a cubic centimetre) can roughly be estimated by an empirical relationship[47] (eqn (1)).

$$[B] = 8.44 \times 10^{30} \times \exp(-0.048 \times W) \ (\text{cm}^{-3}). \tag{1}$$

W is the wavenumber corresponding to the peak of the Lorentzian component of the Raman spectra at 400–600 cm^{-1}. The estimated concentration of boron doping in the BDD electrode is found to be 5.66 × 10^{20} cm^3.

We further explored the impact of electrolysis conditions (pH and current density) on the electrooxidation using a divided flow cell (compartment separation was achieved using a Nafion 324 cation exchange membrane, see Fig. S2†) and an acetic acid/sodium acetate solution as both catholyte and anolyte. The results are shown in Fig. 3. The FE is relatively independent of the pH and current density. Methanol is the dominant product, while the consecutive reaction to methyl acetate is most pronounced at low current densities and moderate pH values (pH 5). Under constant-flow conditions (60 mL min^{-1}), the highest accumulative FE towards methanol was achieved at 100 mA cm^{-2}. The high selectivity towards methanol is attributed to the high concentration of weakly adsorbed OH

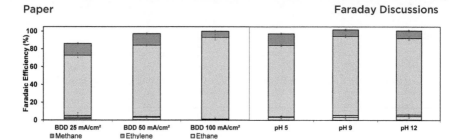

Fig. 3 The effect of current density (left) on the electrooxidation of 1 M acetic acid/sodium acetate on BDD in a flow cell at pH 5 and the pH effect (right) on FE towards methanol and methyl acetate at 50 mA cm^{-2} in a flow cell at 60 mL min^{-1} (catholyte and anolyte flow).

radicals, which promote oxidation of concentrated acetic acid/sodium acetate to methanol.

When decreasing the concentration of acetic acid to 0.5 M, the FE towards methanol remains dominant (Fig. S7†). Thus, the oxidation is not limited by the hydroxyl radical concentration or autoinhibition[39] at 0.5 M acetic acid concentration. Interestingly, in earlier reports, methyl acetate was not detected during acetic acid oxidation on BDD electrodes,[28] so here we report the full product distribution of acetic acid oxidation at BDD electrodes under different conditions for the first time.

Platinised electrodes

To alter the selectivity from the formation of methanol to Kolbe electrolysis, *i.e.*, ethane formation, and to investigate the behaviour of the Pt thin films and nanoparticles on a well-characterized and stable substrate, BDD was functionalised with thin platinum layers by sputtering (5, 20, 50 and 100 nm Pt) and Pt nanoparticles by electrodeposition. For thin film samples, the ECSA normalized to the mass deposited (m^2 g$_{pt}^{-1}$) was obtained by integration of the H$_{upd}$ (hydrogen underpotential deposition) region from the CVs, assuming a monolayer hydrogen charge of 210 µC cm$_{Pt}^{-2}$ (Fig. S8b†). The ECSA normalized to the loading of the amount of platinum deposited decreases in order of increasing platinum layer thickness (Fig. S8c†). This is in contrast to the behaviour observed for platinum nanoparticles (Fig. S16†), but in agreement with the literature.[48] Importantly, the inhomogeneous BDD might influence the nucleation and film growth,[49] likely leading to a non-uniform coating of the substrate. Indeed, non-normalized ECSA data revealed that the surface area (cm^2) and roughness factor increase from the 5 nm sputtered layers up to the 20 nm thin films and only small differences are observed for thick coatings (Fig. S9b and c†).

To investigate the potential-dependent product distribution of the Kolbe electrolysis of acetic acid/sodium acetate solution in real-time, an ECMS investigation was performed with BDD and platinised BDD (5 nm sputtered Pt) electrodes. ECMS is a modern technology compared to conventional DEMS (differential electrochemical mass spectroscopy), as discussed in ESI section 27.† The mass spectrometry data recorded during the CV are shown in Fig. 4. Generally, the transient (*m/z*) signals confirm the presence of (non-)Kolbe products.

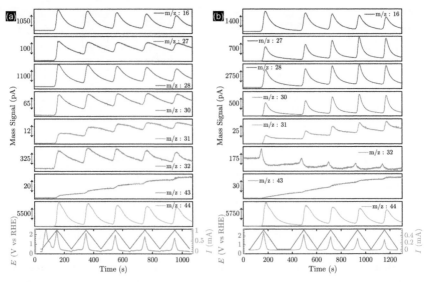

Fig. 4 Electrochemical mass spectrometry (ECMS) measurement of Kolbe electrolysis and indirect oxidation using (a) BDD and (b) platinised BDD electrodes. Cyclic voltammetry has been performed in 1 M acetic acid/sodium acetate at pH 5 in both cases. The recorded signals correspond to methane ($m/z = 16$), ethane ($m/z = 27$ and 30), ethylene ($m/z = 27$ and 28), the by-product CO_2 ($m/z = 44$), the competitive reaction product O_2 ($m/z = 32$) and the indirect oxidation/Hofer–Moest products methanol ($m/z = 31$) and methyl acetate ($m/z = 43$).

As shown in Fig. 4a, the onset of the OER for bare BDD is observed at 2.136 V_{RHE}. Considering that for BDD electrodes the onset of the OER is usually observed at potentials >2.3 V,[50] it is speculated that oxygen evolution proceeds *via* decomposition of hydrogen peroxide produced by OH radical recombination, as depicted in eqn (2)–(4).

$$H_2O \xrightarrow{+H, -e} {}^{\bullet}OH \qquad (2)$$

$$^{\bullet}OH + {}^{\bullet}OH \rightarrow H_2O_2 \qquad (3)$$

$$2H_2O_2 \rightarrow O_2 + 2H_2O \qquad (4)$$

Subsequently, decarboxylation products are detected. Methanol and methyl acetate are accumulated in the liquid phase during cyclic voltammetry, due to their low vapor pressure. Compared to the signal intensities of CO_2, the main Kolbe product, ethane, is only observed in minor amounts, in agreement with the above presented batch and flow cell experiments. Importantly, here we reveal that the onset of the decarboxylation reaction and oxygen evolution and the transient product signals are similar.

This is in contrast to the thin-film-coated Pt/BDD electrodes (Fig. 4b), for which oxygen evolution and the decarboxylation reaction are well-separated in potential. The onset potential for the OER is 1.98 V_{RHE} on Pt/BDD, and only at potentials >2.1 V_{RHE} are Kolbe and indirect oxidation products detected.

Importantly, at the onset of Kolbe product formation, the OER is inhibited, being in strong contrast to the profiles observed for bare BDD. In addition, oxygen evolution is decreasing during successive cycling, while Kolbe-product formation is more pronounced, as revealed by the intensity of ethane ($m/z = 30$). Thus, likely a barrier layer builds up[7] during successive cycling, enabling a shift in selectivity towards Kolbe products. Finally, it is apparent that Kolbe products are primarily produced on Pt/BDD, as revealed by the higher currents of the respective signals. Nevertheless, the presence of the methanol and methyl acetate suggests that the BDD surface is not fully covered with Pt and exposed BDD facets enable OH formation and alcohol and ester formation.

Constant-current electrolysis of acetic acid was subsequently performed at 25 mA cm^{-2} using a variable thickness of Pt (Fig. 5). For 5 nm Pt on BDD electrodes, initially a high selectivity towards OER is observed and only during extended polarisation is product formation transitioning to the Kolbe product formation (Fig. S11†). This is likely due to the formation of (110) and (100) facets of Pt (Fig. S9a†), which are not present during the CV recorded before electrolysis. After 50 minutes of constant polarisation, the selectivity of the reaction for the 5 nm-thick Pt film shifted primarily to ethane, but is still limited to a FE of ~59% (Fig. S10†). CVs recorded in 1 M acetic acid/sodium acetate show the presence of an inflection zone at 2.5 V$_{RHE}$ for 20, 50 and 100 nm Pt layers, and the peak oxidation current of 33.5 mA cm^{-2} was achieved with a 5 nm Pt layer. The absence of an inflection zone for the 5 nm Pt layer shows the dominancy of the OER, which is also displayed in the product distribution, where up to 80% faradaic efficiency reflects the selectivity of the reaction towards the OER as the electrolysis starts.

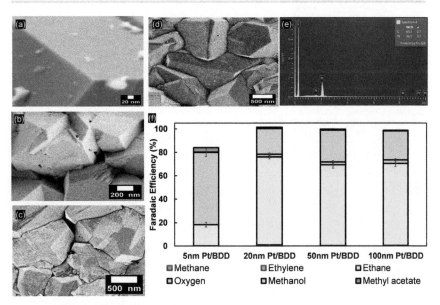

Fig. 5   SEM images of Pt-sputtered BDD electrodes: (a) 5 nm, (b) 20 nm, (c) 50 nm and (d) 100 nm Pt; (e) EDX of 20 nm Pt/BDD showing the presence of Pt and carbon from diamond; (f) averaged FE towards (non-)Kolbe electrolysis with different platinum-layer thicknesses on BDD, obtained during electrolysis at 25 mA cm^{-2} in 1 M acetic acid/sodium acetate (pH 5).

For all other electrodes, *i.e.* with Pt films >20 nm, instantaneous Kolbe-product formation is observed at FE >70% (Fig. S11c†). This is still lower than that obtained for Pt foil, where an FE towards ethane formation of ~90% is reproducibly observed (Fig. S12†). Simultaneously, methyl acetate is detected, allowing the electron balance to be closed. Thus, it appears that methyl radical recombination is still less favourable for BDD-supported Pt, likely due to an insufficient surface radical concentration. Instead, the exposed BDD surface (see Fig. S13†) allows for formation of methanol (Fig. 5f). Likely, some uncovered BDD surface is maintained, in between BDD crystals as suggested by SEM analysis (Fig. 5).

Considering the interesting product formation transients observed for 5 nm Pt on BDD, and in an attempt to study lower Pt loadings, BDD was functionalised with Pt nanoparticles obtained by electrodeposition (see Materials and methods).

The electrodeposited mass of Pt on the BDD is variable due to different deposition times and applied potentials. The FE and production rate of the products of each type of electrodeposited nanoparticle on the BDD were compared to evaluate the selectivity towards Kolbe electrolysis. For porous platinum nanoflowers (ED-A; for SEM see Fig. 6a), an average FE toward ethane of

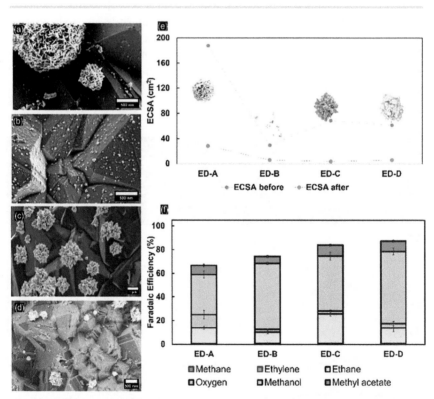

Fig. 6 SEM images of Pt electrodeposition on BDD electrodes: (a) ED-A nanoflowers, (b) ED-B dispersed nanoparticles, (c) ED-C nano-thorns, and (d) ED-D nanocrystals; (e) ECSA (cm^2) in 1 M H$_2$SO$_4$ before and after (non-)Kolbe electrolysis; (f) average FE towards (non-) Kolbe electrolysis/indirect oxidation products from different geometries of platinum nanoparticles on BDD at 25 mA cm^{-2} in 1 M acetic acid/sodium acetate (pH 5). The electron balance was not closed, which could be due to the formation of H$_2$O$_2$ or other side products (not detected).

around 14% (±1.134) was obtained. Initially, the OER dominated with faradaic efficiencies up to 50% (Fig. S17†). Nevertheless, during extended electrolysis, the OER becomes negligible, similar to the trends observed for 5 nm Pt thin films. Post-electrolysis liquid analysis revealed that methanol and methyl acetate were obtained at faradaic efficiencies of 33% ± 2.9 and 7.8% ± 1.0, respectively. For freestanding, well-dispersed platinum nanoparticles (ED-B; for SEM see Fig. 6b), the FE towards Kolbe products (9.2% ± 1.5) is low throughout the electrolysis. The OER is suppressed but the FE towards methanol (55.7% ± 1.8) shows that, despite the coverage, the dispersed nanoparticles are not highly selective for Kolbe electrolysis and instead the decarboxylation reaction predominantly occurs at the BDD substrate. With platinum nano-thorns (ED-C; for SEM see Fig. 6c), an ethane selectivity of up to 43% ± 1.9 was observed, similar to the reported high selectivity towards Kolbe electrolysis when deposited on carbon fibre paper.[14] With extended electrolysis time, the FE decreased to 20% and remained stable. In agreement with the high Kolbe-product selectivity, low FEs towards methanol and methyl acetate (46.4% and 9.4%, respectively) were observed. Finally, the lowest Kolbe-product selectivity was observed for platinum nanocrystals (ED-D; for SEM see Fig. 6d). Clearly, the structure and shape of the Pt particles is of high relevance for Kolbe electrolysis or OER activity, as revealed by the high selectivity towards the OER on Pt nanoflower structures (ED-A). For ED-A electrodes, a mass-normalised ECSA of 12.4 $m_{Pt}^2\ g_{Pt}^{-1}$ and roughness factor of 35.4 were determined. Considering that for ED-C a similar mass-normalized ECSA (13.7 $m_{Pt}^2\ g_{Pt}^{-1}$) at a significantly lower roughness factor (11.6) has been determined, and that for ED-C samples the electrochemical surface area and roughness factor are 5.31 $m^2{}_{Pt}\ g_{Pt}^{-1}$ and 5.6, respectively, it is likely that a high selectivity for Kolbe products is achieved for a fine-balance of the Pt-surface-to-volume ratio and abundance of low-coordinated Pt sites that contribute to an enhanced electrocatalytic activity.[51] To support this hypothesis, BDD modified with platinum nanoflowers was prepared using even higher Pt loadings. Therefore, ED-A electrodeposition was carried out for 30 minutes, which results in a fully covered surface without any exposed BDD facets (Fig. S18a†). The determined ECSA increased to 21.5% at a roughness factor of 60.5, being stable after electrolysis. Here, OER is the predominant reaction (FE 97%, Fig. S19†).

Eliminating the effect of the substrate, the complete inhomogeneous coverage fails to promote Kolbe electrolysis due to exceedingly high roughness factors and porosity, which omits formation of a homogenous acetate barrier layer. It also proves that for promoting the Kolbe reaction, flat surfaces are more favourable. Finally, it is important to note that the loss in activity observed during Kolbe electrolysis is independent of the precise shape of the platinum nanoparticles. This is also in agreement with the ECSA data obtained before and after electrolysis (Fig. 6e) that reveal a loss of 80–90% of nanoparticles during electrolysis. Stabilisation of Pt nanoparticles might be achieved by selective modification of the (111) facets of BDD, as reported in the literature.[52] Facet-selective modification will be evaluated in ongoing research.

To disclose the importance of the substrate used for Pt nanoparticle deposition, electrodeposition of nanoflower-like Pt particles (ED-A) has been repeated on different substrates (graphite, FTO and nickel foam). The particles retain the same shape when deposited on graphite and FTO substrates, but differences in surface coverage were noticeable (Fig. 7a–d). On NiF, the same electrodeposition

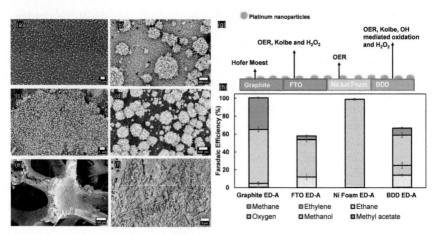

Fig. 7 SEM images of Pt nanoflowers electrodeposited on different substrate electrodes: (a and b) FTO, (c and d) graphite, and (e and f) nickel foam; (g) illustration of predominant reactions on platinum nanoparticles on different substrates; (h) average FE towards (non-)Kolbe electrolysis on platinum nanoflower particles with different substrates at 25 mA cm^{-2} in 1 M acetic acid/sodium acetate (pH 5).

method resulted in the formation of a Pt layer, as confirmed by the EDX analysis shown in Fig. S20†. Kolbe electrolysis performed under the same conditions as used for the Pt/BDD samples revealed that only for the Pt/FTO electrodes was (non-)Kolbe product formation observed (Fig. 7g). As shown above, bare FTO samples are inactive for Kolbe electrolysis, yet after Pt-nanoparticle deposition, ethane was detected at a FE of up to 11% at the expense of the OER. Moreover, traces of H_2O_2 were detected. For Pt-modified graphite electrodes, the product distribution is also slightly influenced by the presence of Pt and the OER was increased to 5% at similar methanol and methyl acetate formation rates. This shows that the exposed surface of the substrate is greatly influencing the selectivity of Kolbe electrolysis. Finally, Pt was not effectively influencing the efficiency and the OER was taking place exclusively. In comparison with bulk platinum electrodes, not only is the tailoring of selectivity towards (non-)Kolbe electrolysis greatly affected by the size, shape or facets of the nanoparticles, but also the substrate is greatly influencing the reaction that is predominantly occurring. Therefore, it is important to find a correlation between the coverage of platinum nanoparticles, their shape, and the substrate to design an electrode that efficiently promotes Kolbe electrolysis.

Conclusions

(Non-)Kolbe/indirect oxidation was performed on platinum and BDD electrodes in 1 M acetic acid/sodium acetate at pH 5. It is shown that BDD can be used as a standalone electrode for alcohol and ester production from carboxylic acids, or as an ideal substrate for surface functionalization. The decarboxylation reaction on bare BDD is OH-radical mediated, and results in methanol, methyl acetate and ethane production as shown by ECMS. The reaction is slightly limited by the concentration of acetic acid and autoinhibition for concentrated electrolytes (above 1 M acetic acid solution). The reaction selectivity can be shifted by tuning the surface of the BDD with a thin layer of platinum or by using electrodeposited platinum nanoparticles. For thicker Pt layers (>20 nm), the reaction is highly selective to Kolbe products and less selective to indirect oxidation products. The use of nanoparticles can alter the selectivity towards (non-)Kolbe electrolysis, but low faradaic efficiencies are still retained, which decrease over time. The complex shapes and structures of platinum nanoparticles are decisive factors for their stability on the substrate and selectivity towards Kolbe electrolysis. Other electrode substrates (FTO, Ni and graphite) are not suitable for the extreme conditions necessary, and generally promote side reactions instead of Kolbe electrolysis.

Author contributions

Talal Ashraf: conceptualization, methodology, experimentation, data analysis and writing. Bastian Timo Mei: conceptualization, methodology, supervision, data analysis, writing and reviewing. Ainoa Paradelo Rodriguez: EC-MS experiments and discussions. Guido Mul: supervision, reviewing and editing.

Conflicts of interest

There are no conflicts to declare.

Acknowledgements

The authors would like to thank and acknowledge support from the European Union's Horizon 2020 Research and Innovation Program, which provided valuable financial support for the EBio project under grant agreement no. 101006612. The authors extend their thanks to Professor Cees Otto and Aufrid Lenferink from the Medical Cell Biophysics group at the University of Twente for conducting the Raman spectroscopy experiments. Additionally, the authors express their gratitude to the EBio consortium for their support and to Tobias Graßl from Condias GmbH Germany for producing the boron-doped-diamond electrodes.

References

1 I. G. O. Črnivec, P. Djinović, B. Erjavec and A. Pintar, *RSC Adv.*, 2015, **5**, 54085–54089.
2 A. Ong, J. Pironon, P. Robert, J. Dubessy, M.-C. Caumon, A. Randi, O. Chailan and J.-P. Girard, *Geofluids*, 2013, **13**, 298–304.
3 R. V. Shende and J. Levec, *Ind. Eng. Chem. Res.*, 1999, **38**, 3830–3837.
4 C. Stang and F. Harnisch, *ChemSusChem*, 2016, **9**, 50–60.
5 F. J. Holzhäuser, J. B. Mensah and R. Palkovits, *Green Chem.*, 2020, **22**, 286–301.
6 E. Klocke, A. Matzeit, M. Gockeln and H. J. Schäfer, *Chem. Ber.*, 1993, **126**, 1623–1630.
7 S. Liu, N. Govindarajan, H. Prats and K. Chan, *Chem Catal.*, 2022, **2**, 1100–1113.
8 A. K. Vijh and B. E. Conway, *Chem. Rev.*, 1967, **67**, 623–664.
9 I. Markó and F. Chellé, in *Encyclopedia of Applied Electrochemistry*, ed. G. Kreysa, K. Ota and R. F. Savinell, Springer, New York, NY, 2014, pp. 1151–1159.

10 M. O. Nordkamp, B. Mei, R. Venderbosch and G. Mul, *ChemCatChem*, 2022, **14**, e202200438.

11 I. Sekine and H. Ohkawa, *Bull. Chem. Soc. Jpn.*, 1979, **52**, 2853–2857.

12 Z. Yu, *Int. J. Electrochem. Sci.*, 2021, 210240.

13 Y. Qiu, J. A. Lopez-Ruiz, G. Zhu, M. H. Engelhard, O. Y. Gutiérrez and J. D. Holladay, *Appl. Catal., B*, 2022, **305**, 121060.

14 G. Yuan, L. Wang, X. Zhang, R. Luque and Q. Wang, *ACS Sustainable Chem. Eng.*, 2019, **7**, 18061–18066.

15 G. Yuan, C. Wu, G. Zeng, X. Niu, G. Shen, L. Wang, X. Zhang, R. Luque and Q. Wang, *ChemCatChem*, 2020, **12**, 642–648.

16 Y. Peng, Y. Ning, X. Ma, Y. Zhu, S. Yang, B. Su, K. Liu and L. Jiang, *Adv. Funct. Mater.*, 2018, **28**, 1800712.

17 S. Xu, X. Niu, G. Yuan, Z. Wang, S. Zhu, X. Li, Y. Han, R. Zhao and Q. Wang, *ACS Sustainable Chem. Eng.*, 2021, **9**, 5288–5297.

18 J. Ranninger, P. Nikolaienko, K. J. J. Mayrhofer and B. B. Berkes, *Chemsuschem*, 2022, **15**, e202102228.

19 T. Zhang, Z. Xue, Y. Xie, G. Huang and G. Peng, *RSC Adv.*, 2022, **12**, 26580–26587.

20 B. Zribi and E. Scorsone, in *Proceedings of Eurosensors 2017*, MDPI, Paris, France, 2017, p. 452.

21 X. Lyu, J. Hu, J. S. Foord, C. Lou and W. Zhang, *J. Mater. Eng. Perform.*, 2015, **24**, 1031–1037.

22 X. Du, Z. Mo, Z. Li, W. Zhang, Y. Luo, J. Nie, Z. Wang and H. Liang, *Environ.Int.*, 2021, **146**, 106291.

23 J. M. Freitas, T. da C. Oliveira, R. A. A. Munoz and E. M. Richter, *Front. Chem.*, 2019, **7**, 190.

24 Y. Tian, X. Chen, C. Shang and G. Chen, *J. Electrochem. Soc.*, 2006, **153**, J80.

25 K. Wenderich, B. A. M. Nieuweweme, G. Mul and B. T. Mei, *ACS Sustainable Chem. Eng.*, 2021, **9**, 7803–7812.

26 M. Panizza, G. Siné, I. Duo, L. Ouattara and Ch. Comninellis, *Electrochem. Solid-State Lett.*, 2003, **6**, D17.

27 J. D. Wadhawan, F. J. Del Campo, R. G. Compton, J. S. Foord, F. Marken, S. D. Bull, S. G. Davies, D. J. Walton and S. Ryley, *J. Electroanal. Chem.*, 2001, **507**, 135–143.

28 A. Arts, M. T. de Groot and J. van der Schaaf, *Chem. Eng. J. Adv.*, 2021, **6**, 100093.

29 A. Kapałka, B. Lanova, H. Baltruschat, G. Fóti and C. Comninellis, *J. Electrochem. Soc.*, 2009, **156**, E149.

30 T. Kashiwada, T. Watanabe, Y. Ootani, Y. Tateyama and Y. Einaga, *ACS Appl. Mater. Interfaces*, 2016, **8**, 28299–28305.

31 N. Kurig, J. Meyers, F. J. Holzhäuser, S. Palkovits and R. Palkovits, *ACS Sustainable Chem. Eng.*, 2021, **9**, 1229–1234.

32 K. Arai, K. Watts and T. Wirth, *ChemistryOpen*, 2014, **3**, 23–28.

33 D. E. Collin, A. A. Folgueiras-Amador, D. Pletcher, M. E. Light, B. Linclau and R. C. D. Brown, *Chem.–Eur. J.*, 2020, **26**, 374–378.

34 G. Lyu, S. Wu and H. Zhang, *Front. Energy Res.*, 2015, **3**, 28.

35 M. Fryda, Th. Matthée, S. Mulcahy, A. Hampel, L. Schäfer and I. Tröster, *Diamond Relat. Mater.*, 2003, **12**, 1950–1956.

36 T. L. Read and J. V. Macpherson, *J. Visualized Exp.*, 2016, 53484.

37 W. Zhang, L. Zhang, G. Zhang, P. Xiao, Y. Huang, M. Qiang and T. Chen, *New J. Chem.*, 2019, **43**, 6063–6068.

38 P. R. Sajanlal, T. S. Sreeprasad, A. K. Samal and T. Pradeep, *Nano Rev.*, 2011, **2**, 5883.

39 A. Kapałka, G. Fóti and C. Comninellis, *J. Electrochem. Soc.*, 2008, **155**, E27.

40 M. Steimecke, G. Seiffarth, C. Schneemann, F. Oehler, S. Förster and M. Bron, *ACS Catal.*, 2020, **10**, 3595–3603.

41 Z. Zhang, Q. Zhang, G. Li, M. Liu and J. Gao, *Chin. J. Chem. Eng.*, 2016, **24**, 1584–1599.

42 N. Wächter, C. Munson, R. Jarošová, I. Berkun, T. Hogan, R. C. Rocha-Filho and G. M. Swain, *ACS Appl. Mater. Interfaces*, 2016, **8**, 28325–28337.

43 N. Dubrovinskaia, L. Dubrovinsky, N. Miyajima, F. Langenhorst, W. A. Crichton and H. F. Braun, *Z. Naturforsch. B*, 2006, **61**, 1561–1565.

44 J. V. Macpherson, *Phys. Chem. Chem. Phys.*, 2015, **17**, 2935–2949.

45 J. Xu, Y. Yokota, R. A. Wong, Y. Kim and Y. Einaga, *J. Am. Chem. Soc.*, 2020, **142**, 2310–2316.

46 A. C. Ferrari and J. Robertson, *Phys. Rev. B: Condens. Matter Mater. Phys.*, 2001, **63**, 121405.

47 Y. Chen, X. Gao, G. Liu, R. Zhu, W. Yang, Z. Li, F. Liu, K. Zhou, Z. Yu, Q. Wei and L. Ma, *Funct. Diam.*, 2021, **1**, 197–204.

48 S. Taylor, E. Fabbri, P. Levecque, T. J. Schmidt and O. Conrad, *Electrocatalysis*, 2016, **7**, 287–296.

49 P. Schmitt, V. Beladiya, N. Felde, P. Paul, F. Otto, T. Fritz, A. Tünnermann and A. V. Szeghalmi, *Coatings*, 2021, **11**, 173.

50 A. Kapałka, H. Baltruschat and C. Comninellis, in *Synthetic Diamond Films*, John Wiley & Sons, Ltd, 2011, pp. 237–260.

51 Y. Yang, M. Luo, W. Zhang, Y. Sun, X. Chen and S. Guo, *Chem*, 2018, **4**, 2054–2083.

52 I. González-González, E. R. Fachini, M. A. Scibioh, D. A. Tryk, M. Tague, H. D. Abruña and C. R. Cabrera, *Langmuir*, 2009, **25**, 10329–10336.

Faraday Discussions

PAPER

Utilisation and valorisation of distillery whisky waste streams *via* biomass electrolysis: electrosynthesis of hydrogen

Robert Price, ⓘ *ab Lewis MacDonald, ⓘ a Norman Gillies,b Alasdair Day,b Edward Brightman ⓘ a and Jun Li ⓘ a

Received 1st May 2023, Accepted 2nd June 2023

DOI: 10.1039/d3fd00086a

Fuel-flexible hydrogen generation methods, such as electrochemical conversion of biomass, offer a route for sustainable production of hydrogen whilst valorising feedstocks that are often overlooked as waste products. This work explores the potential of a novel, two-stage electrolysis process to convert biomass-containing solid (draff/spent barley) and liquid (pot ale and spent lees) whisky co-products, from the Isle of Raasay Distillery, into hydrogen, using a phosphomolybdic acid ($H_3[PMo_{12}O_{40}]$ or PMA) catalyst. Characterisation results for whisky distillery co-products will be presented, including thermogravimetric, differential scanning calorimetric, CHN elemental, total organic carbon and chemical oxygen demand analysis data. The results indicated that the characteristics of these co-products align well with those reported across the Scotch whisky distillation sector. Subsequently, the concept of thermal digestion of each co-product type, using the Keggin-type polyoxometalate PMA catalyst to abstract protons and electrons from biomass, will be outlined. UV-visible spectrophotometry was employed to assess the extent of reduction of the catalyst, after digestion of each co-product, and indicated that draff and pot ale offer the largest scope for hydrogen production, whilst digestion and electrolysis of spent lees is not viable due to the low biomass content of this distillation co-product. Finally, details of electrolysis of the PMA–biomass solutions using a proton-exchange membrane electrolysis cell (PEMEC) will be provided, including electrochemical data that help to elucidate the performance-limiting processes of the PEMEC operating on digested biomass–PMA anolytes.

Introduction

The replacement of fossil fuels with 'green' hydrogen, produced using renewable electricity, provides an alternative route to decarbonization of sectors, such as

aDepartment of Chemical and Process Engineering, University of Strathclyde, James Weir Building, 75 Montrose Street, Glasgow, G1 1XJ, UK. E-mail: robert.price@strath.ac.uk

bIsle of Raasay Distillery, R&B Distillers Ltd., Borodale House, Isle of Raasay, Kyle, Scotland, IV40 8PB, UK

transport and domestic/commercial/process heating.[1] Typically, electrolysis of steam or water can be employed to produce hydrogen for synthesis of feedstock chemicals,[2] sustainable aviation fuel,[3] or to be used as a fuel in electrochemical energy conversion and combustion techniques.[4] Many companies are active in the research, development and supply of electrolysis systems based upon proton-exchange membrane (PEM), alkaline, anion-exchange membrane (AEM)[4] and solid-oxide electrolyte (SOE) electrochemical cells.[2] There is also a growing interest in developing novel, more efficient and fuel-flexible hydrogen generation methods, capable of producing large quantities of fuel gas, using feedstocks that would otherwise go to waste.

Electrochemical conversion of waste biomass offers a fuel-flexible route to sustainable, net-zero greenhouse gas emissions production of hydrogen whilst valorising waste biomass feedstocks. In recent years, the examination of poly-oxometalate (POM) materials as catalysts,[5,6] which can digest biomass, has been undertaken. POM materials comprise large, polyanionic MO_x clusters, in which M = transition metals such as Mo, W and V,[5,7] as well as a variety of cations. Oxygen atoms in the polyanionic clusters are able to act as electron donors (Lewis bases) whilst the unoccupied orbitals of the transition-metal ions are able to act as electron acceptors (Lewis acids).[5] Consequently, these materials are highly redox stable, in addition to being very thermally stable, allowing thermal and electrochemical cycling to take place during the digestion of biomass. In this digestion process, electrons can be transferred from biomass to the unoccupied orbitals of the transition-metal ions to reduce the POM material, with protons also being abstracted to balance the charge. Ultimately, the reduced-POM catalyst can undergo electrolysis either in 'bulk' electrolyte solutions or in electrochemical membrane reactors to produce high-purity hydrogen gas, whilst regenerating the POM catalyst.[8-10]

Electrolysis of biomass–POM solutions gives rise to generation of hydrogen from biomass sources and occurs at significantly lower voltages (<950 mV)[9] than the standard potential of water electrolysis (1230 mV) or 1500–1600 mV in practical commercial PEM electrolysers,[9,11,12] as it avoids the sluggish electrode kinetics of the oxygen evolution reaction, and does not require expensive iridium-based catalysts to function effectively. Consequently, the electricity demand and, thus, the cost per unit volume of hydrogen produced via electrolysis of biomass–POM solutions can be substantially reduced;[10] key factors in the investment and deployment of hydrogen production systems.

Our current research focuses on the demonstration of the potential of a novel two-stage electrolysis process (using PEM electrolysis cells) to convert biomass streams into hydrogen, with a Keggin-type structured POM catalyst known as phosphomolybdic acid ($(H_3[PMo_{12}O_{40}])$ or PMA), whose polyanionic cluster is centred with a phosphorus heteroatom.[5,7] PMA has been successfully used in the depolymerization of lignin,[8,10] degradation of glucose[13] and alcohols,[14] thermal and photo-irradiative digestion of a variety of biomasses[10,15] and subsequent electrolysis in bulk electrolyte solution reactors[9] and PEM flow cells,[10] as well as in electron-coupled proton buffers for energy storage and conversion.[16-18] Therefore, employment of this redox stable and highly water-soluble POM material provides an excellent platform for the evaluation of the potential of whisky co-products as biomass sources for hydrogen production.

Stage one involves thermal digestion of biomass, to reduce the PMA catalyst via removal of protons and electrons from the biomass source. Stage two involves the use of a PEM electrolyzer to reoxidise the reduced PMA catalyst on a carbon felt anode. Upon application of a potential, protons and electrons dissociate from the reduced PMA at the anode. The protons then move through the aqueous PMA solution to the proton-conducting Nafion™ membrane, through which they migrate, whilst electrons flow around an external circuit, both moving towards the cathode. The protons are then reduced by the electrons to form hydrogen at the cathode, with the aid of a platinum catalyst (Pt/C on carbon paper).

In this research, the authors focus on the conversion of whisky co-product streams from the Isle of Raasay Distillery to hydrogen using the previously described two-stage electrolysis process. Three main co-products are generated as a result of the whisky production process, as summarized in Fig. 1. Firstly, draff is a solid co-product of the barley mashing process, the desired product of which is a high sugar-content liquid (known as wort) for fermentation and conversion to alcohol using yeast. Subsequently, this fermented wort (wash) of ~8 volume% (vol%) ethanol undergoes distillation in a wash still to produce low wines (~20 vol% ethanol), in addition to an organic-rich, yeast-laden co-product of pot ale. Finally, a spirit distillation is performed on the low wines (and 'foreshots' and 'feints' from previous spirit distillations) to produce 'new make' spirit (~70 vol% ethanol) for aging (into Scotch whisky), more foreshots and feints (which are distilled along with the following batch of low wines) and a co-product of spent lees.

Initially, characterisation data for whisky co-products will be presented including thermogravimetric, differential scanning calorimetric and carbon–hydrogen–nitrogen analysis of solid (draff/spent barley) and liquid wastes (pot ale and spent lees), in addition to total organic carbon and chemical oxygen demand analyses of liquid samples. Subsequently, the process of thermal digestion of each waste-type, using the Keggin-type polyoxometalate PMA catalyst to abstract protons and electrons from biomass, will be outlined, including assessment of the reduction extent of the PMA catalyst used to digest each biomass source. Finally, details of electrolysis of the PMA–biomass solutions using a PEM flow cell

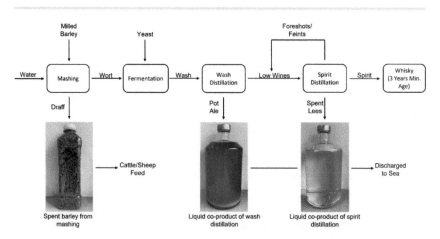

Fig. 1 A basic flow-diagram of the whisky production process and the co-products generated as a result.

will be covered, including electrochemical data (electrochemical impedance spectroscopy, voltage–current measurements and chronoamperometric operation) in order to determine the optimal operating conditions for the process, in addition to the respective amount of hydrogen produced from each biomass source.

Experimental

Characterisation of catalyst material and whisky distillation co-products

Thermogravimetric analysis (TGA), differential scanning calorimetry (DSC) and differential thermal analysis (DTA). Simultaneous thermal analysis (STA), *i.e.* TGA and DSC/DTA, of phosphomolybdic acid ($H_3[PMo_{12}O_{40}] \cdot xH_2O$, Thermo Scientific Chemicals, ACS Reagent) was performed using a Netzsch STA 449 F3 Jupiter® instrument. Data were collected upon heating to 600 °C at a rate of 5.00 °C min^{-1}, using alumina crucibles for both the reference and sample, under a flow of compressed air (50.0 cm^3 min^{-1}) and a flow of N_2 protective gas (60.0 cm^3 min^{-1}). STA of the draff, pot ale and spent lees co-products was performed using a Netzsch STA 449 C Jupiter® instrument. Data were collected upon heating using ramp rates of 1.00 °C min^{-1} to 120 °C and 10.0 °C min^{-1} to 900 °C, under a 20.0 cm^3 min^{-1} flow of compressed air.

Correction runs (using an empty alumina sample crucible) were performed prior to analysis of all samples and were applied to the experimental data, using Netzsch Proteus analysis software.

Carbon–hydrogen–nitrogen (CHN) analysis. Samples of draff, pot ale and spent lees were decanted into evaporating basins and placed into a drying oven, in order to dry the materials prior to CHN analysis. Samples were heated to 120 °C for a period of 12 hours, before cooling to room temperature. Triplicate analysis of samples of dried draff and the residues from the pot ale and spent lees was performed by placing samples into tin capsules and loading into a Flash 2000 Organic Elemental Analyzer (Thermo Scientific) for CHN analysis through dynamic flash combustion at 900 °C. A flow rate of 140 cm^3 min^{-1} of helium carrier gas was used for separation of flash combustion products and subsequent analysis in the thermal conductivity detector (TCD). Prior to the analysis of whisky co-products, an acetanilide standard material was analysed to calibrate the instrument.

Total organic carbon (TOC) analysis and chemical oxygen demand (COD) analyses. Samples of pot ale were diluted by a factor of 10 using de-ionised water (DI H_2O, 15 MΩ cm), whilst the spent lees samples were used undiluted, to ensure that TOC and COD values were in the specified ranges for the test cuvettes employed. Total organic carbon 300–3000 mg L^{-1} C (purging method) LCK387 cuvette test kits (Hach Lange Gmbh) were used to determine the TOC concentration, whilst chemical oxygen demand 1000–10 000 mg L^{-1} O_2 LCK014 cuvette test kits (Hach Lange Gmbh) were employed to determine the COD of pot ale and spent lees co-products, according to procedures specified by the manufacturer.[19,20] Finally, the cuvettes for both TOC and COD were inserted into a Hach DR6000 UV-vis spectrophotometer and analysed to determine the TOC and COD of each liquid co-product. Measurements were performed in triplicate for each of three samples of pot ale and spent lees, for both TOC and COD analysis.

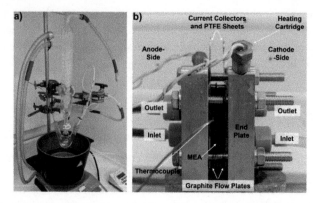

Fig. 2 (a) Image of the setup employed to thermally digest biomass using the PMA catalyst and (b) a labelled image of the PEM flow cell used for electrolysis experiments.

Digestion of biomass samples

Thermal digestion of biomass samples using the PMA catalyst was performed according to the state of the whisky co-product. Samples of solid draff (spent barley) were added to a transparent, yellow-coloured 300 mmol dm^{-3} solution of PMA in deionised water (DI H$_2$O, 15.0 MΩ cm). In contrast, for samples of pot ale and spent lees (both liquid), the PMA catalyst was added directly into 10.0 cm^3 of the whisky co-product and stirred magnetically for 18 hours at 27 °C, until dissolution of the PMA had occurred. Due to the high water of crystallisation content of the PMA catalyst, it is typical to observe a significant increase in the volume of the solution (in this case 5.00 cm^3) as a result of dissolution of the catalyst and release of water. Therefore, only a further 10.0 cm^3 of liquid co-product was added to achieve a 300 mmol dm^{-3} concentration solution of PMA.

25.0 cm^3 of the relevant biomass–PMA solution was decanted into a three-neck, round bottom flask. The flask was attached to a reflux condenser and the side necks were used to introduce a K-type thermocouple (embedded in a sealed bung), to monitor the temperature of the reaction mixture, and a purge of argon from the opposite side. Once sealed, the base of the flask was lowered into a silicone oil heating bath and a purge of argon was initiated to perform the digestion under anaerobic conditions. The oil bath was magnetically stirred at a rate of 800 revolutions per minute (rpm) and was heated to achieve a temperature of 100 °C in the reaction mixture using a hotplate stirrer. The argon purge was halted once the reaction temperature had been reached and the purge line was diverted to a gas syringe in order to collect any gaseous products evolved during the reaction, which proceeded for 4 hours. An image of the thermal digestion apparatus is presented in Fig. 2a.

Electrochemical analysis and UV-visible spectrophotometry of PMA catalyst

Cyclic voltammetry and chronoamperometry. A solution of 300 mmol dm^{-3} PMA catalyst was analysed by cyclic voltammetry (CV), using a carbon felt working electrode, a platinised titanium mesh counter electrode and a Ag/AgCl reference electrode (conditioned in a 100 mmol dm^{-3} KCl solution). A BioLogic SP300 potentiostat was used to sweep the voltage positively at a rate (v) of 10 mV s^{-1} to 1200 mV, before sweeping the voltage negatively to -200 mV at the same rate, whilst recording the current response of the electrolyte.

Chronoamperometry was employed to produce PMA solutions exhibiting varying extents of reduction (with respect to the two-electron process observed at 580 mV). A glass H-cell setup was employed for this electrochemical treatment using a Nafion™ 115 (Fuel Cell Store) proton-exchange membrane to separate the working-electrode compartment (containing 30.0 mmol dm^{-3} PMA in DI H$_2$O) from the counter-electrode compartment (containing 1.00 mol dm^{-3} phosphoric acid (H$_3$PO$_4$)). The volume of each solution employed was 12.0 cm^3. The same working, counter and reference electrodes as described in the aforementioned CV analysis were employed along with the BioLogic SP300 potentiostat. Initially, the PMA solution was reduced by passing 69.5 C of charge at a potential of 250 mV ($vs.$ Ag/AgCl), in order to fully reduce the PMA to provide a point of reference. Then the PMA electrolyte solution was oxidised (in 20% increments) at a potential of 850 mV ($vs.$ Ag/AgCl) to produce a series of solutions that could be used to construct a calibration curve.

UV-visible spectrophotometry and construction of a calibration curve. Solutions produced during chronoamperometry of the PMA electrolyte were diluted from a concentration of 30.0 mmol dm^{-3} to 120 μmol dm^{-3} using DI H$_2$O (in order to prevent saturation of the detector during UV-visible spectrophotometry). PMA solutions of 0%, 20%, 40%, 60%, 80% and 100% reduction (with respect to the two-electron process observed at 580 mV $vs.$ Ag/AgCl) were transferred to a quartz cuvette with a 10 mm path length and inserted into a Varian Cary 5000 UV-vis spectrophotometer. Spectra were collected using a scan range of 300 nm to 800 nm and absorbance at 700 nm for each solution was taken to produce a calibration curve (according to previous work by Liu $et\ al.$).[10] A plot of absorbance $vs.$ reduction extent (of PMA solution) was produced and data were subjected to linear fitting to produce a calibration curve. Thermally digested biomass–PMA samples were then analysed spectrophotometrically under the same conditions and reduction extents were determined using this calibration curve.

Electrochemical analysis of PEM electrolyser operating on H–PMA solutions

PEM setup. A labelled image of the complete test setup employed for electrolysis testing is provided in Fig. 2b. Digested biomass–PMA solutions (anolytes) and sulfuric acid sweep solutions (catholytes) were transferred to separate fluid reservoirs and were continuously circulated to (and returned from) the anode and cathode compartments, respectively, using silicone tubing and Masterflex™ L/S™ 16 Tygon™ tubing via a Masterflex™ L/S™ Digital Miniflex Dual-Channel peristaltic pump. Off-gases evolved in the anode compartment (assumed to be CO$_2$) were collected in a graduated gas syringe connected to the anode fluid reservoir, whilst off-gases evolved in the cathode compartment (assumed to be H$_2$) were diverted to an inverted measuring cylinder filled with DI H$_2$O in a water bath. The membrane electrode assembly (MEA) consisted of a Nafion™ 115 proton-exchange membrane (PEM), with a 1 mm thick carbon felt anode and a carbon paper cathode screen-printed with Pt/C catalysts (0.937 mg cm^{-2} Pt). This was housed between two graphite flow plates and sealed with silicone gaskets, before gold-coated copper current collectors, PTFE insulating sheets and aluminium end plates were stacked either side of these components. The setup was compressed to a torque of 900 mN m using a torque screwdriver and was heated using 230 V heating cartridges inserted into the aluminium end plates, controlled by a K-type thermocouple introduced into the graphite flow plates. Nafion™ 115 PEMs employed for H-cell and flow-cell testing were pre-treated by heating in 500 mmol dm^{-3} $H_2SO_{4(aq)}$ for 1 hour, followed by heating in 3.00% $H_2O_{2(aq)}$ for 1 hour, both at 80 °C. Although a sulfuric acid catholyte was employed for flow-cell testing, as opposed to a phosphoric acid catholyte for H-cell testing, it is not expected to have a significant impact on performance due to its function of sweeping away gases evolved in the cathode compartment (*i.e.* hydrogen).

Electrochemical impedance spectroscopy (EIS), voltage–current (V–I) measurements and chronoamperometric operation. A Metrohm Autolab PGSTAT30, with a BSTR10A booster, or PGSTAT204, with a FRA32M frequency response analyser, was connected to the cell and circulation of the biomass–PMA solution and sulfuric acid solution at a rate of 50.0 cm^3 min^{-1} was initiated. EIS of the PEMECs were collected between 100 kHz and 100 mHz, using an excitation amplitude of 10.0 mV. Firstly, 12.0 cm^3 of each digested biomass–PMA solution was subjected to EIS, at 850 mV, and V–I measurements, between 0 mV and 1200 mV, at 50 °C, 60 °C, 70 °C and 80 °C in the PEMEC. Note that voltage–current plots display voltage on the y-axis and current on the x-axis, for comparability to typical V–I curves for PEM water electrolyzers. Upon conclusion of this electro-chemical evaluation experiment, a second 12.0 cm^3 portion of each solution was run in chronoamperometric mode (850 mV or 900 mV) at 80 °C, with EIS and V–I curves being taken before and after, in order to monitor volumes of anode and cathode off-gas evolved during the experiment.

Results and discussion

Characterisation of phosphomolybdic acid catalyst and whisky distillation co-products

Fig. 3a displays the TGA and DSC data for the PMA catalyst material up to 600 °C in a flow of compressed air. Two distinct mass-loss events are observed in the DSC

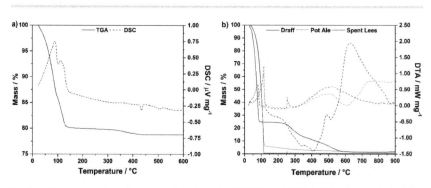

Fig. 3 (a) Simultaneous thermal analysis plots (STA) for $H_3[PMo_{12}O_{40}]\cdot xH_2O$, indicating that the material contains 25.1 moles of water of crystallisation, and (b) STA plots for draff, pot ale and spent lees whisky co-products from the Isle of Raasay Distillery.

trace: (i) between 23 °C and 100 °C and (ii) between 100 °C and 151 °C, representing loss of water from the crystal structure (endothermic peaks). The mass loss observed up to 151 °C (19.9%) corresponds to a molar ratio of 1 : 25.1 PMA : H_2O, *i.e.* 25.1 moles of water of crystallisation are present in the structure of this particular batch of catalyst. Therefore, the formula was determined to be: $H_3[PMo_{12}O_{40}]\cdot 25.1H_2O$. This aligns well with the literature, indicating that PMA may typically contain between 10 and 29 moles of water of crystallisation.[21,22] The additional mass loss (1.38%) between 151 °C and 434 °C represents the loss of hydrogen from the molecular structure of PMA as water of constitution.[22]

Fig. 3b shows plots of TGA and DTA data for draff, pot ale and spent lees collected from the Isle of Raasay Distillery, measured under a flow of compressed air between room temperature and 900 °C. Heating to 130 °C causes evaporation of water and removal of volatile species from all three samples, indicated by the sharp endothermic processes in this region of the DTA trace. According to previous residual alcohol analysis, commissioned by the distillery,[23] the pot ale and spent lees contain 0.079 vol% and 0.068 vol% ethanol, respectively. Therefore, this small alcohol fraction will also evaporate upon heating above 78.35 °C (the boiling point of ethanol).[24] The corresponding mass losses are summarized in Table 1, along with mass losses after 350 °C and 800 °C. After heating (*i.e.* drying) to 130 °C, 75.3% of the mass of draff was lost. The literature suggests that as much as 80 weight% (wt%) of draff comprises residual water from the barley-mashing process at a distillery,[25–27] therefore the sample of draff from the Isle of Raasay Distillery contains a similar mass percentage of water. Combustion of the primarily lignocellulosic matrix of the draff is indicated by the strongly exothermic processes between approximately 130 °C and 550 °C, before any remaining residues undergo thermal decomposition between 550 °C and 900 °C (strongly endothermic peak).

The remaining masses of pot ale and spent lees, at 130 °C, are 5.6 wt% and 0.30 wt%, respectively. It is expected that pot ale contains between 4 and 7 wt% yeast fraction (from the fermentation process),[25] therefore the 5.6 wt% fraction remaining in the distillery's pot ale agrees well with literature indications.[25,28] However, very little residue remains in the spent lees at all. Subsequently, the dead yeast fraction and any remaining organic material in the pot ale and spent lees, respectively, combust between approximately 130 °C and 340 °C (slightly exothermic peaks) followed by thermal decomposition of the residues above 400 °C.

CHN (ultimate) analysis of the residues of the three whisky co-products was also performed to provide insight into the weight fraction of C, H and N, as well as the H/C ratio of these co-product streams. Table 2 provides a summary of the mean CHN weight fractions and H/C ratios for the three co-products. The weight

Table 1　A summary of mass loss data from STA of draff, pot ale and spent lees samples from the distillery

Co-product	Mass loss at 130 °C/%	Mass loss at 350 °C/%	Mass loss at 800 °C/%
Draff	75.3	86.6	98.4
Pot ale	94.4	97.2	99.1
Spent lees	99.7	99.6	99.7

Table 2 A summary of ultimate analysis results for draff, pot ale and spent lees samples from the distillery

Co-product	Draff	Pot ale	Spent lees
Mean C content/wt%	47.4	46.6	32.9
Mean H content/wt%	6.4	6.4	4.4
Mean N content/wt%	3.3	3.8	1.5
Mean O content/wt%	42.9	43.2	61.2
Mean H/C ratio	1.61	1.63	1.58

percentage balance in each material is expected to represent oxygen (O) content; sulfur content was not determined but is expected to be negligible. The draff and pot ale residues give rise to similar weight fractions of C (~47 wt%), H (~6 wt%) and N (3–4 wt%), however the spent lees appears to contain ~18 wt% more O than its counterparts. All three materials give rise to similar H/C ratios of ~1.60. However, it must be noted that although this information is particularly useful in determining the composition of the draff (~25 wt% biomass by weight according to Table 1), the pot ale and, in particular, spent lees both mainly consist of water. Therefore, evaporation of this water (and volatile species) to produce a 'dry' biomass would involve an energy penalty and may not yield sufficient quantities of biomass (in a usable form) to warrant such pre-treatment of co-product streams, prior to biomass digestion.

Subsequently, analysis of the TOC concentration and COD of the liquid co-products, *i.e.* pot ale and spent lees, was performed in order to compare the composition of these co-products to those expected across the industry, as well as to determine the usable fraction of biomass (for digestion and electrolysis) in solution. Fig. 4a and b display bar graphs of mean TOC and COD values, respectively, along with error bars representing standard deviation ($n = 3$).

The TOC concentration of pot ale (~18.8 g L^{-1}) is two orders of magnitude higher than that of spent lees (0.573 g L^{-1}), which is to be expected given the 5.6 wt% total solids fraction (mainly comprising yeast) present in the pot ale. Considering that the spent lees also contains 0.068 vol% ethanol (equivalent to

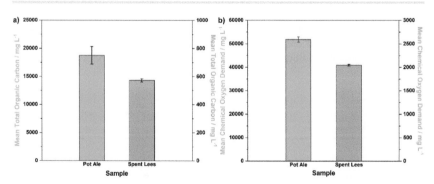

Fig. 4 (a) Mean TOC concentrations and (b) mean COD for pot ale and spent lees collected from the Isle of Raasay Distillery, with error bars representing standard deviation ($n = 3$).

0.536 g L^{-1}), it can be surmised that a large proportion of the TOC concentration of this sample originates from residual alcohol (with the balance most likely comprising volatile fatty and organic acids).[25] The potential for spent lees to act as a biomass source in the digestion and electrolysis process is, therefore, low in comparison to draff or pot ale. The mean COD of pot ale ($\sim 51.9 \text{ g L}^{-1}$) is accordingly higher than that of spent lees (2.05 g L^{-1}), with both values falling at the lower end of the range specified by Bennett et al., for these whisky co-products.[29] Overall, it can be concluded that the compositional parameters of draff, pot ale and spent lees examined in this research are typical of those expected in the wider whisky distillation sector in Scotland.[25,29–31]

Thermal digestion of whisky co-products and assessment of catalyst reduction

Based upon the molecular mass calculated for $H_3[PMo_{12}O_{40}] \cdot 25.1H_2O$ ($\sim 2.28 \text{ kg mol}^{-1}$ or $2277.54 \text{ g mol}^{-1}$), a 300 mmol dm^{-3} solution of PMA in DI H_2O was prepared for electrochemical analysis. Initially, CV was performed to identify the potentials (vs. a Ag/AgCl reference electrode) of the reduction and oxidation processes that the catalyst exhibits. Fig. 5a shows the CV data obtained for this PMA solution. Three reduction–oxidation peak sets are visible over this voltage range, each of which, nominally, represents a two-electron charge-transfer process.[18,32] However, as described by Lewera et al., the first (580 mV vs. Ag/AgCl) and second (290 mV vs. Ag/AgCl) two-electron processes appear to be reversible, whilst the third process (50 mV vs. Ag/AgCl) becomes slightly more complex when the PMA is dissolved in a liquid electrolyte (as opposed to the solid-state response obtained when PMA is deposited onto the surface of a glassy-carbon electrode).[32] The complexity may arise from the facile hydrolysis of the phosphomolybdic acid, in comparison to phosphosilicotungstic acid, for example.[33] It is only the first two-electron process that is expected be utilised in this research to reduce/oxidise the PMA catalyst, either electrochemically or via biomass electrolysis.

In order to reduce PMA by two-electrons (converting the $[PMo_{12}O_{40}]^{3-}$ ion to $[PMo_{12}O_{40}]^{5-}$ in solution), exposure to a potential of 250 mV vs. Ag/AgCl was

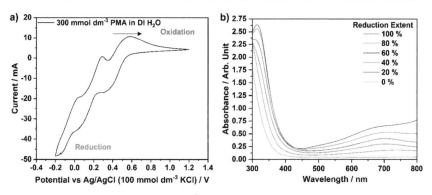

Fig. 5 (a) CV plot of a 300 mmol dm^{-3} solution of $H_3[PMo_{12}O_{40}] \cdot 25.1H_2O$ (in DI H_2O), collected between −200 mV and 1200 mV vs. a Ag/AgCl reference electrode and (b) UV-visible spectra of PMA solution reduced by 0, 20, 40, 60, 80 and 100%, with respect to the first two-electron process, through chronoamperometric experiments.

required, whilst to reoxidise this species, a potential of 850 mV *vs.* Ag/AgCl was employed. Chronoamperometric experiments were, therefore, subsequently performed in order to produce a series of solutions with different extents of reduction. Solutions of 30.0 mmol dm^{-3} PMA were held at a potential of 250 mV *vs.* Ag/AgCl by 69.5 C in order to fully reduce the first redox process by two-electrons. Solutions were then oxidised from this 'reference' point at a potential of 850 mV *vs.* Ag/AgCl until the required amount of charge had been passed to allow solutions with reduction extents of 0, 20, 40, 60 and 80% to be created. UV-visible spectrophotometric analysis of each solution (diluted from 30.0 mmol dm^{-3} to 120 μmol dm^{-3} to prevent saturation of the detector) was performed, yielding the graph in Fig. 5b. As the extent of reduction increases from 0% towards 100%, the absorbance peak at 700 nm becomes more intense. Therefore, the absorbance at this wavelength was chosen to create a calibration curve (absorbance at 700 nm *vs.* percentage reduction), in line with previous work by Liu *et al.*[10]

In order to assess the ability of the three whisky co-products to reduce the PMA catalyst in solution, each was digested at 100 °C for 4 hours under an argon atmosphere to ensure anaerobic conditions were maintained. In each case, no off-gas was evolved or collected in the gas syringe during the digestion process. After the digestion process was complete and the solutions were allowed to cool, a Buchner filtration was performed to remove any solid residues of biomass that may interfere with UV-visible spectrophotometry. Samples were then diluted with DI H$_2$O to a concentration of 120 μmol dm^{-3} (as used for the electrochemically reduced PMA solutions) before undergoing spectrophotometric analysis. The calibration curve (with accompanying line of best fit and equation) produced as a result of the aforementioned chronoamperometric and spectrophotometric analyses is shown in Fig. 6a. Based upon the absorbance value of each of the thermally digested biomass solutions at 700 nm, the equation of the line of best fit was employed to determine the reduction extent of each sample. Data points for each sample are overlain on Fig. 6a and the corresponding percentage reductions are summarised in Table 3.

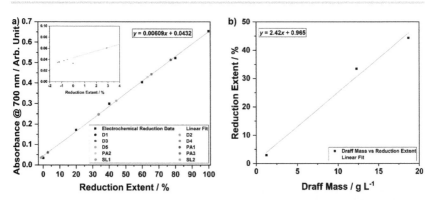

Fig. 6 (a) A calibration curve (absorbance at 700 nm *vs.* percentage reduction of PMA) with overlain data points for thermally digested whisky co-products, inset: magnified region showing the negligible extent of reduction of digested draff and pot ale samples, and (b) a plot of percentage reduction of PMA *versus* draff concentration (dry basis), used to determine that 40.9 g L^{-1} of 'dried' draff should be required to reduce PMA by two electrons.

Table 3 A summary of absorbance at 700 nm and corresponding percentage reduction of PMA for the thermally digested draff, pot ale and spent lees samples

Co-product	Biomass or solids loading/g L^{-1}	Absorbance at 700 nm/arbitrary units	Reduction extent/%
Draff (D1)	0.123	0.0356	−1.25
Draff (D2)	0.123	0.0381	−0.837
Draff (D3)	1.23	0.0607	2.87
Draff (D4)	12.3	0.248	33.6
Draff (D5)	18.7	0.314	44.5
Pot ale (PA1)	44.8	0.514	77.3
Pot ale (PA2)	44.8	0.429	63.4
Pot ale (PA3)	44.8	0.443	65.6
Spent lees (SL1)	2.40	0.0347	−1.40
Spent lees (SL2)	2.40	0.0399	−0.542

The data in Table 3 and Fig. 6a show that digestion of samples of spent lees, which have a dry basis solids loading of 2.40 g L^{-1} and a TOC content of 0.456 g L^{-1} (after dilution by a factor of 1.25 due to release of water of crystallisation from PMA dissolution), and samples of draff, containing 0.123 g L^{-1} of (dry) biomass, results in negligible reduction of the PMA catalyst. This is indicated by the negative values for reduction extent, which are likely to be within experimental error for this calibration curve, due to the small amount of charge transferred from the biomass to the catalyst. As the PMA catalyst is directly dissolved into the spent lees liquid (which has been shown to comprise mostly water and volatiles, including ~0.429 g L^{-1} ethanol at this dilution factor), the biomass loading is fixed. Therefore, the potential of spent lees to act as a hydrogen and electron source is inherently low. In contrast, as solid draff can be added to PMA solutions (in DI H$_2$O) in any quantity desired, a greater extent of reduction could be easily achieved by increasing the mass of concentration of draff. This is demonstrated by samples draff #3 (1.23 g L^{-1}), 4 (12.3 g L^{-1}) and 5 (18.7 g L^{-1} biomass), which give rise to percentage reductions of 2.87%, 33.6% and 44.5%, respectively. It should be noted that the biomass loadings stated for draff have been normalised based upon the water content of the samples determined from TGA (i.e. biomass concentrations recorded in Table 3 are 24.7 wt% of the total mass of draff added). Digestion of the pot ale, with a dry basis solids loading of 44.8 g L^{-1} and 15.0 g L^{-1} TOC (after dilution by a factor of 1.25), leads to reduction extents between 63.4% and 77.3%.

Given that the concentration of draff was the most readily altered of the three co-product types, a series of draff–PMA solutions of varying concentration were prepared and digested in order to estimate the mass loading of draff that would be required to fully reduce a 300 mmol dm^{-3} solution of PMA in DI H$_2$O. Fig. 6b shows a plot of percentage reduction (with respect to the first two-electron process presented in Fig. 5a) versus draff concentration, along with linear fitting data. Based upon this data, it is estimated that 40.9 g L^{-1} of 'dry' draff (or ~166 g L^{-1} 'as received' draff) would be required to fully reduce PMA by two electrons. Given that the thermal digestion of pot ale (with a 44.8 g L^{-1} solids loading) yielded between 63.4% and 77.3% reduction of PMA, this implies that draff has a higher potential for hydrogen production, based upon the 40.9 g L^{-1} of dry basis draff that is predicted to give rise to 100% PMA reduction. Experiments are currently underway to confirm this prediction.

Electrolysis of reduced-PMA solutions using PEMECs

Electrochemical evaluation of PEMEC performance. 12.0 cm^3 portions of each digested solution were subjected to electrochemical performance evaluation using the PEMEC flow-cell setup. V–I curves and EIS were collected at 50 °C, 60 °C, 70 °C and 80 °C for both the digested pot ale and draff-containing solutions. In comparison, due to the low extent of reduction of the digested spent lees-containing solution, measurements were only performed at 80 °C due to the rapid reoxidation of the PMA catalyst.

Fig. 7 displays the V–I curves for the PEMECs operating using a draff (D4)–PMA anolyte (a) and pot ale (PA2)–PMA or spent lees (SL2)–PMA anolytes (b). Between 0 mV and 600 mV, a region of activation polarisation is observed before ohmic losses dominate the V–I curve and current increases with voltage in a linear manner above ~600 mV. The operating voltage was limited to 1200 mV in order to minimise performance contributions from water electrolysis, which can theoretically occur at the thermodynamic potential of 1230 mV.[34] However, the stability of the current response begins to reduce above ~800 mV, possibly arising due to issues with the replenishment of anolyte and catholyte in the respective compartments as a result of the action of the peristaltic pump employed. Furthermore, due to the low reduction extent of the spent lees-PMA anolyte, poor performance of the PEMEC is observed, therefore exacerbating the aforementioned mass-transport issues and showing instability of the current density throughout the measurement. In general, the maximum current density achieved in each test increases as a function of operating temperature, due to the accompanying increase in the conductivity of the PEM and thermal activation of charge-transfer processes.

Fig. 8 displays temperature-sweep complex-plane EIS for PEMECs operating on digested draff (D4)–PMA solution (a) and digested pot ale (PA2)–PMA/spent lees (SL2)–PMA solutions (b), whilst Fig. 9 displays Bode-format EIS for the same anolytes. It should be highlighted that although EIS were collected between 100 kHz and 100 mHz, data points below 12 Hz were excluded from plots due to the

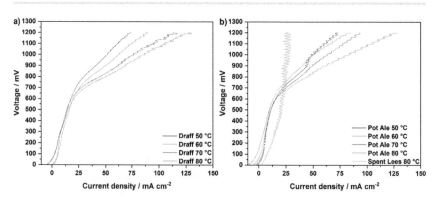

Fig. 7 Temperature-sweep V–I curves for PEMECs operating with (a) a digested draff–PMA anolyte solution and (b) digested pot ale–PMA and spent lees–PMA anolyte solutions, collected at 850 mV using H_2SO_4 as a catholyte.

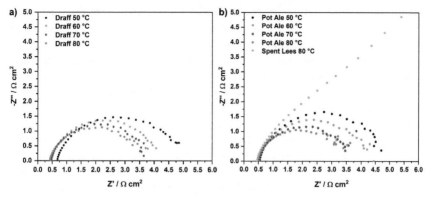

Fig. 8 Temperature-sweep complex-plane EIS for PEMECs operating with (a) a digested draff (D4)–PMA anolyte solution and (b) digested pot ale (PA2)–PMA and spent lees (SL2)–PMA anolyte solutions, collected at 850 mV using H_2SO_4 as a catholyte.

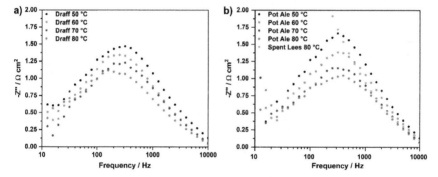

Fig. 9 Temperature-sweep Bode-format EIS for PEMECs operating with (a) a digested draff (D4)–PMA anolyte solution and (b) digested pot ale (PA2)–PMA and spent lees (SL2)–PMA anolyte solutions, collected at 850 mV using H_2SO_4 as a catholyte.

excessive noise in this frequency region. Considering the complex-plane EIS, series resistances (R_s) of the PEMECs decrease with increasing temperature, as expected, due to the thermal activation of proton conductivity in Nafion™ membranes.[35] Slade *et al.* examined the membrane area resistance of a variety of Nafion™ membranes in 1 mol dm^{-3} H_2SO_4 at 25 °C, indicating that a R_s value of 110–130 mΩ cm^2 should be expected.[36] The R_s values observed here (420–670 mΩ cm^2) are significantly higher than these nominal values and this is most likely due to poor contact at the electrode/electrolyte interface. Hot-pressing or calendering is commonly used to produce well-adhered and highly interconnected electrode/ electrolyte interfaces in MEAs, however, it was not possible to carry out during the current research campaign, particularly when using thick (1 mm) carbon felt anodes. Therefore, employment of commercial, pre-treated MEAs may yield higher performance in future experiments.

As a consequence of the elevated R_s recorded, the polarisation resistances (R_p) and, therefore, area-specific resistances (ASR) of each PEMEC are also likely to be proportionally larger than expected. The shape of the spectra for the PEMECs

operating on draff and pot ale-based solutions are similar, with the total polarisation resistance also showing thermal activation (*i.e.* a decrease in resistance as
a function of increasing operating temperature).

Equivalent circuit fitting (ECF) of the spectra included in Fig. 8 was performed,
using circuit models presented in Fig. 10a, and the polarisation resistance values
obtained were used to construct an Arrhenius-style plot in Fig. 10b. All resistance
values and activation energies (E_A) are summarised in Table 4, along with information on the PEMEC used for each experiment. The spectra recorded as
a function of temperature for PEMEC1 operating on draff- and pot-ale-based
anolytes display three processes: (i) a high-frequency process, R_{p1}, with
a frequency maximum (f_{max}) = 2000–1200 Hz; (ii) a mid-frequency process, R_{p2},
(f_{max} = 400–300 Hz); and (iii) a low-frequency process, R_{p3}, (f_{max} = 90–40 Hz). The
R_{p1} process for PEMEC1, operating on digested draff–PMA and pot ale–PMA
anolytes, show temperature independence with negligible E_A values. Siracusano
et al. observed a similar type of low-resistance, temperature-independent process
during water electrolysis experiments. Although the experimental conditions were
significantly different in comparison to PMA electrolysis (potential = 1500 mV,
using PEMECs comprising Aquivion® membranes, IrRuOx anode catalysts and

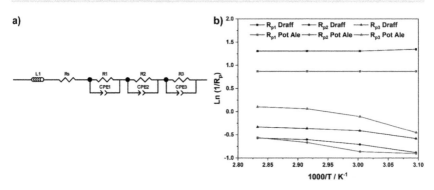

Fig. 10 (a) Equivalent circuit model employed to determine fitting parameters for the EIS
presented in Fig. 8 and (b) Arrhenius-style plots of ln(1/R_p) *versus* 1000/T for resistance
data obtained from ECF of EIS.

Table 4 A summary of ECF parameters determined from analysis of EIS for PEMECs 1 and
2 operating on digested draff–PMA, pot ale–PMA and spent lees–PMA anolytes

Parameter	Anolyte								
	Draff D4 (12.3 g L^{-1})				Pot Ale PA2 (44.8 g L^{-1})				Spent Lees SL2 (2.40 g L^{-1})
Temperature	50 °C	60 °C	70 °C	80 °C	50 °C	60 °C	70 °C	80 °C	80 °C
Inductance (L)/μH	1.34	1.43	1.37	1.51	1.53	1.65	1.94	1.52	0.825
R_s/Ω cm^2	0.67	0.47	0.47	0.42	0.54	0.51	0.5	0.48	0.46
R_{p1}/Ω cm^2	0.26	0.27	0.27	0.27	0.42	0.42	0.42	0.42	0.85
R_{p2}/Ω cm^2	2.42	2.03	1.83	1.77	2.47	2.37	1.95	1.75	4.08
R_{p3}/Ω cm^2	1.79	1.51	1.44	1.39	1.57	1.11	0.94	0.90	170.7
ASR/Ω cm^2	5.14	4.28	4.01	3.85	5.00	4.41	3.81	3.55	176.1
PEM cell ID	PEMEC1				PEMEC1				PEMEC2

Pt/C cathode catalysts),[37] similar charge-transfer processes at electrode–electrolyte interfaces will occur in PEMECs, therefore, this process may relate to charge transfer.[37,38] As PEMECs containing Pt/C-based cathodes were utilised in this research, as well as the aforementioned reports, it is possible that this common characteristic arises from charge transfer in the cathode during hydrogen evolution. Due to the elevated ohmic resistance, and proportionally higher polarisation resistance values, recorded whilst operating at 50 °C with the digested draff–PMA anolyte, this data point was omitted during linear fitting of the Arrhenius-style plots. The R_{p2} and R_{p3} arcs both exhibit thermal activation giving rise to average E_A values of 12.6 kJ mol^{-1} and 10.8 kJ mol^{-1}, respectively. Given the f_{max} of the mid-frequency process, R_{p2}, it may tentatively be assigned to charge transfer at the carbon felt anode, whilst the low-frequency, R_{p3}, process most likely relates to reactant diffusion limitations. This is exemplified by the fact that the analogous process in the EIS collected for PEMEC2, operating on the digested spent lees–PMA anolyte, exhibits a resistance of 170.7 Ω cm^2, due to the low extent of reduction of the spent lees-based anolyte and, consequently, the low concentration of $[PMo_{12}O_{40}]^{5-}$ in solution in comparison to the other anolytes.

Hydrogen production from reduced PMA solutions under chronoamperometric operation. Separate 12.0 cm^3 portions of each anolyte solution characterised using EIS and V–I curves were subjected to gas evolution experiments at 80 °C, with EIS and V–I curves being taken before and after chronoamperometric operation. Fig. 11a displays the changes in current density of PEMECs 2, 3 and 4 operating on PMA solutions reduced by thermal digestion of (i) spent lees (SL2), (ii) 12.3 g L^{-1} draff (D4), (iii) 18.7 g L^{-1} draff (D5), (iv) pot ale (PA2) and (v) pot ale (PA3), whilst Fig. 11b shows flow-rate dependent EIS of PEMEC3 operating on a digested draff–PMA solution (D5, 18.7 g L^{-1} concentration), indicating that spectrum quality and coherence in the low-frequency domain may be improved by lowering the flow rate to 25.0 cm^3 min^{-1}. A summary of anolytes that were reoxidised using the PEM flow cell, including the anode and cathode off-gas volumes, actual and theoretical charges passed (Q) and associated test durations, is provided in Table 5.

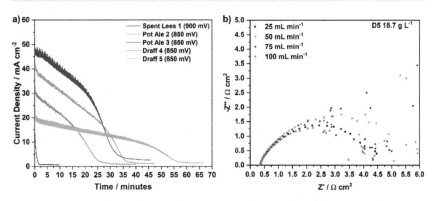

Fig. 11 (a) Chronoamperometric data for the PEMECs operating on digested draff–PMA, pot ale–PMA (850 mV) and spent lees–PMA (900 mV) anolyte solutions during gas evolution experiments and (b) EIS collected for PEMEC3, operating on a draff (D5)–PMA anolyte, illustrating the effect of peristaltic pump flow rate on the spectrum quality.

Table 5 A summary of data obtained from electrolysis of digested co-product–PMA anolytes using PEM flow cells

Parameter	Anolyte				
	Spent lees SL2	Draff D4	Draff D5	Pot ale PA2	Pot ale PA3
Biomass or solids loading/g L^{-1}	2.40	12.3	18.7	44.8	44.8
Reduction extent/%	−0.542	33.6	44.5	63.4	65.6
Equivalent charge/C	—	233	308	440	455
Expected H_2 evolved/cm^3	—	30	39	56	58
Actual charge/C	3.52	112	170	207	256
Cathode off-gas (H_2) evolved/cm^3	1	12	14	23	26
Anode off-gas (CO_2) evolved/cm^3	3	8	4	10	—
PEM ID	PEMEC2	PEMEC2	PEMEC3	PEMEC2	PEMEC4

Firstly, the spent lees–PMA anolyte showed negligible reduction according to the calibration curve in Fig. 6a, therefore, chronoamperometry at 900 mV (50 mV higher than for the draff and pot ale samples) resulted in a rapid degradation of the current response to a steady-state level after less than 2 minutes. This confirmed that the spent lees co-product offers very little potential for hydrogen production (generating 1 cm^3 of cathode off-gas) and that the slightly higher operating voltage made little difference to the discharge, given the low extent of reduction. In contrast, both the draff and pot ale-based anolyte samples that were reoxidised in the PEM flow cell had much higher reduction extents and evolved more cathode off-gas (assumed to be hydrogen) during chronoamperometry at 850 mV. In particular, it is possible to see more clearly the 'discharge' profile of the PEMECs operating on these anolytes in Fig. 10a. Initially a steady-state degradation of the current density is observed, relating to the depletion of $[PMo_{12}O_{40}]^{5-}$ and enrichment of $[PMo_{12}O_{40}]^{3-}$ in solution, accompanied by hydrogen production in the cathode compartment and, possibly, CO_2 in the anode compartment. As the concentration of $[PMo_{12}O_{40}]^{5-}$ becomes critically low, a steep decline in current density is observed before the current stabilises at 2–3 mA cm^{-2}, signalling the completion of the reoxidation experiment.

Interestingly, the reaction details summarized in Table 5 indicate that the theoretical amount of charge required to be passed to reach the observed reduction extents of PMA represent approximately double the charge that was actually passed during the experiment. Taking draff (D4) and pot ale (PA2) as examples, 33.6% (233 C) and 63.4% (440 C) reduction of PMA, respectively, was achieved, whereas only 112 C and 207 C were passed during electrolysis of the anolytes, respectively. For the draff (D4) anolyte, 12 cm^3 of hydrogen was captured and recorded, whilst 23 cm^3 was recorded for the pot ale (PA2) anolyte, however, based upon the reduction extents of PMA achieved, production of 30 cm^3 and 56 cm^3 from the digested draff–PMA and digested pot ale–PMA anolyte solutions should have been achieved.

Experiments are currently underway to identify the origin of this mismatch in theoretical *vs.* observed hydrogen yields and charge passed. Notably, after reoxidation of the biomass–PMA anolytes a distinct brown colour was observed within the predominantly transparent yellow coloured solution. This may imply that corrosion and removal of the carbon felt anode material and/or the graphite flow plate could be occurring, which might explain the reduced charge passed during

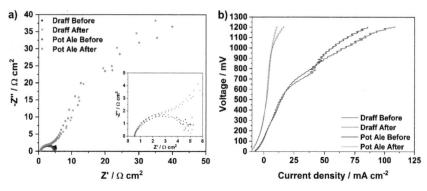

Fig. 12 Complex-plane EIS for PEMEC2 operating with digested draff (D4)–PMA and pot ale (PA2)–PMA anolyte solutions, collected at 850 mV before and after chronoamperometry experiments, and (b) V–I curves for PEMEC2 operating with the same anolyte solutions, before and after chronoamperometry experiments at 80 °C using H_2SO_4 as a catholyte.

electrochemical production of hydrogen (if a chemical side reaction was occurring in the anode compartment) and may also explain the elevated levels of anode off-gas (possibly CO_2) if oxidation of the carbon components of the flow cell was occurring. It is acknowledged that further experimentation and gas analysis is required to verify these theories.

Fig. 12a displays impedance spectra collected before and after chronoamperometric measurements were performed on PEMEC2, operating on the digested draff (D4)–PMA and pot ale (PA2)–PMA anolytes at 80 °C. For both anolytes, the initial spectra exhibit similar ASRs to each other, though these are higher than ASRs observed during EIS of equivalent anolytes employed for PEMEC1 (possibly relating to slight differences in compression and therefore contacting in each cell). However, after the chronoamperometric measurements, a high-resistance arc evolves in the low-frequency region for both samples, most likely relating to mass-transport limitations arising from depletion of the $[PMo_{12}O_{40}]^{5-}$ in solution, which is to be expected and serves as confirmation that reoxidation of the catalyst has occurred. This is further reinforced by the characteristics of the V–I curves (Fig. 12b) collected before and after chronoamperometry for PEMEC2 operating on these anolyte solutions. Despite the noise observed in the higher voltage region of the V–I curves collected before chronoamperometry (due to the pulsating flow generated by the peristaltic pump), collection of EIS at 850 mV falls into the pseudo-linear region of these curves, which are dominated by ohmic losses within the PEMEC. However, after reoxidation of the anolytes, an EIS collected at the same voltage would result in perturbation of the cell closer to the concentration polarisation (mass-transport limited) region of the V–I curve, explaining the observation of a high-resistance arc in the low-frequency region of the post-chronoamperometry EIS.

Conclusions

Whisky co-products (namely draff, pot ale and spent lees) collected from the Isle of Raasay Distillery were characterised by simultaneous thermal analysis (STA) and carbon–hydrogen–nitrogen (CHN) analysis, as well as being analysed for total organic carbon concentration (TOC) and chemical oxygen demand (COD). These analyses indicated that the characteristics of the three co-products agreed well with those reported for the Scotch whisky distillation sector. Importantly, solids content (0.30 wt%) and total organic carbon concentration (0.573 g L^{-1}) of the spent lees liquid was determined to be too low to offer significant potential as a biomass source for thermal digestion and electrolysis with a $H_3[PMo_{12}O_{40}]\cdot 25.1H_2O$ (PMA) catalyst. In comparison, the pot ale liquid yielded a 5.60 wt% solids loading, containing 18.8 g L^{-1} of total organic carbon, primarily due to the presence of a dead yeast fraction from fermentation, whilst the draff (containing 75.3 wt% water), offered a lignocellulose-rich biomass source whose concentration (in the acid catalyst solution) could be more easily altered than those of the liquid wastes.

Thermal digestion of each co-product using a 300 mmol dm^{-3} solution of PMA (either dissolved directly into the liquid co-products or into deionised water for addition of solid draff) was performed under an inert atmosphere at 100 °C for 4 hours. Subsequent determination of reduction extent of the PMA catalyst, using UV-visible spectroscopy, found negligible reduction with respect to the first two-electron process identified during cyclic voltammetry of PMA, when employed to digest the spent lees co-product. Digestion of pot ale, containing 44.8 g L^{-1} of dry matter or 15.0 g L^{-1} TOC (assuming pot ale to have a density of 1.00 kg L^{-1} and taking account of the dilution factor of 1.25 employed during digestion) resulted in reduction extents between 63.4% and 77.3%, whilst a draff loading (dry basis) of 18.7 g L^{-1} resulted in a reduction extent of 44.5%. Based upon a plot of reduction extent *versus* draff loading (dry basis), it is anticipated that a draff concentration of 40.9 g L^{-1} should give rise to 100% reduction of the PMA catalyst. This suggests that there is a higher potential for hydrogen production from draff, than pot ale, through the thermal digestion and electrolysis method. Experiments are currently underway to verify this estimation.

Finally, electrolysis experiments using a proton-exchange membrane (PEM) flow-cell setup were performed on each digested biomass–PMA anolyte to determine the yields of off-gas from the anode (carbon dioxide) and cathode (hydrogen). This confirmed that the volume of gas collected from the cathode compartment increased as a function of the extent of reduction of the PMA in each anolyte, in this case: pot ale > draff > spent lees. However, by increasing the draff concentration to 40.9 g L^{-1} (to achieve the theoretical 100% PMA reduction), it is expected that the hydrogen production potential would be greatest for the draff co-product stream. Electrochemical analysis of PEM electrolysis cells (PEMECs) operating on the digested biomass–PMA anolytes indicated that three rate-limiting polarisation processes could be identified within the electrochemical impedance spectra collected: (i) a high-frequency process ($\sim$2000 Hz) possibly relating to charge transfer during hydrogen evolution at the cathode, (ii) a mid-frequency process ($\sim$400 Hz) tentatively assigned to charge transfer at the anode and (iii) a low-frequency process (90–40 Hz) most-likely relating to mass transport and reactant diffusion issues in the anolyte. This information provides valuable indications as to where electrolysis performance can be enhanced to improve hydrogen yields and efficiency, particularly given the fact that only 47–56% of the theoretical charge was passed during the electrolysis experiments.

Conflicts of interest

There are no conflicts of interest to declare.

Acknowledgements

The authors wish to thank Dr Gavin S. Peters (University of St Andrews) for performing TGA and DSC measurements of the whisky co-products, Orfhlaith McCullough (London Metropolitan University) for carrying out CHN analysis, Dr Mara Knapp and Marcella McIlroy (University of Strathclyde) for assistance in the preparation and analysis of TOC and COD samples, as well as Liam Kirkwood and Cameron Gemmell for modification and commissioning of flow-cell test setups. In addition, many thanks go to Dr Ian Heywood and the KTP West of Scotland Centre for support. Knowledge Transfer Partnerships (KTPs) aim to help businesses improve their competitiveness and productivity through better use of knowledge, technology and skills within the UK knowledge base. This KTP was co-funded by UKRI through Innovate UK and R&B Distillers Ltd.

References

1 V. Masson-Delmotte, P. Zhai, H.-O. Pörtner, D. Robert, J. Skea, P. R. Shukla, A. Pirani, W. Moufouma-Okia, C. Péan, R. Pidcock, S. Connors, J. B. R. Matthews, Y. Chen, X. Zhou, M. I. Gomis, E. Lonnoy, T. Maycock, M. Tignor and T. Waterfield, Global warming of 1.5 °C. An IPCC Special Report on the impacts of global warming of 1.5 °C above pre-industrial levels and related global greenhouse gas emission pathways, in *The Context of Strengthening the Global Response to the Threat of Climate Change, Sustainable Development, and Efforts to Eradicate Poverty*, IPCC, 2018.

2 J. B. Hansen, *Faraday Discuss.*, 2015, **182**, 9–48.

3 Department for Transport, *Sustainable Aviation Fuels Mandate*, 2021.

4 M. Götz, J. Lefebvre, F. Mörs, A. McDaniel Koch, F. Graf, S. Bajohr, R. Reimert and T. Kolb, *Renewable Energy*, 2016, **85**, 1371–1390.

5 S. S. Wang and G. Y. Yang, *Chem. Rev.*, 2015, **115**, 4893–4962.

6 N. L. Z. Z. Adil, T. S. T. Saharuddin, L. N. Ozair and F. W. Harun, *IOP Conf. Ser.: Mater. Sci. Eng.*, 2021, **1173**, 012073.

7 D. L. Long, R. Tsunashima and L. Cronin, *Angew. Chem., Int. Ed.*, 2010, **49**, 1736–1758.

8 X. Du, H. Zhang, K. P. Sullivan, P. Gogoi and Y. Deng, *ChemSusChem*, 2020, **13**, 4318–4343.

9 H. Oh, Y. Choi, C. Shin, T. V. T. Nguyen, Y. Han, H. Kim, Y. H. Kim, J.-W. Lee, J.-W. Jang and J. Ryu, *ACS Catal.*, 2020, **10**, 2060–2068.

10 W. Liu, Y. Cui, X. Du, Z. Zhang, Z. Chao and Y. Deng, *Energy Environ. Sci.*, 2016, **9**, 467–472.

11 E. Fabbri, A. Habereder, K. Waltar, R. Kötz and T. J. Schmidt, *Catal. Sci. Technol.*, 2014, **4**, 3800–3821.

12 S. Song, H. Zhang, X. Ma, Z. Shao, R. T. Baker and B. Yi, *Int. J. Hydrogen Energy*, 2008, **33**, 4955–4961.

13 Y. Li, W. Liu, Z. Zhang, X. Du, L. Yu and Y. Deng, *Commun. Chem.*, 2019, **2**, 67.

14 F. Sheng, Q. Yang, D. Cui, C. Liu, Y. Sun, X. Wang and W. Su, *Energy Fuels*, 2020, **34**, 10282–10289.

15 M. Li, T. Wang, M. Zhao and Y. Wang, *Int. J. Hydrogen Energy*, 2022, **47**, 15357–15369.

16 L. G. Bloor, R. Solarska, K. Bienkowski, P. J. Kulesza, J. Augustynski, M. D. Symes and L. Cronin, *J. Am. Chem. Soc.*, 2016, **138**, 6707–6710.

17 L. MacDonald, B. Rausch, M. D. Symes and L. Cronin, *Chem. Commun.*, 2018, **54**, 1093–1096.

18 M. D. Symes and L. Cronin, *Nat. Chem.*, 2013, **5**, 403–409.

19 Hach Lange Gmbh, *Working Procedure: LCK387: Total Organic Carbon*, 2017.

20 Hach Lange Gmbh, *Working Procedure: LCK014 COD*, 2019.

21 P. Singh, K. Kumari and R. Patel, *J. Pharm. Appl. Chem.*, 2017, **3**, 53–56.

22 A. Micek-Ilnicka, *J. Mol. Catal. A: Chem.*, 2009, **308**, 1–14.

23 A. Day, Personal communication.

24 R. L. Brown and S. E. Stein, in *NIST Chemistry WebBook, NIST Standard Reference Database Number 6*, ed. P. J. Linstrom and W. G. Mallard, National Institute of Standards and Technology, Gaithersburg MD, 2023.

25 J. C. Akunna and G. M. Walker, in *The Alcohol Textbook*, ed. G. M. Walker, C. Abbas, W. M. Ingledew and C. Pilgrim, Ethanol Technology Institute, Duluth, Georgia, 6th edn, 2017, pp. 529–537, ch. 34.

26 J. Bennett, PhD thesis, University of Abertay Dundee, 2013.

27 S. I. Mussatto, G. Dragone and I. C. Roberto, *J. Cereal Sci.*, 2006, **43**, 1–14.

28 J. S. White, K. L. Stewart, D. L. Maskell, A. Diallo, J. E. Traub-Modinger and N. A. Willoughby, *ACS Omega*, 2020, **5**, 6429–6440.

29 J. Bennett, G. M. Walker, D. Murray, J. C. Akunna and A. Wardlaw, in *Distilled Spirits – Future Challenges, New Solutions: Proceedings of the Worldwide Distilled Spirits Conference*, ed. I. Goodall, R. Fotheringham, D. Murray, R. A. Speers and G. M. Walker, Context, Packington, 2015, pp. 303–312.

30 C. Edwards, C. C. McNerney, L. A. Lawton, J. Palmer, K. Macgregor, F. Jack, P. Cockburn, A. Plummer, A. Lovegrove and A. Wood, *Resour., Conserv. Recycl.*, 2022, **179**, 106114.

31 J. Graham, B. Peter, G. M. Walker, A. Wardlaw and E. Campbell, in *Distilled Spirits – Science and Sustainability: Proceedings of the Worldwide Distilled Spirits Conference*, ed. G. M. Walker, I. Goodall, R. Fotheringham and D. Murray, Nottingham University Press/The Institute of Brewing and Distilling, Nottingham, 2012, pp. 1–7.

32 A. Lewera, M. Chojak, K. Miecznikowski and P. J. Kulesza, *Electroanalysis*, 2005, **17**, 1471–1476.

33 M. Sadakane and E. Steckhan, *Chem. Rev.*, 1998, **98**, 219–237.

34 S. Wang, A. Lu and C. J. Zhong, *Nano Convergence*, 2021, **8**, 4.

35 K. D. Kreuer, M. Schuster, B. Obliers, O. Diat, U. Traub, A. Fuchs, U. Klock, S. J. Paddison and J. Maier, *J. Power Sources*, 2008, **178**, 499–509.

36 S. Slade, S. A. Campbell, T. R. Ralph and F. C. Walsh, *J. Electrochem. Soc.*, 2002, **149**, A1556–A1564.

37 S. Siracusano, S. Trocino, N. Briguglio, V. Baglio and A. S. Arico, *Materials*, 2018, **11**, 1368.

38 J. C. Garcia-Navarro, M. Schulze and K. A. Friedrich, *J. Power Sources*, 2019, **431**, 189–204.

Faraday Discussions

ROYAL SOCIETY
OF CHEMISTRY

DISCUSSIONS

Electrofuels: general discussion

Talal Ashraf, (iD) Mickaël Avanthay, (iD) Belen Batanero, (iD)
Christoph Bondue, (iD) Dylan G. Boucher, Rokas Gerulskis, (iD)
Alexander Kuhn, (iD) Shelley D. Minteer, (iD) Andrew Mount, (iD)
Zachary A. Nguyen, Robert Price, (iD) Shahid Rasul, (iD) Naoki Shida, (iD)
Eniola Sokalu (iD) and Jeannie Z. Y. Tan (iD)

DOI: 10.1039/d3fd90041j

Dylan G. Boucher opened the discussion of the paper by Talal Ashraf: What does the surface termination look like after electrolysis? BDD surface chemistry can change quite substantially during the process.

Talal Ashraf responded: Since we are using boron-doped diamond with a high concentration of boron (2000–5000 ppm boron atoms in the diamond lattice), the high doping is known to facilitate more oxygen-related species. We can pretreat the BDD electrochemically to convert it to a hydrogen-terminated surface, but during electrolysis of acetic acid/sodium acetate, the high oxidation potential converts the H-terminated surface to an O-terminated surface. The reaction pathway can be changed from OH-mediated oxidation to Kolbe coupling if the surface retains its H termination (hydrophobic), thus avoiding water oxidation and favouring the coupling reaction.

Eniola Sokalu remarked: For the synthesis of your platinum nanoparticles (https://doi.org/10.1039/d3fd00066d) you used different platinum-containing precursors for different nanoparticle shapes, namely chloroplatinic acid (H_2PtCl_6) and potassium hexachloroplatinate (K_2PtCl_6), where the only difference between the precursors is the counterion. Can you explain how and why you chose these different platinum-containing precursors?

Talal Ashraf answered: Platinum salts with different cations have different reduction potentials (higher reduction potentials may result in slower reduction rates, allowing more time for crystal growth and favouring the formation of larger nanoparticles). The larger cations can also cause overcrowding in the double-layer region and can form non-covalent interactions (weak non-covalent interaction: K^+ Cs^+; strong non-covalent interaction: Na^+, Li^+). This can affect the adsorption of platinum species and the nucleation and growth of nanoparticles. We wanted to test the differently shaped platinum nanoparticles and therefore we referred to the literature for the electrodeposition method. It might be a good investigation for the future to use scanning electrochemical microscopy (SECM) to determine why/how nanoparticles nucleate differently in the presence of different cations.

Shelley D. Minteer commented: What is the difference between differential electrochemical mass spectroscopy (DEMS) and your *in situ* mass spectrometry systems? What are the advantages of your systems *versus* DEMS for electrosynthesis applications?

Talal Ashraf answered: Electrochemical mass spectrometry (ECMS) is the latest commercial technology compared to DEMS, which is more often homemade; our ECMS system is advantageous in terms of sensitivity, reproducibility, electrolyte saturation, time resolution and quantification, and dynamic range. ECMS provides high-turnover resolution for gaseous analytes with 100% collection efficiency and fast control of dissolved gases. ECMS membrane chips are precisely made by accurate and high-quality microfabrication in Nanolab, providing high reproducibility due to a well-defined flux of molecules. Porous Teflon™ DEMS membranes are naturally highly variable in their characteristics. DEMS runs in flow mode, which can cause the reaction products to be lost in flow conditions; the water evaporation rate in DEMS is higher. Measuring the small number of products with decent time resolution with DEMS is impossible. Our ECMS cells operate in a stagnant cell without differential pumping. It is also possible to quantify the sub-monolayer concentrations.

Christoph Bondue said: I wonder about your approach to deposit a metal on the semiconductor. At the interface, a Schottky barrier or an ohmic contact might form, which might affect the charge transfer from the electrode to the electrolyte. Might there be an effect on the distribution?

Talal Ashraf responded: Thank you for your question. We wanted to test if the Pt nanoparticles are active for Kolbe coupling and stability and found it unstable at high oxidation potential. Therefore, we didn't conduct an investigation on the charge-transfer resistance. Since we used highly-doped BDD (2000–5000 ppm boron atoms in the diamond lattice), we have a high concentration of available charge carriers (electrons) in the BDD. Highly-doped BDD has a Fermi level closer to the conduction band due to electron abundance. I think that in the BDD with platinum nanoparticles, the Fermi level alignment can be more favourable, leading to a lower effective Schottky barrier height. We will investigate it more in the future if we find a way to stabilise platinum nanoparticles on BDD.

Belen Batanero asked: Did you apply these electrode conditions to decarboxylate larger acidic molecules?

If yes: was the olefin formation (after a further oxidation of the radical) never a competitive pathway towards the radical coupling?

Talal Ashraf answered: Thank you for the question. I didn't use a large molecular acid for decarboxylation, which is a goal for the upcoming months. At acidic pH on the platinum surface, we have found that Kolbe coupling is favourable and hence no further oxidation of the radical takes place. The olefin and alcohol formation are more favourable at alkaline pH (in the presence of carbonate and bicarbonates, the overcrowding effect in the double layer region leads to overoxidation of the radical).

Shahid Rasul asked: As an engineer, considering the commercialization and scale-up process of BDD (boron-doped diamond) electrodes, what electrode materials would be suitable for scaling up under harsh conditions? Are titanium-based alloys and hard carbons viable options? Any recommendations?

Talal Ashraf responded: Thank you for the question. Metal oxide-coated titanium electrodes (also known as dimensionally stable anodes (DSA)) find extensive use in electro-oxidation applications. However, their practicality as electrode is impeded by the limited availability of crucial mixed metal oxide (MMO) materials, notably ruthenium, a scarce component typically found during the platinum group metals extraction. Titanium, when exposed to high oxidation potentials, undergoes oxide formation and the oxide acts as an insulating layer, rendering it unsuitable for prolonged application.

Lead-based anodes also participate in electro-oxidation reactions; nonetheless, their limitations are evident. Graphite, glassy, and pyrolytic carbon encounter challenges when exposed to high potentials within aqueous environments.

An alternative with proven utility is the boron-doped diamond (BDD) anode, acknowledged for its prolonged operational lifespan, even under extreme conditions. This technology has successfully transitioned into commercial and industrial domains, serving roles in applications such as H_2O_2 synthesis and wastewater treatment. However, the preparation of BDD anodes demands substantial energy input, rendering them relatively expensive. In an effort to address this cost factor, companies are actively exploring the integration of renewable energy sources to power their manufacturing processes, which holds potential for reducing production expenses.

Alexander Kuhn remarked: You seem to have major problems with the stability of the Pt deposits on your BDD electrodes and therefore I wonder whether a Pt/Ir alloy could show a better behaviour in terms of stability. We recently saw that even with a very low concentration of around 5% Ir there is a very significant improvement in terms of stability,[1] so it's maybe worth trying because such a low amount of Ir should not affect the selectivity of the electrode.

1 S. Butcha, S. Assavapanumat, S. Ittisanronnachai, V. Lapeyre, C. Wattanakit and A. Kuhn, *Nat. Commun.*, 2021, **12**, 1314, DOI: 10.1038/s41467-021-21603-8.

Talal Ashraf replied: Thank you for the suggestions. We haven't tried bimetallic electrocatalysis for the Kolbe reaction; iridium is not active for Kolbe, but it is also prone to dissolution at high oxidation potentials (Cherevko's group's work on fuel cells and electrolysers[1]). As the paper you mentioned shows that Ir enhances the stability of Pt, it will be worth trying for our reaction, and therefore we are considering approaching this strategy; thanks for the very insightful paper and suggestions.

1 S. Cherevko, S. Geiger, O. Kasian, A. Mingers and K. J. J. Mayrhofer, *J. Electroanal. Chem.*, 2016, **773**, 69–78, DOI: 10.1016/j.jelechem.2016.04.033.

Alexander Kuhn commented: I understand that iridium is not active for Kolbe reaction, but the amount of iridium I have in mind is typically very small (only a few percent). So maybe you can try with very small amounts of 1/2/3%?

Talal Ashraf replied: Thank you for your suggestions. I will try it in the upcoming month following the preparation method reported in your paper for a Pt/Ir alloy *via* electrodeposition on BDD or by solid-state dewetting.

Naoki Shida remarked: The uniqueness of BDD electrodes for generating radical species is fascinating. When radical species such as hydroxy radicals or methyl radicals (*via* a Kolbe reaction) are generated, do these radicals exist as molecular radical species? Or do they have any interaction with the surface of the BDD electrode?

Talal Ashraf responded: Methyl radicals from the carboxylic acid scavenge the OH radicals, but we also observed a slight faradaic efficiency towards H_2O_2, which is coming from recombination of OH radicals. Due to the inertness of BDD, the OH radicals are very weakly adsorbed and the concentration of OH radicals decreases from the distance of the electrode in the presence of acetate (observed during our upcoming modelling results).

Naoki Shida remarked: Considering the structure of BDD is mainly composed of sp^3 carbons, BDD electrodes seem to be chemically inert. Is there any known electrolytic activity?

Talal Ashraf replied: It depends on the quality of the BDD; high-quality BDD will be extremely inert.

The electrocatalytic activity from BDD can also be caused by the graphitised carbon during manufacturing or corrosion of BDD by the radicals, which scavenge OH radicals weakly adsorbed on the surface.[1] Giving an example, we acquired BDD electrodes from Condias (the leading manufacturer) and our Raman and SEM results show them to be of high quality; we found it very electrocatalytic inert by indirect (OH-mediated) oxidation of acetate to methanol. We also tried BDD from another company, Neocoat, and we found that the quality was not great, and thus we observed a side reaction (methyl acetate formation with high faradaic efficiency (FE)), which comes from electrochemical oxidation of adsorbed acetate on graphitized carbon.

1 T. Kashiwada, T. Watanabe, Y. Ootani, Y. Tateyama and Y. Einaga, *ACS Appl. Mater. Interfaces*, 2016, **8**, 28299–28305, DOI: 10.1021/acsami.5b11638.

Zachary A. Nguyen opened the discussion of the paper by Robert Price: What scale of biomass are you planning to process? What is the specific composition of this biomass mix and which of them acts as the proton source?

Robert Price responded: The potential of spent lees to act as a proton (and electron) source is quite low and so it is not being considered for this process. The draff and pot ale offer significantly higher potentials. The draff comprises up to 80 weight % (wt%) water, with the remaining 20 wt% including cellulose, hemi-cellulose, lignin, protein and starch (please see ref. 25 in the corresponding manuscript (https://doi.org/10.1039/d3fd00086a)). It is likely that the cellulose, lignin and hemi-cellulose act as the main proton and electron sources here, though the mechanisms for abstraction (and, therefore, degradation of this biomass) have not yet been elucidated in our work. Approximately 400 000 kg of draff is produced at the distillery per annum.

The pot ale contains a significant fraction of dead yeast (5.62 wt%) along with other organic acids, amino acids and protein (please see ref. 28 in the corresponding manuscript), which will be the most likely sources for protons and electrons. Approximately 1.5 million litres of pot ale are produced at the distillery per annum.

Rokas Gerulskis remarked: It seems that you are adding substrate stoichiometrically in an initial reaction, and then regenerating it in a subsequent electrochemical reaction. Couldn't you add only catalytic amounts if you ran these reactions concurrently? What specific catalytic requirements preclude doing so?

Robert Price responded: Initially, we used a fixed concentration of phosphomolybdic acid (PMA, 0.3 M) to ensure that plenty of catalyst was available to be able to abstract protons and electrons from the biomass. This helped us to understand the potential of each waste stream to reduce the PMA and, in all cases, more biomass was required to do so. Therefore, we could certainly reduce the concentration of PMA used for the pot ale and spent lees streams, which have fixed biomass loadings.

In the case of draff, as it is a solid, the required amount to reduce that specific concentration of PMA could be added and controlled much more easily. This would, therefore, ensure that all of the PMA in the reaction mixture would act as a catalyst in future experiments.

The digestion and electrolysis steps couldn't be run concurrently in our present setup as the residues from digestion would need to be filtered off to prevent blockages and pressure build-up in the PEM electrolyser. Also, considering that biomass would continually need to be added to the digestion vessel to reduce the PMA, this wouldn't be feasible for the liquid wastes as addition would continually dilute the catalyst in the digestion mixture. It is rather a process constraint than a catalytic limitation.

Mickaël Avanthay asked: Once the draff has been digested and the hydrogen extracted, how do you dispose of the residue?

Robert Price responded: Ideally, the concentration of draff added to the phosphomolybdic acid (PMA) catalyst solution would be optimised so that full degradation of the organic material would occur and the residue would be minimised. I would expect that any remnants would be washed to recover as much PMA as possible, before the remaining material is disposed of as chemical waste.

Alexander Kuhn commented: I wonder why you have chosen PMA for these experiments because molybdenum-based Keggin ions are known to be not very stable with respect to hydrolysis, especially in neutral pH. This might also explain the partial irreversibility that you observe in your electron transfer process.

Robert Price responded: We chose phosphomolybdic acid because thermally it is very stable (required for the initial digestion step) and the first two-electron process in the cyclic voltammogram exhibits reversibility (during electro-chemical H-cell cycling); this is the process expected to be accessed during the thermal digestion of biomass (pH = 2).

I agree that, certainly, the second and third two-electron processes exhibit some irreversibility, most likely due to hydrolysis, and that this could also be the case when performing reduction *via* biomass digestion (rather than electro-chemical reduction in a H-cell). Other polyoxometalate catalysts that exhibit higher stability could also be trialled to see if this irreversibility can be avoided.

Andrew Mount remarked: This work presents an interesting alternative for "green" hydrogen generation compared to water electrolysis. In terms of economics, is there already a likely energy benefit identified when using this alternative approach (for example, there appears to be an additional thermal energy requirement. Is this more than offset by the reduced voltage/electrical power requirements)?

Robert Price answered: According to previous literature reports for similar research, though using other native biomass sources, a power consumption as low as 0.69 kW h per Nm^3 could be achieved for the electrolysis step (please see ref. 10 in the accompanying manuscript), in comparison to *e.g.* ~4.3 kW h Nm^3 for commercial, pressurised alkaline water electrolysers. Although the power consumption would likely increase when scaling this technology to a system level, due to the balance of plant required, there could still be a significant reduction in power consumption. Ideally, the thermal digestion step would use waste heat (in this case, from the distillery), therefore the power consumption of the first step is expected to be minimal. Alternatively, the step could be performed at lower temperature *via* light irradiation.

Andrew Mount asked: The Spent Lees EIS data in Fig. 8b of your paper (https://doi.org/10.1039/d3fd00086a) in particular appears to show a 45° line in the Nyquist plot characteristic of diffusion; indeed this is consistent with the comment "the low-frequency, R_{p3}, process most likely relates to reactant diffu-sion limitation". Diffusion is typically represented by a Warburg element in EIS, *e.g.*, in the established Randles circuit for analysing diffusion to, and charge transfer at/across an interface. The equivalent circuit used (Fig. 10a of your paper) does not have such an element. It would have been good to have goodness of fit data provided or fit lines shown on Fig. 8 of your paper to enable the equivalent circuit fit to these data to be assessed. However, assuming these to be good, it is worth noting that CPE(3) could also provide this 45° feature if $n = 0.5$ (this cannot be confirmed as no Y_0 or n parameters have been provided for the three CPEs in the circuit). Given a CPE would normally be used to model some form of dispersion rather than diffusion, how confident are the authors that the equiva-lent circuit used in Fig. 10a of your paper has been established to be the most appropriate model of the EIS response?

Robert Price answered: The data points in the electrochemical impedance spectra (EIS) presented in Fig. 8b of our paper have been excluded below 12 Hz (as mentioned in the text) due to poor data quality, in part because of non-optimal contacting/compression of the cell and the action of the peristaltic pump. Consequently, the data set for the spent lees does indeed appear to give a 45° line that resembles a Warburg element. For reference, the n parameters for constant phase elements (CPEs), determined during equivalent circuit fitting (ECF) of the spent lees EIS: R_{p1}, R_{p2} and R_{p3} are 0.95, 0.83 and 0.75, respectively.

However, the EIS relating to PEMEC2 operating on digested draff–PMA and pot ale–PMA solutions collected after electrolysis (Fig. 12a of our paper) illustrate this analogous low-frequency process more clearly, owing to improved data quality. When the concentration of $[PMo_{12}O_{40}]^{5-}$ in solution becomes critically low (as is the case for the spent lees sample), a high-resistance partial arc, which is approaching its frequency maximum, dominates the low-frequency region of the spectrum, rather than a Warburg element with a 45° line. For this reason, a CPE was chosen to model the low-frequency process. It is acknowledged that this process may be better described as a phenomenon relating to high reactant utilisation, rather than diffusion (*e.g.*, similar to the commonly observed 'gas conversion' processes in EIS of solid oxide fuel cells/electrolysers at high fuel utilisation, particularly when testing stacks and systems).

Shelley D. Minteer asked: You can now send your waste streams to the sea, but if you do chemistry (in this case electrochemistry) on them, do you end up with more regulatory issues to deal with the waste streams?

Robert Price responded: The discharge of the liquid (pot ale and spent lees) waste streams to sea is carried out in line with guidelines from the Scottish Environmental Protection Agency. In the case of deployment of a 'thermal digestion and electrolysis of biomass' system that utilises the phosphomolybdic acid (PMA) catalyst, there are likely to be more regulatory procedures to be followed to ensure that the surrounding environment is not exposed to the PMA catalyst and that the whisky/gin production process is also not impacted by the presence of the catalyst on-site. The residues left over from the digestion and electrolysis process will, most likely, have to be treated and disposed of as 'chemical waste'.

Jeannie Z. Y. Tan addressed Robert Price and Talal Ashraf: In the literature, researchers proposed the use of pulsed potential to enhance the product selectivity, yield and durability of the system. Have the presenters applied pulse potential to their systems? If yes, what are their observations?

Robert Price answered: We have not yet considered the use of a pulsed potential as there are quite a few areas of the flow cell electrolyser that first need to be optimised. For example, reducing the amount of anode off-gas produced, ensuring that the theoretical charge corresponds to the actual charge passed and improving materials compatibility for use with phosphomolybdic acid. It could be explored in the future once a good baseline performance has been achieved.

Talal Ashraf responded: Phil Baran recently investigated the application of pulsed electrolysis during the Kolbe reaction published in *Science*[1] and observed a significant difference. In our upcoming work, we will also apply the same strategy to different acids' decarboxylation with pulsed electrolysis.

1 Y. Hioki, M. Costantini, J. Griffin, K. C. Harper, M. P. Merini, B. Nissl, Y. Kawamata and P. S. Baran, *Science*, 2023, **380**, 81–87, DOI: 10.1126/science.adf4762.

Conflicts of interest

There are no conflicts to declare.

Faraday Discussions

PAPER

Alkene reactions with superoxide radical anions in flow electrochemistry†

Rojan Ali, [ID] Tuhin Patra [ID] ‡ and Thomas Wirth [ID] *

Received 22nd February 2023, Accepted 6th March 2023

DOI: 10.1039/d3fd00050h

Alkenes were cleaved to ketones by using dioxygen in an electrochemical flow set-up. The pressurised system allowed efficient gas–liquid mixing with a stabilised flow. This mild and straightforward approach avoids the use of transition metals and harsh oxidants.

Introduction

Significant progress has been made in the electrochemical reduction of oxygen to provide hydrogen peroxide for wastewater treatment or as a bleaching agent.[1] The coupling of electrochemical oxygen reduction with preparative organic chemistry and the set-up of a flow system for such paired synthetic transformations is, however, a largely unexplored area. Flow electrochemistry has been a focus area of research recently.[2] We have developed electrochemical flow reactors, which also have been commercialised.[3] Biphasic reactions in microreactors have already attracted a lot of interest due to increased mixing abilities.[4] Even in electrochemical microreactors such biphasic flow systems have been investigated and analysed.[5] The gas/liquid biphasic reaction system we propose to use in this research will allow an intense mixing. As oxygen is used as a gas, the precise control of the potential in the electrochemical flow reactor will result in a convenient generation of the superoxide radical anion (O_2·$^-$), which can be used for proton uptake and hydrogen atom abstraction.[6] However, the use of a nonconductive gas as the second phase in such biphasic mixtures might require additional optimisation in the flow set-up.

Results and discussion

Here we describe how a biphasic gas/liquid solution can be employed in electrochemical flow reactors to effect oxygen activation and perform a direct reaction

School of Chemistry, Cardiff University, Park Place, Main Building, Cardiff, CF10 3AT, UK. E-mail: wirth@cf. ac.uk

† Electronic supplementary information (ESI) available. See DOI: https://doi.org/10.1039/d3fd00050h

‡ Present address: School of Basic Sciences, Indian Institute of Technology Bhubaneswar, Argul, Odisha 752050, India.

with alkenes as exemplary substrates. Electrochemically generated superoxide radical anions $(O_2{}^{\cdot-})$ have been reported previously and used in reactions with esters.[7] The electrochemical flow experiments were stimulated *via* the corresponding batch electrochemical procedure which has been described by Chiba and co-authors.[8] Due to the small dimensions of the electrochemical microreactor, the use of oxygen does not pose any safety hazards as the total amount of hazardous material is low. A back-pressure regulator was employed to maintain a pressurised system resulting in an efficient gas–liquid mixing, while allowing a stable flow rate at the same time. The electrochemical flow experiments were performed at room temperature in an undivided, commercially available electrochemical flow reactor.[3] The flow of the oxygen gas was controlled by use of a mass flow controller.[9] The resulting biphasic reaction mixture was pumped into the electrochemical reactor with a peristaltic pump.[10] This pump controlled the combined flow rate of the biphasic reaction mixture and can be used to monitor the back-pressure created *via* a tuneable back-pressure regulator (BPR) positioned after the electrochemical reactor, as shown in the set-up in Fig. 1.

For the flow experiments, platinum electrodes with an active surface area of 12 cm^2 each were separated by a 0.5 mm fluorinated ethylene propylene (FEP) spacer, creating a 0.6 mL volume channel in the reactor. The reactions were performed under constant current electrolysis. 1,1-Diphenylethylene **1a** was investigated to determine the optimal conditions required for its electrolysis to benzophenone **2** (Table 1). Initially, the reaction was performed in acetonitrile using platinum

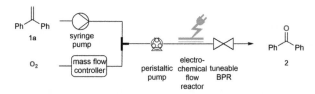

Fig. 1 Experimental set-up for the electrochemical flow reactions with oxygen.

Table 1 Optimisation studies for the oxidative cleavage of 1,1-diphenylethylene **1a**[a]

Entry	Solvent	Combined flow rate (mL min^{-1})	Charge (F)	Current (mA)	Back-pressure (bar)	Yield 2 (%)
1	MeCN	0.1	3	12	2.8	25
2	^{i}PrCN	0.1	3	12	2.8	36
3	^{i}PrCN	0.2	2	16	2.8	53
4	^{i}PrCN	0.2	2	16	2.2	51
5	**^{i}PrCN**	**0.2**	**2**	**16**	**1.1**	**51**
6	^{i}PrCN	0.2	2	16	0	41
7	^{i}PrCN	0.2	2.5	20	1.1	46
8	^{i}PrCN	0.2	3	24	1.1	35

[a] Standard reaction conditions: undivided flow cell, Pt anode and cathode (active surface area: 12 cm^2 each), interelectrode distance: 0.5 mm, **1a** (0.025 M) and $^nBu_4NClO_4$ (0.05 M) in solvent. Yield determined by gas chromatography-flame ionization detection (GC-FID) using benzonitrile as an internal standard.

electrodes. The alkene solution and oxygen gas were pumped with a flow rate of 0.05 mL min^{-1} each resulting in a combined flow rate of 0.1 mL min^{-1}. The molar ratio of oxygen : alkene was at least 4.5 : 2.5 (no back-pressure). With an applied charge of 3 F and a back-pressure of 2.8 bar, the desired product 2 was obtained in 25% yield (Table 1, entry 1). GC-MS analysis of the crude reaction mixture showed several undesired side products, which were generated due to the solvent participating in the reaction. The formation of these side products was suppressed by changing acetonitrile to isobutyronitrile (^{i}PrCN) allowing 2 to be obtained in 36% yield (Table 1, entry 2) (see ESI†). On increasing the flow rate to 0.2 mL min^{-1}, and decreasing the applied charge to 2 F, the yield of 2 was increased to 53% (Table 1, entry 3). The yield remained unaffected at lower back-pressures of 2.2 and 1.1 bar (Table 1, entries 4 and 5), while a lack of back-pressure (Table 1, entry 6) reduced the yield to 41%. Finally, on increasing the charge applied from 2 F to 2.5 F and 3 F, the yield of 2 decreased (Table 1, entries 7 and 8).

Using the optimised reaction conditions, several di- and tri-substituted alkenes were converted into their corresponding carbonyl derivatives (Fig. 2). Starting from the disubstituted alkene 1,1-diphenylethylene 1a, benzophenone 2 was obtained in 45% yield. Tri-substituted olefins 1b and 1c produced 2 in 33% and 27% yields, respectively. The reaction of α-methylstyrene 1d resulted in the formation of acetophenone 3 in 39% yield. Additionally, indene and 4-bromo- and 4-fluorostyrene were investigated in this reaction. Although these starting materials were fully consumed, the corresponding aldehydes were not obtained. This might not be completely surprising as it has been reported that superoxide radical anions can decompose aromatic aldehydes through a base-catalysed process.[11]

The proposed reaction mechanism is shown in Fig. 3. Initially, oxygen undergoes a one-electron reduction to form the corresponding superoxide radical anion ($O_2^{\bullet-}$).[12] An aprotic solution containing the superoxide radical anion has been reported to be stable at room temperature for several hours.[13] Therefore,

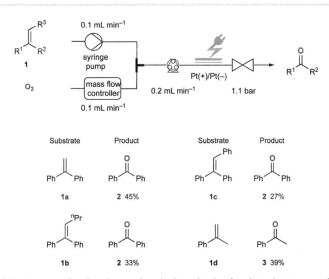

Fig. 2 Substrate scope for the electrochemical synthesis of carbonyl compounds from di- and tri-substituted alkene derivatives in flow.

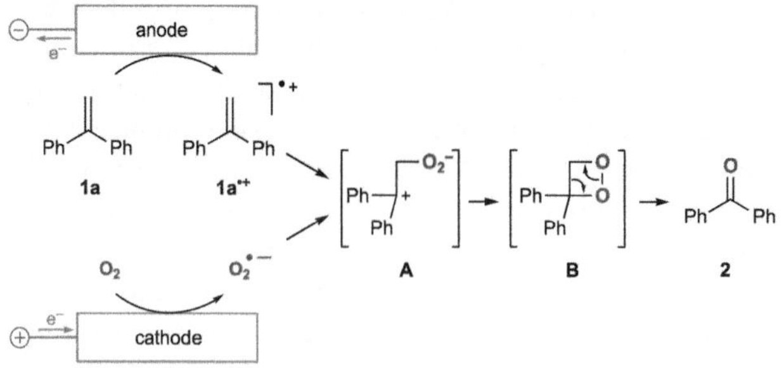

Fig. 3 Proposed reaction mechanism.

$O_2{}^{\cdot-}$ is stable enough to capture the radical cation of the alkene ($1a^{\cdot+}$) generated by the one-electron oxidation of the alkene. This could form the intermediate **A** followed by the generation of dioxetane intermediate **B**, which decomposes to the corresponding carbonyl compound.[14]

Conclusions

In conclusion, a continuous biphasic electrochemical flow method was developed to efficiently generate superoxide radical anions and oxidatively cleave di- and trisubstituted alkenes into their corresponding carbonyl compounds. This is a mild and straightforward approach that uses molecular oxygen as a green oxygen source in combination with flow electrochemistry as an alternative to transition metals and harsh oxidants. Moreover, an intrinsic advantage of flow chemistry is that the procedures can be easily scaled up. Finally, this method allowed the synthesis of several ketones in moderate yields.

Author contributions

The authors confirm contributions to the paper as follows: study conception and design: T. Wirth; experiments and data collection: R. Ali and T. Patra; analysis and interpretation of results: all authors; draft manuscript preparation: T. Patra, R. Ali and T. Wirth. All authors reviewed the results and approved the final version of the manuscript.

Conflicts of interest

There are no conflicts of interest to declare.

Acknowledgements

The authors acknowledge the financial support provided by the Welsh Government (SMART Expertise grant).

References

1 (*a*) Y. Pang, H. Xie, Y. Sun, M.-M. Titirici and G.-L. Chai, *J. Mater. Chem. A*, 2020, **8**, 24996; (*b*) N. Wang, S. Ma, P. Zuo, J. Duan and B. Hou, *Adv. Sci.*, 2021, **8**, 2100076.

2 (*a*) N. Tanbouza, T. Ollevier and K. Lam, *iScience*, 2020, **23**, 101720; (*b*) M. Elsherbini and T. Wirth, *Acc. Chem. Res.*, 2019, **52**, 3287; (*c*) T. Noël, Y. Cao and G. Laudadio, *Acc. Chem. Res.*, 2019, **52**, 2858; (*d*) N. Kockmann, P. Thenée, C. Fleischer-Trebes, G. Laudadio and T. Noël, *React. Chem. Eng.*, 2017, **2**, 258; (*e*) M. Elsherbini, B. Winterson, H. Alharbi, A. A. Folgueiras-Amador, C. Génot and T. Wirth, *Angew. Chem., Int. Ed.*, 2019, **58**, 9811; (*f*) D. Pletcher, R. A. Green and R. C. D. Brown, *Chem. Rev.*, 2018, **118**, 4573.

3 Ion electrochemical reactor, https://www.vapourtec.com/products/flow-reactors/ion-electrochemical-reactor-features/, accessed February 2023.

4 M. J. Hutchings, B. Ahmed-Omer and T. Wirth, in *Microreactors in Organic Chemistry and Catalysis*, ed. T Wirth, Wiley-VCH, Weinheim2013, p. 197.

5 Y. Cao, N. Padoin, C. Soares and T. Noël, *Chem. Eng. J.*, 2022, **427**, 131443.

6 M. Hayyan, M. A. Hashim and I. M. AlNashef, *Chem. Rev.*, 2016, **116**, 3029.

7 (*a*) F. Magno and G. Bontempelli, *J. Electroanal. Chem.*, 1976, **68**, 337; (*b*) D. Bauer and J. P. Beck, *J. Electroanal. Chem.*, 1972, **40**, 233.

8 Y. Imada, Y. Okada, K. Noguchi and K. Chiba, *Angew. Chem., Int. Ed.*, 2019, **58**, 125.

9 Gas Mass flowmeter and controllers, https://www.omega.co.uk/pptst/FMA5400A_5500A.html, accessed February 2023.

10 SF-10 reagent pump, https://www.vapourtec.com/products/sf-10-pump-features/, accessed February 2023.

11 M. J. Gibian, D. T. Sawyer, T. Ungermann, R. Tangpoonpholvivat and M. M. Morrison, *J. Am. Chem. Soc.*, 1979, **101**, 640.

12 (*a*) D. L. Maricle and W. G. Hodgson, *Anal. Chem.*, 1965, **37**, 1562; (*b*) Q. Li, C. Batchelor-McAuley, N. S. Lawrence, R. S. Hartshorne and R. G. Compton, *J. Electroanal. Chem.*, 2013, **688**, 328.

13 J. M. McCord and I. Fridovich, *J. Biol. Chem.*, 1969, **244**, 6049.

14 A. A. Frimer, *Chem. Rev.*, 1979, **79**, 359.

Faraday Discussions

ROYAL SOCIETY
OF CHEMISTRY

PAPER

Spatio-temporal detachment of homogeneous electron transfer in controlling selectivity in mediated organic electrosynthesis†

Jack W. Hodgson, [iD] Ana A. Folgueiras-Amador, [iD] Derek Pletcher, David C. Harrowven, [iD] Guy Denuault [iD] * and Richard C. D. Brown [iD] *

Received 3rd May 2023, Accepted 31st May 2023

DOI: 10.1039/d3fd00089c

In electrosynthesis, electron transfer (ET) mediators are normally chosen such that they are more easily reduced (or oxidised) than the substrate for cathodic (or anodic) processes; setting the electrode potential to the mediator therefore ensures selective heterogeneous ET with the mediator at the electrode, rather than the substrate. The current work investigates the opposite, and counter intuitive, situation for a successful mediated electroreductive process where the mediator (phenanthrene) has a reduction potential that is negative to that of the substrate, and the cathode potential is set negative to both ($E_{ele} < E_M < E_s$). Simulations reveal a complex interplay between mass transport, the relative concentrations of the mediator and substrate as well as the heterogeneous and homogeneous rate constants for multiple steps, which under suitable conditions, leads to separation of the homogeneous chemistry in a reaction layer detached from the electrode. Reaction layer detachment is a spatio-temporal effect arising due to opposing fluxes of the mediator radical anion $M^{•-}$ and the substrate 1, which ultimately prevents 1 from reaching the electrode, thereby affording a different reaction pathway. Simulations representative of unstirred batch (1D) and flow (2D) electrolysis are presented, which qualitatively reproduce the experimental selectivity outcomes for mediated and unmediated electroreductive cyclisation of aryl iodide 1. The potential to use highly reducing homogeneous ET agents, possessing reduction potentials beyond those of the substrates, offers exciting opportunities in mediated electrosynthesis.

Introduction

Aryl radical cyclisation is an established approach for the synthesis of carbocyclic and heterocyclic compounds, applied in research laboratories around the world

School of Chemistry, University of Southampton, Highfield, Southampton SO17 1BJ, UK. E-mail: rcb1@soton. ac.uk

† Electronic supplementary information (ESI) available. See DOI: https://doi.org/10.1039/d3fd00089c

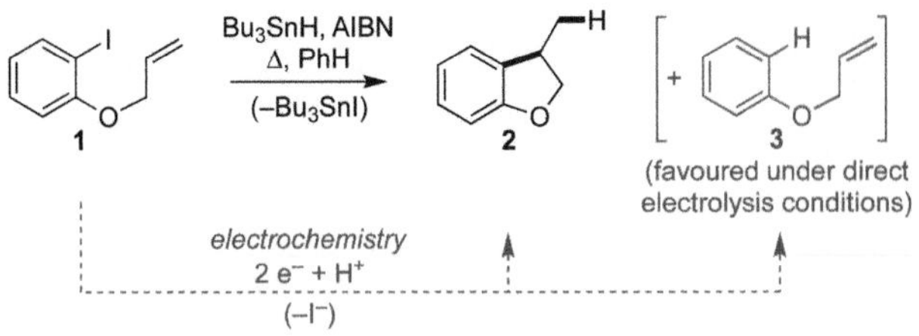

Scheme 1 Classical Bu₃SnH mediated radical cyclisation of aryl iodide **1** and electro-reductive approach.

towards a plethora of interesting and useful target molecules. A classical method involves heating an aryl bromide or iodide, such as **1**, in the presence of Bu₃SnH and a suitable initiator (*e.g.* AIBN) to give the cyclised product **2** and the corresponding tributyltin halide (Scheme 1). Although radical cyclisation using tin hydrides provides a reliable synthesis with broad scope, its appeal from environmental and safety perspectives is diminished somewhat due to the toxicity of organotin compounds, hazards associated with initiators, and the high reaction temperatures often required. Significant effort has been devoted to replacing Bu₃SnH with other hydrides, developing conditions that are substoichiometric in organotin, and efficient methods for removing organotin impurities, but in spite of these developments the use of Bu₃SnH in radical cyclisation often remains the method of choice for laboratory applications. Recent years have seen vigorous interest in alternative methods to access aryl radicals from a variety of precursors, particularly, using photochemical activation.[1–3]

Electrochemistry offers a more environmentally acceptable method to produce aryl radicals by reduction of aryl halides.[4] However, a limitation is that direct cathodic electrolysis of aryl halides favours hydrogenolysis products (*e.g.* **3**) over radical cyclisation (*e.g.* **2**) for all but a limited sub-set of substrates (Scheme 1).[5,6] The electrochemical mechanism involves electron transfer (ET) to aryl iodide **1**, giving rise to a frangible radical anion **1**·⁻ which very rapidly loses iodide ion to afford an aryl radical **4** (Scheme 2).[7] The relative rates of the onward reactions of **4** affect the selectivity for either formal hydrogenolysis product **3** or cyclised product **2**, and as will be discussed below, a crucial factor is the spatial proximity of the aryl radical to the cathode due to the highly favourable ET to **4** giving the aryl anion **5**.[8]

The use of outer sphere electron transfer mediators in electrosynthesis is attracting increasing interest from organic chemists.[9] In this context, the mediator (**M**) behaves as an electron shuttle between electrode surface and substrate such that ET to/from the substrate occurs as a homogeneous process. It is important to recognise that heterogeneous ET – focussing here on reduction of a mediator to its radical anion – occurs within molecular distances from the electrode, whereas homogeneous ET between mediator radical ions and substrate occurs anywhere both species meet, with the proviso that the reduction potential of **M** is sufficiently negative.[9] The ability to move ET away from the electrode provides an opportunity for altered and useful selectivity, for example, in situations where ET is followed by a very fast chemical reaction leading to an intermediate that can either undergo a second ET (ECE pathway) or a chemical

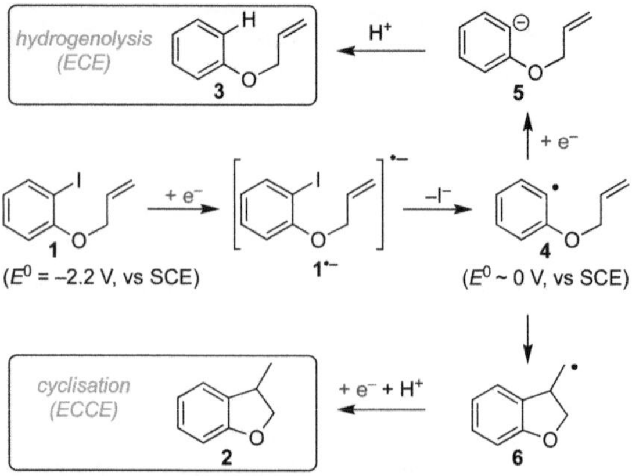

Scheme 2 Competing hydrogenolysis and cyclisation pathways observed for direct electroreduction of aryl iodide 1. In the pathway acronyms (ECE or ECCE), E stands for a heterogeneous ET at the electrode while C stands for a homogeneous process in solution.

reaction (ECCE pathway). An illustrative case is the cathodic reduction of aryl halides highlighted below, where direct electrolysis favours hydrogenolysis (ECE), whereas, in the presence of suitable mediators the radical intermediate can undergo selective transformations such as cyclisation onto pendant unsaturation (ECC).[7]

Grimshaw *et al.* reported that direct electrolysis of aryl chloride 7 at a Hg cathode in a divided cell gives the hydrogenolysis product 9, whereas electrolysis of the same substrate in the presence of stilbene (10) affords the cyclised indolene 8 in good yield (Scheme 3a).[5d] In another study, Mitsudo *et al.* demonstrated that

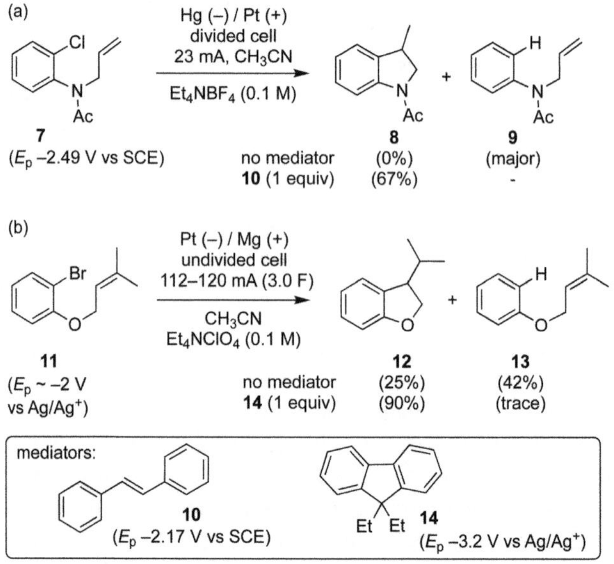

Scheme 3 Direct and mediated electroreductions of aryl halides 7 and 11.

very high selectivity for the cyclised dihydrobenzofuran **12** is achieved from electrolysis of aryl bromide **11** conducted in the presence of 1 equivalent of diethylfluorene **14** in an undivided cell equipped with a sacrificial electrode.[10] In the absence of a mediator, the selectivity is reversed yielding mainly hydrogenolysis product **13** (**12** : **13** ∼ 1 : 1.7). At this juncture, it is worth highlighting that in the former case the mediator **10** has a reduction potential that is positive to the respective substrate, whereas the reverse is true for the reaction mediated by diethylfluorene (*i.e.* diethylfluorene is actually harder to reduce at the cathode than the aryl bromide substrate **11**). Other examples of mediated reductive electrocyclization of aryl halides *via* the intermediacy of aryl radicals have been reported.[11,12] The studies described thus far highlight the interplay of the two different reaction manifolds shown in Scheme 2, which are dependent on a number of factors, including homogeneous and heterogeneous ET rate constants and mass transport.[13] The preferential formation of hydrogenolysis products in direct electrochemical reduction of aryl halides is well established (Scheme 2), arising because loss of halide ion from frangible radical anions such as **1**$^{\cdot-}$ is so rapid that the aryl radical **4** is formed close to the cathode.[7] Aryl radicals exhibit significantly more positive reduction potentials ($E^0_{\text{Ph}^\cdot/\text{Ph}^-} \sim 0$ V *vs.* SCE) compared to aryl halides; they are thus readily reduced at the potential required to form the parent radical anion, and therefore the ECE pathway prevails. On the other hand, when a suitable mediator is present, ET from **M**$^{\cdot-}$ to the aryl halide **1** can occur wherever both radical anion and substrate are found in the electrolyte solution. When ET takes place away from the electrode, the ensuing rapid ejection of halide ensures that the aryl radical **4** is also formed away from the electrode.

A fast chemical reaction step (*e.g.* cyclisation to **6**) outpaces reduction of the aryl radical, as the latter requires either mass transport to the cathode, or a homogeneous ET with **M**$^{\cdot-}$ (or ArI$^{\cdot-}$, disproportionation). Therefore, in order for the radical cyclisation pathway (ECCE) to predominate, the aryl radical **4** should not be present close to the cathode where heterogeneous ET becomes the most favourable process.

In texts discussing mediated electrosynthesis it is stated that the mediator (**M**) must be more easily reduced or oxidised than the substrate in order to favour ET between the mediator and electrode, rather than the substrate and electrode, to allow continuous regeneration of the mediator *in situ*.[9] In the context of mediated electroreduction, this follows a common observation that in situations where two or more electroactive species are present, that with the less negative reduction potential is reduced preferentially at the electrode, assuming sufficiently different reduction potentials. However, this neglects the important influence of mass transport, which limits the overall rate of electrochemical reactions involving very fast ET and chemical steps. Interestingly, several papers describe reductive cyclisation of aryl halides in the presence of mediators with reduction potentials that are negative to that of the substrate (*i.e.* the mediator is harder to reduce than the substrate). For example, phenanthrene ($E^0 = -2.5$ V *vs.* SCE), and fluorene ($E_p = -3.5$ V *vs.* Ag/Ag$^+$) are reported as mediators for cyclisations of aryl halides ($E^0 \sim -2$ V *vs.* SCE) (see Scheme 3b).[10,11b] Consequently, under the potential required to drive the electrochemical reduction of the mediator, direct cathodic reduction of the substrate is a highly favourable process. The authors did not propose a satisfactory explanation to account for the observed selectivity for cyclised

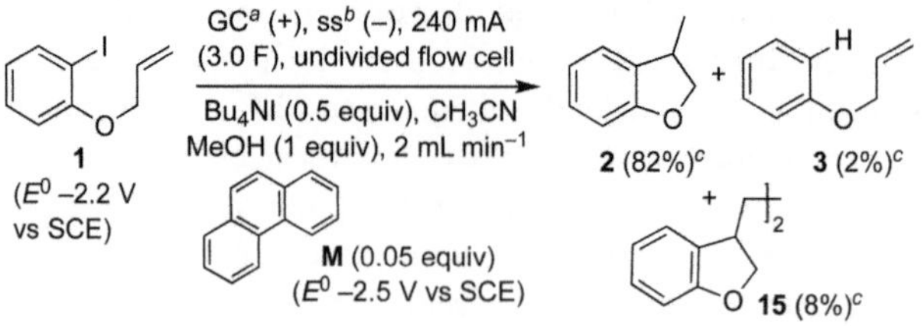

Scheme 4 Selective electroreductive cyclisation of aryl iodide **1** in an undivided electrochemical flow cell, using a highly reducing catalytic mediator. [a] Glassy carbon anode. [b] Stainless steel cathode. [c] Yields are estimated using calibrated GC.

products in the presence of mediators, and a rationale based only on reduction potentials is neither obvious nor sufficient.

Recently, we reported an electroreductive cyclisation of aryl halides, including **1**, in the presence of substoichiometric amounts of a highly reducing mediator in an undivided flow cell (Scheme 4).[14] Voltammetry clearly shows that the mediator, phenanthrene (**M**), has a reduction potential negative to that required to electrochemically reduce the substrate **1**. When mediator is present, the major product is dihydrobenzofuran **2** (82%) with radical dimerisation product **15** (8%) and hydrogenolysis product **3** (2%) as minor by-products. In the absence of mediator, selectivity is reversed and **3** is the major product (48%, **2** : **3** : **15** ~ 1 : 2 : 0). We proposed that, in the mediated process, homogeneous ET takes place in a reaction layer (or zone) that is detached from the cathode surface, and that the substrate **1** is prevented from reaching the electrode surface by the flux of highly reducing mediator radical anion $M^{\cdot-}$ diffusing outwards from the cathode and balancing the inward flux of aryl halide. In the work described here, we present simulations to illustrate and highlight the important role of mass transport in achieving selectivity in the mediated electroreduction of ArI **1** as a consequence of a time-dependent detachment of the homogeneous reaction layer, and discuss the factors influencing detachment in unstirred batch and laminar flow regimes. Significantly, we show that the reduction potential of the mediator does not need to be positive to that of the substrate to achieve a selective mediated electrosynthesis using catalytic mediator *in situ*. We believe that a better understanding and appreciation of the interplay between mass transport, electron-transfer and chemical steps will offer great opportunities in electrosynthesis, and will also account for other outcomes that cannot be rationalised by electrochemical mechanisms alone.

Simulations

1D simulations were carried out using DigiElch v.8 and COMSOL Multiphysics v.6.0 whereas 2D simulations were performed with COMSOL Multiphysics v.6.0. The reaction steps (1)–(11) described in Schemes 5–7, were used in all cases, unless stated otherwise. In the 1D simulations, diffusion perpendicular to the electrode was the only form of mass transport considered. In contrast, 2D simulations involved convection (flow) and diffusion parallel to the electrode as well as diffusion normal to the electrode. For the 1D simulations, the cathode (1 cm²) was located 0.5 mm away from an inert wall.[15] For the 2D simulations, the

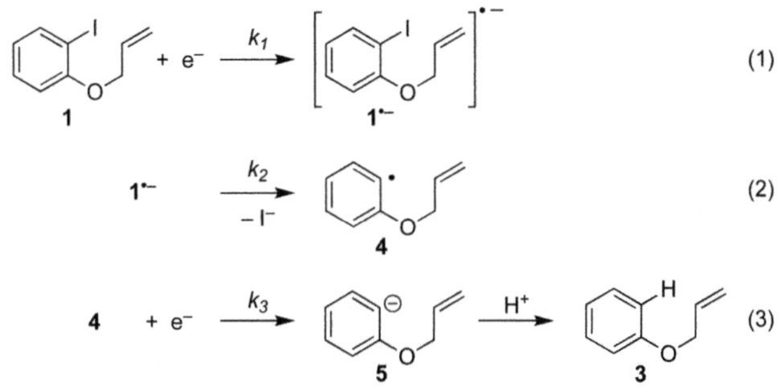

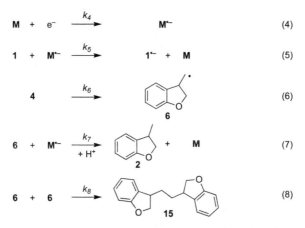

Scheme 5 ECE(C) pathway: reaction steps involved in direct electrochemical reduction (hydrogenolysis) of ArI **1**.

cathode (0.95 mm long, 2 mm deep) was located on the floor of a narrow channel flow cell (0.5 mm high, 1 mm long, 2 mm deep).[15] For simplicity, solution velocities within the channel were computed assuming laminar flow. With Dig-iElch, adaptive meshing was used to allow the numerical procedure to track the rapidly changing concentration profiles. In COMSOL, an expanding grid with an adaptive mesh environment was used to confine a high density of mesh elements near the electrode surface where narrow distributions of species were expected. Without this, the numerical solver is not able to compute highly localised concentration gradients. However, the optimum mesh is difficult to predict as the reaction layer thickness and its distance away from the electrode depend on the rate constants of the homogeneous processes and on the concentrations and diffusion coefficients of the species involved.[16] In COMSOL, the mesh was first optimised by comparing the results of the 1D simulations with those obtained with DigiElch. For the 2D simulations, the optimised 1D mesh was then used to constrain the size and growth of mesh elements along the normal to the electrode. The mesh was further refined near the upstream electrode edge where high current densities were expected. Along the channel, larger mesh elements were selected as concentrations were not expected to vary rapidly.

Scheme 6 Reaction steps in the mediated electroreductive cyclisation of aryl iodide **1**, including dimerisation of alkyl radical **6**.

$$6 + e^- \xrightarrow[+ H^+]{k_9} 2 \qquad (9)$$

$$4 + M^{\cdot-} \xrightarrow{k_{10}} 5 + M \xrightarrow{H^+} 3 + M \qquad (10)$$

$$1^{\cdot-} + 4 \xrightarrow{k_{11}} 1 + 5 \xrightarrow{H^+} 1 + 3 \qquad (11)$$

Scheme 7 Additional reaction steps (9)–(10) included in simulations.

Mechanism of direct and mediated electrochemical reduction of aryl iodide 1

The present work concerns simulations of direct and mediated electroreduction of aryl iodide **1**, for which we have previously disclosed electroanalytical data and preparative results in batch and flow electrolysis reactors.[14] A proposed mechanism for direct electrochemical reduction of **1** leading to hydrogenolysis product **3** incorporates reactions (1)–(3) (Scheme 5).[7] Much effort has been devoted by others to the study of these mechanisms, including determination of kinetic and thermodynamic parameters for a variety of aryl halides that are not directly accessible by electrochemical techniques.[7] Where necessary, the present work uses literature values for rate constants and reduction potentials from related compounds/intermediates where we were unable to determine them.

The rate of ET of aryl iodide **1** at the electrode is primarily determined by the standard rate constant for ET, k_1, which reflects the extent of reorganisation energy required to form the transition state, and by the electrode potential which sets the ET driving force. The latter is high because the electrode potential of -2.8 V $vs.$ SCE, set in the simulation, is more negative than the E^0 for aryl iodide **1**; as a result, **1** is readily reduced at the electrode surface where its concentration is driven to zero. The standard potential for **1** ($E^0 = -2.2$ V $vs.$ SCE) and rate constant ($k_1 = 5 \times 10^{-3}$ cm s^{-1}) for heterogeneous electron transfer were obtained from simulated voltammograms fitted to the experimental voltammetry.[14,17] The characteristic features of the voltammogram show a single irreversible wave due to a rate limiting electron transfer and very fast fragmentation of ArI$^{\cdot-}$ giving aryl radical **4** and I$^-$, the former undergoing facile ET and protonation. A rate constant (k_2) of 10^{10} s^{-1} or greater is expected for reaction (2) based upon published experimental and theoretical values for fragmentation of radical anions of aryl iodides and bromides.[18] As discussed above, the heterogeneous ET in reaction step (3) is also assumed to be rapid as the electrode potential necessary for reduction of aryl halide **1** is well negative to that required for reduction of Ar$^{\cdot}$ **4** ($E^0_{\text{Ph}\cdot/\text{Ph}^-} \sim 0.0$ V $vs.$ SCE).[8a] An estimation of the heterogeneous electron transfer rate constant ($k_3 \sim 0.03$ cm s^{-1}) for step (3) was taken from Andrieux et al.[8a] The final protonation of the aryl anion **5** by the solvent (CH$_3$CN) is incorporated within step (3) as protonation of this highly basic species is under diffusion control, and in any case, anion **5** is not involved in any other reaction steps.

The electrochemical and chemical steps (4)–(8) are involved in the mediated reductive cyclisation of aryl iodide **1** giving rise to the experimentally observed cyclisation and radical dimerisation products **2** and **15**, respectively (Scheme 6). Reactions (5) and (7) involve homogeneous ET, and their rate constants ($k_5 = 4.0 \times 10^5$ M^{-1} s^{-1}, and $k_7 = 1 \times 10^9$ M^{-1} s^{-1}) are approximated from literature values of related processes.[19,20] Step (4) is a heterogeneous ET to the mediator and values for the ET rate constant ($k_4 = 3 \times 10^{-2}$ cm s^{-1}), and $E^0_{M/M^{\bullet-}}$ (-2.5 V vs. SCE) were obtained from the simulated voltammetry fitted to our experimental data.[14,17] The rate constant ($k_6 = 8 \times 10^9$ s^{-1}) for cyclisation step (6) is taken from experimental values estimated for the same reaction.[21] As protonation of highly basic alkyl anionic species by a component of the electrolysis medium is rapid, and has no bearing on onwards reactions, it is incorporated in step (7).[22] Previous studies in deuterated solvents provide strong evidence that dihydrobenzofuran is formed principally by step (7) rather than through H-atom abstraction. Therefore, abstractions of a hydrogen atom from solvent by aryl and alkyl radicals have not been included in the present simulations. The rate constant ($k_8 = 10^9$ M^{-1} s^{-1}) for dimerisation of alkyl radical 6 (step (8)) is approximated to that reported for the 2-phenylethyl radical.[23]

Additional reactions included in simulations are heterogeneous reduction of the cyclised radical 6 (step (9)), homogeneous reduction of aryl radical 4 by M$^{\bullet-}$ (step (10)) or through disproportionation step (11) with 1$^{\bullet-}$ (Scheme 7). While all the products of these reactions are observed experimentally, heterogeneous reduction of the cyclised radical is only expected to be important in the direct mechanism. Bimolecular reactions (10) and (11) are unlikely to contribute significantly toward the formation of 3 as both radical anion 1$^{\bullet-}$ and aryl radical 4 are consumed in very fast unimolecular reaction steps (2) and (6).

Diffusion coefficients for the substrate (1, 3.3×10^{-5} cm^2 s^{-1}) and phenan-threne (M, 2.0×10^{-5} cm^2 s^{-1}) were determined using voltammetry.[14] The diffusion coefficients for intermediates and mediator radical anions are approximated to those of their respective parent compounds.

Results

Reaction product profiles

Electrolysis of ArI 1 was first simulated in one dimension (1D) combining the electrochemical steps (1)–(11) using COMSOL Multiphysics. The results were validated with simulations performed in DigiElch for identical conditions. This allowed optimisation of the COMSOL mesh elements along the electrode normal. Subsequently, the same mechanism was simulated in COMSOL in the presence of fluid flow (see below). The 1D simulations, where mass transport is only through diffusion perpendicular to the electrode, are representative of unstirred batch electrolysis conditions, whereas the 2D flow simulations are representative of channel flow cells as found in the Ammonite 8 electrochemical reactor,[24] where mass transport involves convection and diffusion parallel to the electrode surface as well as diffusion perpendicular to the electrode. The fluid dynamics models do not include migration of charged species under the influence of electric field, which will be a topic of further work. Fig. 1a shows percentage conversions for electrolysis of ArI ($t = 0$ s, [1] = 0.025 M) from the 1D simulation over 60 s in the absence of the mediator ([M] = 0 M) with the electrode potential set to -2.8 V vs. SCE, a value that is negative to both substrate and phenanthrene. As conversion of ArI proceeds, cyclisation and hydrogenolysis products (2 and 3, respectively) are both formed with a modest selectivity for 3 (2 : 3 $\sim$ 1 : 1.2), which qualitatively is comparable with the experimental results for direct electrolysis in a stirred batch cell (2 : 3 $\sim$ 1 : 2).[14] In the unmediated reaction, the balance between

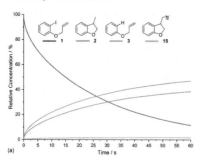
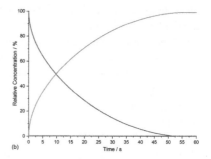

Fig. 1 1D simulation (representative of unstirred batch conditions) showing relative concentrations of ArI **1**, and products **2**, **3** and **15**, respectively, as a function of time for: (a) direct electrolysis of ArI **1**; (b) electrolysis of **1** in the presence of phenanthrene (1 equiv.). Starting concentration of ArI **1** is 0.025 M. Electrolysis time = 0 to 60 s under constant potential of −2.8 V vs. SCE.

hydrogenolysis and cyclisation pathways is influenced by the relative rates of steps (3) and (6). Cyclisation of aryl radical **4** is a 5-*exo-trig* process with a high rate constant $(k_6 = 8 \times 10^9 \text{ s}^{-1})$,[21] whereas the rate of heterogeneous ET step (2) is determined by both the standard rate constant, k_2, and the electrode potential. At conditions of high cathode overpotential, heterogeneous ET to Ar˙ $(E^0 \sim 0 \text{ V } vs.$ SCE) is a very fast process, giving rise to the observed selectivity.

The product **15** from the dimerisation of alkyl radical **4** (step (8)) is seen in small amount (2%) under the simulation conditions where the cathode potential is set to −2.8 V vs. SCE. The influence of cathode potential on product selectivity is discussed below. Repeating the simulation in the presence of 1 equivalent of phenanthrene ([**M**] = 0.025 M) shows a dramatic switch in the selectivity (Fig. 1b), now strongly in favour of cyclised product **2**, with minor amounts of hydrogenolysis and dimerisation products. After 60 s, the simulated selectivity profile $(\mathbf{2}:\mathbf{3}:\mathbf{15} \sim 99:0.5:0.5)$ again displays a similar trend to the experimental outcome from our batch electrolysis (**2** (74%):**3** (3%):**15** (1%) $\sim 95:4:1$). Another interesting feature of the simulation in the presence of phenanthrene is that very little ArI remains after 50 s, compared to ∼10% at 60 s when the mediator is absent. The various factors influencing the selectivity outcome of the electrolysis are discussed in the ensuing sections.

Time dependence of concentration profiles of the reaction components

Defining the spatial distribution of reactive species is central to this work and is presented here using concentration profiles with respect to distance from the cathode for the 1D simulations. Consumption of species at the cathode gives rise to concentration gradients between the electrode surface and the inert wall, and these are seen to evolve in a time dependent fashion. In the absence of phenanthrene, concentration profiles for ArI **1** show its concentration to be highest near the inert wall and dropping to zero at the cathode surface where it is consumed by heterogeneous ET step (1) to give Ar˙⁻ (Fig. 2a). Concentration profiles for **1** flatten from $t = 1$–60 s as its total concentration decreases by its progressive conversion to products **2** and **3**. Simulated profiles for **3** and **2** show concentrations that are highest at the cathode, where they are formed by the

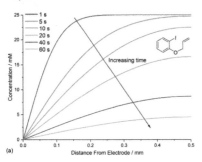
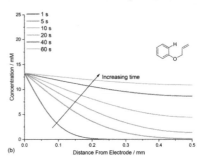

Fig. 2 Time dependent profiles for direct electrolysis of ArI 1 showing: (a) ArI 1; (b) hydrogenolysis product 3. Starting concentration of 1 is 0.025 M. Electrolysis time = 0 to 60 s under constant potential of −2.8 V vs. SCE.

direct mechanism steps (1)–(3), or (1), (2), (6) and (9), respectively (Fig. 2b, only the profile for 3 is shown as the one for 2 behaves similarly). The steepest concentration profile gradients are at short reaction times (1 s) and close to the electrode where conversion is driven by fast heterogeneous ET and chemical reactions, and relatively little diffusion has occurred. Profiles for the reactive intermediates $1^{\cdot-}$ and 4 are not shown as these species do not accumulate, or escape the proximity of the cathode in the direct process, due to their consumption in very fast onward reaction steps (2) and (3)/(6), respectively.

The simulated profiles for electrolysis of 1 in the presence of phenanthrene (1 equiv.) at a cathode potential of −2.8 V vs. SCE reveal a more interesting time dependent behaviour, particularly in the early stages of the process (Fig. 3). Under the conditions $E_{ede} < E^0_M < E^0_1$, both 1 and M are electrochemically reduced (steps (1) and (4)) at the onset of electrolysis, with their concentrations tending to zero at the cathode (Fig. 3a and b). However, as $M^{\cdot-}$ diffuses away from the electrode it reacts homogeneously with 1 thereby accelerating the depletion of 1 near the electrode. Even after a short time ($t < 1$ s), 1 is sufficiently depleted in the region close to the cathode for M to act as the charge shuttle and the mechanism for reduction of 1 switches from heterogeneous (direct) step (1) to homogeneous (mediated) step (5). As time progresses, the region depleted in 1 grows from the cathode as 1 is consumed by the strongly reducing mediator radical anion $M^{\cdot-}$ as it diffuses outwards. The concentration profiles for $M^{\cdot-}$ show that its flux from the cathode surface overcomes the inwards flux of ArI 1 (Fig. 3c). Considering hydrogenolysis product 3, its profiles at the beginning of electrolysis show concentration highest close to the cathode (Fig. 3d), which arises from the initial direct (ECE) process. As electrolysis time progresses, the concentration of 3 at the cathode decreases, tending towards a constant concentration across the cell showing that no further 3 is being produced and the direct mechanism is no longer contributing significantly. Thus, the hydrogenolysis product 3 that is present in the reaction solution is only produced during the very early stages of the process, highlighting the switch from direct to mediated mechanism. Formation of 3 by homogeneous reactions (10) and (11) does not appear to contribute significantly as already discussed above. Direct visualisation of the detached reaction layer is challenging as intermediates $1^{\cdot-}$ and Ar$^{\cdot}$ 4 do not accumulate due to very rapid onward reactions. However, the concentration

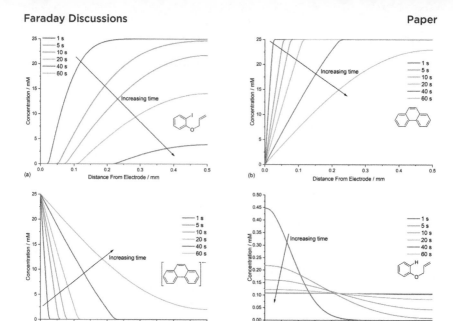

Fig. 3 Time dependent concentration profiles for electrolysis of ArI **1** in the presence of phenanthrene (1 equiv.) showing: (a) ArI **1**. (b) Phenanthrene (**M**). (c) Phenanthrene radical anion (**M**$^{\cdot-}$). (d) Hydrogenolysis product **3**. Starting concentration of ArI **1** is 0.025 M. Electrolysis time = 0 to 60 s under constant potential of −2.8 V *vs*. SCE.

profile for alkyl radical **6** – formed from **1**$^{\cdot-}$ by fragmentation and cyclisation steps (5) and (6) – provides insight into the zone where **1**$^{\cdot-}$ is produced (and consumed). In Fig. 4 the maximum, albeit low, concentration of **4** can be seen to move progressively outwards from the cathode with time, which is a manifestation of homogeneous ET step (5) occurring away from the cathode, *i.e.* reaction layer detachment.

It is important to emphasise that under the simulated conditions in the presence of phenanthrene, the process starts as direct plus mediated before

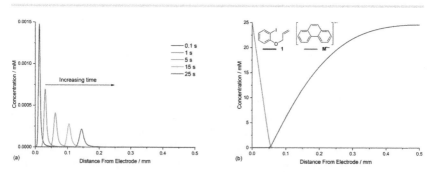

Fig. 4 Illustrative visualisation of reaction layer detachment showing simulated concentration profiles for: (a) the cyclised radical **4** between 0.1 s and 25 s of electrolysis, moving progressively away from the cathode. (b) **1** and **M**$^{\cdot-}$ after 5 s of electrolysis intersecting where their fluxes are matched.

becoming fully mediated. The time taken to establish the mediated pathway depends on several parameters, including, heterogeneous and homogeneous rate constants, diffusion coefficients, and bulk concentrations. The influence of these, and other variables, will be explored through simulations presented and discussed in the ensuing sections.

Diffusion coefficient of mediator

The diffusion coefficients for phenanthrene (2.0×10^{-5} cm^2 s^{-1}) and aryl iodide **1** (3.3×10^{-5} cm^2 s^{-1}) were established experimentally from voltammetry, and values for reaction intermediates are assumed to be the same as their parent molecules.[14,17] Simulations of concentration profiles allowed us to visualise the influence of varying diffusivity of mediator **M** using three different values (1, 2, and 4×10^{-5} cm^2 s^{-1}), and in all cases time dependent reaction layer detachment was observed (Fig. 5). Detachment progresses more quickly for a mediator with a higher diffusion coefficient, and at $t = 5$ s the maximum concentration of cyclised intermediate **6** occurs further from the cathode (Fig. 5a). Greater diffusivity of **M/M$^{•-}$** supports an increased flux of reducing species perpendicular from the cathode, pushing homogeneous ET outwards into solution, which is also indicated by the depletion of **1** at an increased distance from the electrode (Fig. 5b). These results are consistent with previous work using microelectrodes.[16] It is also apparent that the larger diffusion coefficient also increases the rate of conversion, indicated from the higher concentration of cyclised intermediate (integral of peak area from Fig. 5a).

Mediator stoichiometry

Opposing fluxes of mediator radical anion and ArI **1** play a key role in determining the region where homogeneous ET reaction (5) takes place, and a sufficient flux of **M$^{•-}$** is required for detachment of the reaction layer. As the flux is the product of the diffusion coefficient and concentration gradient, the stoichiometry of mediator is an important parameter. Fig. 6a–c show the effect of varying amounts of mediator (phenanthrene, 1.0, 0.5, and 0.05 equiv.) relative to the substrate **1** (0.025 M). Under the conditions of the simulation, detachment of the reaction layer is observed even with catalytic electron transfer mediator (0.05 equiv.) at

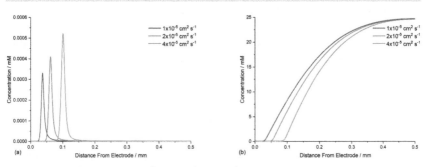

Fig. 5 Influence of mediator ($E^0 = -2.5$ V vs. SCE) with different diffusion coefficients (1×10^{-5} cm^2 s^{-1}, 2×10^{-5} cm^2 s^{-1}, and 4×10^{-5} cm^2 s^{-1}) on reaction layer detachment. Concentration profiles of: (a) cyclised radical **6** at $t = 5$ s; (b) ArI **1** at $t = 5$ s.

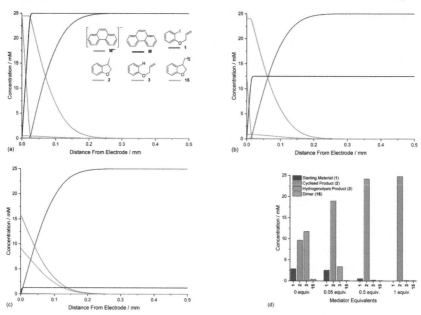

Fig. 6 Influence of mediator (phenanthrene, M) stoichiometry on reaction layer detachment. Concentration profiles for species 1, 2, 3, 15, M and M$^{\bullet-}$ for simulated electrolysis under constant potential of -2.8 V vs. SCE at $t = 1$ s. (a) 1 equiv. of M. (b) 0.5 equiv. of M. (c) 0.05 equiv. of M. (d) Product selectivity profile from simulations with different mediator loadings after electrolysis for 60 s.

short time (1 s of simulation). Detachment is realised more quickly for higher mediator loadings (Fig. 6a and b), which is evident from the smaller amount of hydrogenolysis product (area under the red line). It should be stressed that no further hydrogenolysis product is being produced under any of the mediator loadings at 1 s, and all simulations show that the mediated mechanism has taken over at this time. The simulated product distribution after electrolysis for 60 s shows high selectivity for the cyclised product 2 at all mediator loadings investigated (Fig. 6d), which is in good qualitative agreement with experimental results using catalytic (0.05 equiv.) and stoichiometric (1 equiv.) mediator.[14] As the simulated conditions only include mass transport by diffusion, they cannot be considered truly representative of the experimental conditions in batch (stirred) or flow. None the less, the similar trends in the experimental and simulated results are noteworthy. Simulations of mediated electrolysis under flow conditions will be discussed in later parts of this paper.

Influence of mediator

The rate constants for homogeneous and heterogeneous ET steps (4), (5), (7) and (10) are related to the structure and reduction potential of the mediator. Furthermore, the rate of heterogeneous ET step (4) is affected by the potential of the cathode, and the diffusion coefficient of the mediator also directly influences concentration profiles as has been exemplified above. In view of these complex relationships, and challenges obtaining accurate rate constants for all ET steps,

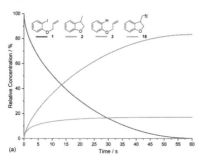
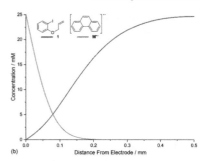

Fig. 7   Influence of mediator M′ ($E^0 = -2.0$ V *vs.* SCE) on product selectivity and reaction layer. (a) Relative concentrations of ArI **1** and **2**, **3** and **15** as a function of time. (b) Concentration profiles for species **1** and $M'^{•-}$ for simulated electrolysis under constant potential of -2.8 V *vs.* SCE at $t = 5$ s. Starting concentration of ArI **1** is 0.025 M. Electrolysis time $= 0$ to 60 s under constant potential of -2.8 V *vs.* SCE.

the present study takes, for the purpose of illustration, a single hypothetical mediator (**M′**) with a standard potential (-2.0 V *vs.* SCE) that is positive to that of the model substrate **1**. Now ET from the mediator to **1** is thermodynamically unfavourable, however, the E^0 for the mediator should remain sufficiently negative such that the rate of homogeneous ET step (5), when followed by a very rapid irreversible reaction (2), leads to a reasonable rate for aryl radical formation. In the simulation, a value of $k_5 = 100$ M^{-1} s^{-1} is used for ET step (5), which is 250 times slower than when using phenanthrene.[19] The ET rate constants for heterogeneous step (4) and homogeneous step (7) are also adjusted accordingly ($k_4 = 7$ cm s^{-1} and $k_7 = 10^8$ M^{-1} s^{-1}), and all other conditions are unchanged from those used in Fig. 3.[20,25]

Simulation of the product selectivity profile with **M′** ($E^0 = -2.0$ V *vs.* SCE) retains selectivity for the cyclised product **2**, even under conditions where the electrode potential is negative to both mediator and substrate (Fig. 7a). However, an increased amount of hydrogenolysis product **3** is produced compared to the simulation with the more strongly reducing mediator, phenanthrene. Inspection of concentration profiles for the reducing species $M'^{•-}$ and **1** show that some of the substrate now reaches the electrode surface due to the slower rate of homogeneous ET step (5), which effectively leads to a broadening of the homogeneous reaction zone such that it does not detach from the electrode completely under the simulated conditions. Although the simulation implies an advantage of using the more strongly reducing mediator phenanthrene, some caution should be exercised due to the uncertainty in the ET rate constants used. It should also be highlighted that, for a selective mediated synthesis using **M′**, the electrode potential would normally be set such that $E^0_1 < E_{ede} < E^0_{M'}$ to decrease the rate of the unmediated reaction.

Electrode potential

The effect of varying electrode potential on the electroreduction of aryl iodide **1** in the presence of phenanthrene (1 equiv.) as mediator is shown in Fig. 8. At cathode potentials that are negative or equal (-2.8 V and -2.5 V *vs.* SCE) to the standard potential for phenanthrene the mediated mechanism occurs already 1 s after electrolysis commences (Fig. 8a and b). The profiles at a cathode potential of −2.8 V are discussed above (see Fig. 3, 4 and 6a), and show reaction layer detachment. When the electrode potential is −2.5 V – close to E^0 for phenanthrene – the mediated mechanism is again in operation, although the detached reaction layer takes longer to establish and is closer to the cathode at corresponding times (Fig. 8b). More hydrogenolysis product 3 is present, formed during a short period at the onset of electrolysis by the direct mechanism. The slower progress of detachment at −2.5 V can be explained by the lower heterogeneous rate constant for ET to the mediator, lowering the concentration gradient and flux of $\mathbf{M^{\bullet-}}$ from the cathode.

A further positive shift of the electrode potential to −2.2 V – now positive to phenanthrene and close to E^0 for ArI – returns the mechanism to the direct one as very little $\mathbf{M^{\bullet-}}$ is produced at the cathode (Fig. 8c). Even after an extended time (25 s) no detachment is seen at an electrode potential of −2.2 V, with the direct mechanism prevailing (Fig. 8d). An interesting feature of the simulated profiles at −2.2 V is the formation of dimer 15, rather than 2, which is explained by the relatively slow heterogeneous reduction of alkyl radical 6 at the more positive electrode potential.[26] This should be investigated experimentally, as other reactions of cyclised alkyl radical 6 that are not included in the simulation may become significant at the more positive potential. Finally, and unsurprisingly, adjustment of the electrode potential to −2.0 V – positive to standard potentials for both ArI and phenanthrene – results in the rate of electrochemical reduction becoming very small indeed (not shown).

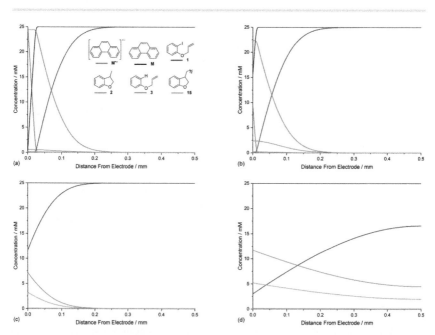

Fig. 8 Influence of electrode potential on reaction mechanisms for electroreduction of 1 in the presence of phenanthrene (M, 1 equiv.). Concentration profiles for species 1, 2, 3, 15, M and $M^{\bullet-}$ at electrode potentials of: (a) −2.8 V vs. SCE at $t = 1$ s. (b) −2.5 V vs. SCE at $t = 1$ s. (c) −2.2 V vs. SCE at $t = 1$ s. (d) −2.2 V vs. SCE at $t = 25$ s.

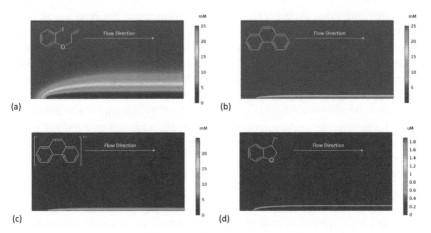

Fig. 9 Flow simulation for electroreduction of **1** in the presence of phenanthrene (M, 1 equiv.). Concentration heat maps for: (a) ArI **1**. (b) Phenanthrene (M). (c) Phenanthrene radical anion (M$^{\cdot-}$). (d) Cyclised radical intermediate **6**. Starting concentration of ArI **1** is 0.025 M. Electrolysis time = 0 to 2.4 s under constant potential of −2.8 V vs. SCE.

Simulations of electroreduction in a laminar flow regime

We recently described electroreductive cyclisation of ArI **1** in the presence of phenanthrene (1 equiv.) in a flow reactor, the Ammonite 8,[14] which possesses a long path length (100 cm) and small interelectrode gap (0.5 mm) (see Scheme 4).[24,27] Although it is beyond the scope of the present work to develop a simulation for mediated electrolysis in the 100 cm long spiral channel of this reactor, preliminary 2D simulations of the process have been performed over a short channel (1 mm) with an interelectrode gap of 0.5 mm using COMSOL Multi-physics and the conditions described above, but with introduction of parabolic laminar flow of the electrolyte solution parallel to the cathode.

In batch simulations (1D), the concentration profiles of the different species only require the dimension perpendicular to the electrode, whereas in the flow reactor the concentrations of the different reaction components also vary along the channel length and it is beneficial here to consider 2D "heat maps" of the concentrations. The heat map in Fig. 9d, showing the spatial distribution of the cyclised radical intermediate **6**, is a surrogate for the zone where the radical anion **1**$^{\cdot-}$ is formed by either heterogeneous ET step (1) or homogeneous ET step (5), as **1**$^{\cdot-}$ is converted to **6** very rapidly.

Furthermore, the radical **6** does not accumulate as it is consumed in onwards bimolecular reactions (7) and (8). The concentration heatmap for **6** shows that it is produced close to the cathode at the start of the electrode segment in the flow channel, but that its formation moves outwards from the electrode as the electrolyte flows downstream. While in unstirred batch reactors the reaction pathway switches from direct to mediated as time progresses, the 2D simulations demonstrate that under flow, the change in pathway from direct to mediated occurs as the solution progresses downstream from the electrode edge, a complex function of time and distance.

Another important aspect is that the flow yields a boundary layer where mass transport evolves from convection far from the electrode to diffusion near the electrode. At higher flow rates (not shown), the boundary layer will be thin and confine the diffusion region closer to the electrode. This will enhance the flux of **1** towards the electrode and it will take longer (both in time and distance) for the reaction pathway to switch from direct to mediated. At low flow rates, the boundary layer will extend further in solution, the diffusion region will widen, the flux of **1** will decrease, and the switch from direct to mediated pathway will occur sooner. A more detailed analysis of effect of the flow rate on the mechanism is the subject of ongoing work.

Discussion

A common logic applied in reductive electrosynthesis using ET mediators is that the redox potential of the mediator should be positive to the potential of the substrate to promote selective electron transfer from the cathode to the redox catalyst.[9] Thus, for selective mediated electroreduction, the conditions $E^0_M > E_{ede} > E^0_S$ would be expected, where ede is the cathode, M is the mediator, and S is the substrate. However, this is not the situation seen in the mediated electroreductive cyclisations in Schemes 3b and 4, where $E^0_S > E^0_M$, and direct reduction of the substrate is expected to be favoured at electrode potentials required to reduce the mediator. We have proposed that the observed selectivity results from spatio-temporal effects where homogeneous ET occurs in a reaction layer that becomes progressively detached from the electrode surface.[14] Under the conditions of detachment, and despite facile reduction of the substrate at the cathode, the substrate is not directly reduced because it does not reach the cathode surface. In view of the complex interplay between mass transport, time, and the rates of multiple steps – including chemical, homogeneous, and heterogeneous ET processes – we have established simulations in order to illustrate the key requirements for reaction layer detachment.

The 1D simulations for the phenanthrene-mediated reduction of aryl iodide **1** support the proposed time dependent detachment of the homogeneous electro-chemistry from the cathode, and are qualitatively consistent with the experimentally observed selectivities. Initially both **1** and **M** are reduced at the cathode, and the direct and mediated mechanisms operate simultaneously. However, as $M^{\cdot-}$ diffuses away from the electrode it reacts homogeneously with **1** thereby accelerating the depletion of **1** near the electrode. After a period of time, substrate **1** no longer reaches the cathode and the process becomes fully mediated where **M** acts as the charge shuttle between the electrode and the detached homogeneous reaction layer, favouring cyclisation product **2**. The time taken to establish the mediated pathway depends on several parameters, primarily, heterogeneous and homogeneous rate constants, diffusion coefficients, and bulk concentrations. It is perhaps convenient to think of the position of the homogeneous reaction layer being determined by opposing fluxes of $M^{\cdot-}$ and **1**, which are the products of their concentration gradients (time dependent) and diffusion coefficients.[28] The thickness of the homogeneous reaction layer is dependent on the homogeneous rate constants and diffusion coefficients in the 1D simulation. Slower rates of homogeneous processes lead to broadening of the homogeneous reaction layer arising from longer lifetimes of reactive intermediates.

The flux of mediator radical anion $M^{\cdot-}$ from the cathode is affected by mass transport, concentration of M, and the rate of heterogeneous ET step (4). Although the latter is limited by the electron transfer rate constant (k_4), and transfer coefficient (α), a sluggish ET is overcome by applying sufficient cathode over-potential (*e.g.* -2.8 V *vs.* SCE in the simulation). As the overall rate, *i.e.* the electrode current, is affected by the rate of ET and the rate of mass transport, the simulations also help to explain how the mediated mechanism prevails under galvanostatic conditions. When the applied current exceeds the rate at which the substrate **1** is replenished at the electrode surface by diffusion, the electrode potential adjusts to a value sufficiently negative to drive the electrochemical reduction of the mediator. This process not only provides the additional rate required to match the applied current but also promotes the mediated reduction of the substrate. Thus, reaction layer detachment can also be achieved under constant current conditions when $E^0_1 > E^0_M$, providing that the current density exceeds the limiting current density for the substrate.

Reaction layer detachment is also seen in 2D simulations where mass transport by convection is included to represent conditions found in laminar flow cells. In the 2D simulations, detachment is seen as a function of distance along the flow channel, and the mechanism of reduction switches from direct plus mediated at the start of the electrode, to purely mediated after a short distance downstream from the electrode edge. Again, the initial results showing detachment are consistent with preparative work showing selectivity for the mediated process leading to the cyclised product **2** (Scheme 4). It should be emphasised that the current density and potential are not constant along the length of the flow channel,[29] and 2D simulations in a longer channel require considerably greater computational power. A more detailed investigation is therefore beyond the scope of the current work and is the subject of ongoing work.

Finally, it should be recognised that our experimental work was conducted under constant current conditions, and without a reference electrode present in the flow cell to monitor the electrode potentials. It is, however, entirely reasonable for the cathode potential to be negative of E^0_M, and indeed, the experimental observation of the mediated pathway combined with the results of the simulation provide indirect support for this. The voltage across the flow cell is typically in the region of 5 to 6 V, and the measured cathode potential for the mediated electrolysis of **1** in a batch cell was -3.5 V *vs.* SCE under steady state conditions. On this basis we are confident that the electrode potential in the batch and flow cells are such that the simulated conditions of $E_{ede} < E^0_M < E^0_1$ are in operation.

Conclusions

Simulations presented herein support the proposed role of reaction layer detachment in electroreductive cyclisation of aryl halides in the presence of strongly reducing ET mediators. It is shown that mediators that are harder to reduce than their substrates – $E^0_M < E^0_1$ – can be employed, providing that the conditions of homogeneous reaction layer detachment are achieved. The resulting highly reducing mediator radical anions promote kinetically accelerated and thermodynamically favourable homogeneous ET with substrates and intermediates, as opposed to classical application of mediators relying on thermodynamically uphill ET followed by irreversible coupled reactions.

Reaction layer detachment is an example of a spatio-temporal effect arising due to opposing fluxes of two species – here the mediator radical anion $M^{\cdot-}$ and the substrate **1** – which undergo very rapid homogeneous ET and onwards reactions. The detachment results from a complex interplay between the relative concentrations and diffusion coefficients of the mediator and substrate as well as the rate constant for the homogeneous process between them. No detachment is observed when the substrate is not sufficiently depleted near the electrode *i.e.*, when the homogeneous reaction between substrate and mediator is slow or when the flux of the substrate is much larger than that of the mediator. When detachment occurs, the width and location of the reaction zone are directly determined by mass transport, concentrations of **M** and **1**, and the heterogeneous and homogeneous rate constants.

Experimental and computational work is underway to further investigate and understand the complex interplay of mass transport, heterogeneous and homogeneous ET and coupled chemistry that may lead to new opportunities in electrosynthesis using powerful homogeneous ET mediators.

Author contributions

JWH performed experiments and simulations. AAF performed experiments. DP, DCH, GD and RCDB conceptualised the project. RCDB, DCH and GD supervised the project. RCDB wrote the manuscript with assistance and contributions from all authors.

Conflicts of interest

There are no conflicts to declare.

Acknowledgements

The authors acknowledge financial support from the EPSRC (Photo-Electro Programme Grant EP/P013341/1 and EP/K039466/1), and the use of the IRIDIS High Performance Computing Facility, and associated support services at the University of Southampton, in the completion of this work.

Notes and references

1 For general reviews of organic radicals and their reactions, see: (*a*) W. P. Neumann, *Synthesis*, 1987, 665–683; (*b*) C. P. Jasperse, D. P. Curran and T. L. Fevig, *Chem. Rev.*, 1991, **91**, 1237–1286; (*c*) *Radicals in Organic Synthesis, Vol. 1 and 2*, ed. P. Renaud and M. P. Sibi, Wiley-VCH, Weinheim, 2001; (*d*) S. Z. Zard, *Radical Reactions in Organic Synthesis*, Oxford University Press, Oxford, 2003; (*e*) J. Lalevée and J. P. Fouassier, in *Encyclopedia of Radicals in Chemistry, Biology and Materials, Vol 2: Synthetic Strategies and Applications*, ed. C. Chatgilialoglu and A. Studer, John Wiley and Sons, Weinheim, 2012.

2 For a perspective on radicals in synthesis, see: M. Yan, J. C. Lo, J. T. Edwards and P. S. Baran, *J. Am. Chem. Soc.*, 2016, **138**, 12692–12714.

3 For selected reviews covering aryl radical chemistry from a variety of precursors, see: (*a*) C. Galli, *Chem. Rev.*, 1988, **88**, 765–792; (*b*) G. Pratsch and H. R. Heinrich, *Top. Curr. Chem.*, 2012, **320**, 33–59; (*c*) F. Mo, D. Qiu, L. Zhang and J. Wang, *Chem. Rev.*, 2021, **121**, 5741–5829; (*d*) N. Kvasovs and V. Gevorgyan, *Chem. Soc. Rev.*, 2021, **50**, 2244–2259; (*e*) F. Juliá, T. Constantin and D. Leonori, *Chem. Rev.*, 2022, **122**, 2292–2352; (*f*) J. H. Lan, R. X. Chen, F. F. Duo, M. H. Hu and X. Y. Lu, *Molecules*, 2022, **27**, 5364.

4 Electroreductive cyclisation of aryldiazonium salts has also been described: F. LeStrat, J. A. Murphy and M. Hughes, *Org. Lett.*, 2002, **4**, 2735–2738.

5 For some examples where selective cyclisations of aryl radicals have been achieved by direct electrolysis, see: (*a*) R. Munusamy, K. Samban Dhathathreyan, K. Kuppusamy Balasubramanian and C. Sivaramakrishnan Venkatachalam, *J. Chem. Soc., Perkin Trans. 2*, 2001, 1154–1166; (*b*) J. Grimshaw, R. J. Haslett and J. Trocha-Grimshaw, *J. Chem. Soc., Perkin Trans. 1*, 1977, 2448–2455; (*c*) S. Donnelly, J. Grimshaw and J. Trocha-Grimshaw, *J. Chem. Soc., Chem. Commun.*, 1994, 2171–2172; (*d*) M. Dias, M. Gibson, J. Grimshaw, I. Hill, J. Trocha-Grimshaw and O. Hammerich, *Acta Chem. Scand.*, 1998, **52**, 549–554; (*e*) R. Gottlieb and J. L. Neumeyer, *J. Am. Chem. Soc.*, 1976, **98**, 7108–7109.

6 For examples of cathodic reductive dehalogenation of aryl halides, see: (*a*) C. P. Andrieux, J. Badoz-Lambling, C. Combellas, D. Lacombe, J. M. Savéant, A. Thiebault and D. Zann, *J. Am. Chem. Soc.*, 1987, **109**, 1518–1525; (*b*) K. Mitsudo, T. Okada, S. Shimohara, H. Mandai and S. Suga, *Electrochemistry*, 2013, **81**, 362–364; (*c*) J. Ke, H. Wang, L. Zhou, C. Mou, J. Zhang, L. Pan and Y. R. Chi, *Chem. – Eur. J.*, 2019, **25**, 6911–6914; (*d*) C. B. Liu, S. Y. Han, M. Y. Li, X. D. Chong and B. Zhang, *Angew. Chem., Int. Ed.*, 2020, **59**, 18527–18531; (*e*) L. Lu, H. Li, Y. Zheng, F. Bu and A. Lei, *CCS Chem*, 2020, **2**, 2669–2675; (*f*) A. A. Folgueiras-Amador, A. E. Teuten, D. Pletcher and R. C. D. Brown, *React. Chem. Eng.*, 2020, **5**, 712–718.

7 The reductive electrochemistry of aryl halides has been reviewed, see: (*a*) J.-M. Savéant, in *Adv. Phys. Org. Chem.*, ed. D. Bethel, Academic Press, New York, 1990, vol. 26, pp. 1–130; (*b*) J. Grimshaw, in *Electrochemical Reactions and Mechanisms in Organic Chemistry*, Elsevier Science, 2000, pp. 89–157; (*c*) Z. Chami, M. Gareil, J. Pinson, J.-M. Savéant and A. Thiebault, *J. Org. Chem.*, 1991, **56**, 586–595.

8 ET to aryl radicals ($E^0_{\text{Ph.}/\text{Ph}^-} \sim 0$ V *vs.* SCE) is a relatively facile process in comparison to the parent halides ($E^0 \sim -2$ V *vs.* SCE). (*a*) C. P. Andrieux and J. Pinson, *J. Am. Chem. Soc.*, 2003, **125**, 14801–14806; (*b*) C. Costentin, M. Robert and J.-M. Savéant, *J. Am. Chem. Soc.*, 2004, **126**, 16051–16057.

9 For selected reviews, see: (*a*) E. Steckhan, *Angew. Chem., Int. Ed. Engl.*, 1986, **25**, 683–701; (*b*) R. Francke and R. D. Little, *Chem. Soc. Rev.*, 2014, **43**, 2492–2521; (*c*) R. Francke, A. Prudlik and R. D. Little, in *Science of Synthesis: Electrochemistry in Organic Synthesis*, ed. L. Ackermann, Thieme: Stuttgart, 2021, vol. 1, pp. 293–324; (*d*) C. Zhu, N. W. J. Ang, T. H. Meyer, Y. Qiu and L. Ackermann, *ACS Cent. Sci.*, 2021, **7**, 415–431; (*e*) L. F. T. Novaes, J. Liu, Y. Shen, L. Lu, J. M. Meinhardt and S. Lin, *Chem. Soc. Rev.*, 2021, **50**, 7941–8002; (*f*) W. Shao, B. Lu, J. Cao, J. Zhang, H. Cao, F. Zhang and C. Zhang, *Chem. – Asian J.*, 2023, **18**, e202201093.

10 K. Mitsudo, Y. Nakagawa, J. Mizukawa, H. Tanaka, R. Akaba, T. Okada and S. Suga, *Electrochim. Acta*, 2012, **82**, 444–449.

11 For selected examples of electroreductive cyclisation using organic mediators, see ref. 5*b* and: (*a*) M. D. Koppang, G. A. Ross, N. F. Woolsey and D. E. Bartak, *J. Am. Chem. Soc.*, 1986, **108**, 1441–1447; (*b*) N. Kurono, E. Honda, F. Komatsu, K. Orito and M. Tokuda, *Tetrahedron*, 2004, **60**, 1791–1801; (*c*) A. Katayama, H. Senboku and S. Hara, *Tetrahedron*, 2016, **72**, 4626–4636.

12 For selected examples of Ni mediated electrochemical radical cyclisations of aryl halides, see: (*a*) S. Ozaki, H. Matsushita and H. Ohmori, *J. Chem. Soc., Perkin Trans. 1*, 1993, 2339–2344; (*b*) S. Olivero, J. C. Clinet and E. Duñach, *Tetrahedron Lett.*, 1995, **36**, 4429–4432; (*c*) E. Duñach, M. J. Medeiros and S. Olivero, *Electrochim. Acta*, 2017, **242**, 373–381; (*d*) C. Déjardin, A. Renou, J. Maddaluno and M. Durandetti, *J. Org. Chem.*, 2021, **86**, 8882–8890.

13 C. Costentin and J. M. Savéant, *Proc. Natl. Acad. Sci. U. S. A.*, 2019, **116**, 11147–11152.

14 A. A. Folgueiras-Amador, A. E. Teuten, M. Salam-Perez, J. E. Pearce, G. Denuault, D. Pletcher, P. J. Parsons, D. C. Harrowven and R. C. D. Brown, *Angew. Chem., Int. Ed.*, 2022, **61**, e202203694.

15 For further details of the simulations, see the ESI†

16 (*a*) G. Denuault, M. Fleischmann, D. Pletcher and O. R. Tutty, *J. Electroanal. Chem.*, 1990, **280**, 243–254; (*b*) O. R. Tutty and G. Denuault, *IMA J. Appl. Math.*, 1994, **53**, 95–109; (*c*) O. R. Tutty, *J. Electroanal. Chem.*, 1994, **377**, 39–51.

17 Experimental and simulated cyclic voltammetry data can be found in the supporting information for ref. 14.

18 C. Amatore, M. A. Oturan, J. Pinson, J. M. Saveant and A. Thiebault, *J. Am. Chem. Soc.*, 1985, **107**, 3451–3459.

19 C. P. Andrieux, C. Blocman, J. M. Dumas-Bouchiat and J. M. Saveant, *J. Am. Chem. Soc.*, 1979, **101**, 3431–3441.

20 D. Occhialini, J. S. Kristensen, K. Daasbjerg and H. Lund, *Acta Chem. Scand.*, 1992, **46**, 474–481.

21 (*a*) A. Annunziata, C. Galli, M. Marinelli and T. Pau, *Eur. J. Org. Chem.*, 2001, 1323–1329; (*b*) A. N. Abeywickrema and A. L. J. Beckwith, *J. Chem. Soc., Chem. Commun.*, 1986, 464–465.

22 Labelling studies provide strong evidence that dihydrobenzofuran is formed principally by step (7) rather than through H-atom abstraction. See ref. 14 and 10.

23 P. Neta, J. Grodkowski and A. B. Ross, *J. Phys. Chem. Ref. Data*, 1996, **25**, 709–1050.

24 R. A. Green, R. C. D. Brown, D. Pletcher and B. Harji, *Electrochem. Commun.*, 2016, **73**, 63–66.

25 A. Russell, K. Repka, T. Dibble, J. Ghoroghchian, J. J. Smith, M. Fleischmann, C. H. Pitt and S. Pons, *Anal. Chem.*, 1986, **58**, 2961–2964.

26 C. P. Andrieux, I. Gallardo and J. M. Saveant, *J. Am. Chem. Soc.*, 1989, **111**, 1620–1626.

27 The Ammonite 8 electrochemical reactor is available from Cambridge Reactor Design. https://www.cambridgereactordesign.com/ammonite/ammonite.html, accessed 30 Apr 2023.

28 For examples of reaction layers, or zones in electrochemistry see ref. 13, 16 and: (*a*) M. E. G. Lyons, D. E. McCormack, O. Smyth and P. N. Bartlett,

Faraday Discuss. Chem. Soc., 1989, **88**, 139–149; (*b*) W. J. Albery and A. R. Hillman, *J. Electroanal. Chem. Interfacial Electrochem.*, 1984, **170**, 27–49. For lead references to reaction-diffusion systems, see: (*c*) A. Comolli, A. De Wit and F. Brau, *Phys. Rev. E*, 2019, **100**, 052213; (*d*) A. Li, R. Chen, A. B. Farimani and Y. J. Zhang, *Sci. Rep.*, 2020, **10**, 3894.

29 R. A. Green, R. C. D. Brown and D. Pletcher, *J. Flow Chem.*, 2015, **5**, 31–36.

Faraday Discussions

PAPER

Primary *vs.* secondary alkylpyridinium salts: a comparison under electrochemical and chemical reduction conditions†

Bria Garcia, [a] Jessica Sampson,[b] Mary P. Watson [*a] and Dipannita Kalyani [*c]

Received 13th June 2023, Accepted 6th July 2023

DOI: 10.1039/d3fd00120b

This report details a systematic comparison of the scope of aryl bromides in nickel-catalyzed, reductive cross-electrophile couplings of primary *vs.* secondary alkylpyridinium salts using both electrochemical and chemical reductants. Facilitated by the use of high-throughput experimentation (HTE) techniques, 37 aryl bromides, including 13 complex, drug-like examples, were investigated. By using primary and secondary substrates differing only by one methylene, we observed that the trends in ArBr scope are similar between the primary and secondary alkylpyridinium salts, although distinctions were observed in isolated cases. In addition, the electrochemical conditions compared favorably to those using chemical reductants, especially among the more complex, drug-like aryl halides.

Introduction

Reductive cross-electrophile couplings are a powerful synthetic approach for the formation of $C(sp^3)$–$C(sp^2)$ bonds, offering broader functional group tolerance than their redox-neutral counterparts.[1,2] These types of methods were originally developed with manganese (Mn^0) or tetrakis(dimethylamino)ethylene (TDAE) reductants.[3–5] More recently, electrochemical conditions have been investigated to replace chemical reductants with an inherently more tunable approach.[6] Electroreductive couplings have been developed for alkyl halides, NHPI esters,[7] and very recently Katritzky pyridinium salts.[8] In these investigations, the reaction outcomes using electrochemical conditions are often compared to those using chemical reductants, with electrochemistry often providing complementary or

[a]*Department of Chemistry & Biochemistry, University of Delaware, Newark, Delaware 19716, USA. E-mail: mpwatson@udel.edu*

[b]*High Throughput Experimentation Facility, Department of Chemistry & Biochemistry, University of Delaware, Newark, Delaware 19716, USA*

[c]*Discovery Chemistry, Merck & Co., Inc., Kenilworth, New Jersey 07033, USA. E-mail: dipannita.kalyani@merck.com*

† Electronic supplementary information (ESI) available. See DOI: https://doi.org/10.1039/d3fd00120b

improved results. However, a "mix and match" approach is often applied in scope studies, making systematic comparisons of the differences (or similarities) between scope for primary (1°) *vs.* secondary (2°) *vs.* tertiary (3°) alkyl substrates difficult. Importantly, the rates of oxidative addition of the aryl bromide and the alkyl electrophile activation must be matched for productive catalysis. Hence it is difficult to predict *a priori* whether the trends in ArBr scope would be the same for various alkyl classes.

The ubiquity and diversity of alkyl amines in the inventories of pharmaceutical companies and the ease of activating them as Katritzky pyridinium salts makes deaminative couplings particularly useful in the context of medicinal chemistry applications. Our previous report demonstrated dramatic differences in the optimal reaction conditions for 1° *vs.* 2° alkylpyridinium salts. This result is consistent with previous reactions of alkylpyridinium salts (Scheme 1A).[5] The difference in stability of 1° *vs.* 2° alkyl radicals impacts both the rate and reversibility of C–N bond cleavage,[9] and different conditions are often required to achieve adequate yields using these two classes of substrates across a wide range of reaction types. In reductive couplings, this difference is exacerbated because the alkylpyridiniums may be directly reduced by the reductant, instead of by single electron transfer (SET) with the Ni catalyst, leading to an imbalance in the rate of alkyl radical formation and the rate of ArBr oxidative addition. Because of these differences in the reactivity between 1° and 2° alkylpyridinium salts, it was unclear if they would have complementary or distinct trends in their ArBr scope, making it challenging to extrapolate ArBr scope studies from one alkylpyridinium class to another.

Herein, we report a systematic comparison of nickel-catalyzed reductive cross-electrophile couplings of primary and secondary alkylpyridinium salts with a diverse set of aryl bromides (Scheme 1B). This study builds upon our recent report on electrochemical reductive coupling of alkyl pyridiniums with aryl halides.[8a] We now uncover relative reactivity trends between primary and secondary alkyl pyridinium substrates, where the primary and secondary substrates differ only by one methylene, making systematic comparison appropriate (Scheme 1B). Investigations were done using three distinct reaction conditions: (1) electrochemical reduction, (2) Mn⁰ as a heterogeneous reductant and (3) TDAE as a homogeneous reductant. Facilitated by the use of high-throughput experimentation (HTE) techniques, including use of the recently developed HTe-Chem reactor,[10] 24 aryl bromides were investigated, along with 13 informer halides, which are complex, drug-like aryl halides from the chemistry informer library pioneered by researchers at Merck & Co., Inc., Rahway, NJ,

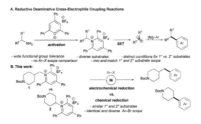

Scheme 1 Comparative strategy for primary and secondary alkylpyridinium salts.

USA.[11] We found that the trends in ArBr scope are fairly consistent between the primary and secondary alkylpyridinium salt, although distinctions were observed in isolated cases. In addition, the electrochemical conditions compared favorably to those using chemical reductants, especially among the more complex informer halides. The translatability of ArBr scope between 1° and 2° alkyl pyridiniums will be particularly useful in the context of medicinal chemistry where predictability of the reaction scope is crucial to provide confidence for applications of cross-couplings using precious drug-discovery program intermediates.

Results and discussion

Investigations with 24 diverse aryl bromides

We selected primary alkylpyridinium **1** and secondary alkylpyridinium **2** as our model substrates (Scheme 1B). Importantly, these substrates differ only by the addition of a methylene, making their results comparable. In addition, they provide mass-active products to facilitate LC/MS analysis of the crude reaction mixtures from the HTE experiments. Aryl bromides **ArBr-1–ArBr-24** were chosen as coupling partners, because they resulted in a wide range of product LC area percents (LCAPs) when used in our previously reported electroreductive coupling of a primary alkylpyridinium salt (Fig. 1).[8a]

To determine suitable conditions for primary alkylpyridinium salt **1**, we initially evaluated conditions from the literature (Scheme 2A). Our previously published electrochemical conditions [NiBr$_2$(DME), pyridine-2,6-bis(carboximidamide) dihydrochloride (**L1**) as ligand, NaI as electrolyte, with a cobalt sacrificial anode and stainless-steel cathode at 60 °C][8a] provided satisfactory product LCAPs. Conditions using Mn^0 and TDAE were chosen based on our previous report of reductive couplings of lysine-derived pyridinium salts.[5f]

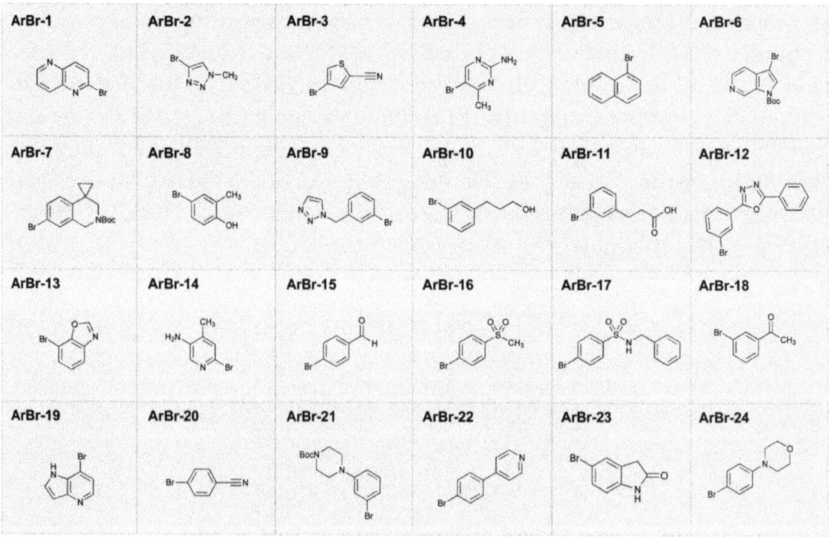

Fig. 1 Structure of 24 aryl bromides used for scope studies.

The results of these HTE campaigns using 1° alkylpyridinium substrate **1** are depicted in Scheme 2A. The electrochemical conditions compared favorably to the reactions using both chemical reductants (Mn and TDAE), with all three sets of conditions providing approximately the same average product LCAP across the 24 ArBr's. The scope of ArBr's for all three sets of conditions roughly follows similar trends. However, consistent with our previous reports, the TDAE conditions are often low-yielding for substrates with protic functional groups, such as **ArBr-8**, **ArBr-11**, and **ArBr-23**. Interestingly, the use of Mn⁰ conditions with **ArBr-11** also gave low LCAP, yet these conditions generally tolerate other protic functional groups.

For the secondary alkylpyridinium salt **2**, we used electrochemical conditions that were optimal in our previous work (Scheme 2B).[8a] The use of electrochemical mediator [Ni(terpy)₂]·2PF₆ was shown to be beneficial for the cross-coupling of secondary alkylpyridiniums and alkyl halides. The anode (Co) and the cathode (stainless steel) were the same as those used for the primary alkylpyridinium salt couplings. The Mn⁰ conditions were inspired by work done by Martin and co-workers, where secondary alkylpyridinium salts were the prime focus.[5a] Because no conditions using TDAE as a reductant have been previously reported for secondary pyridiniums, we evaluated several possible conditions (see ESI†); however, average LCAP was generally poor (<3%). Because optimal reaction conditions have not yet been developed with this substrate class, comparison of the results using TDAE was deprioritized for this study.

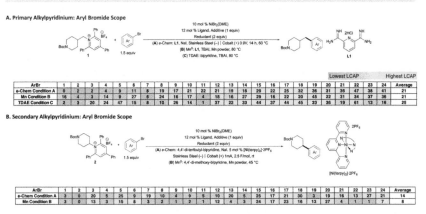

A. Primary Alkylpyridinium: Aryl Bromide Scope

ArBr	1	2	3	4	5	6	7	8	9	10	11	12	13	14	15	16	17	18	19	20	21	22	23	24	Average
e-Chem Condition A	0	2	2	4	8	9	16	19	17	21	22	21	19	18	29	22	25	32	31	36	47	38	41		21
Mn Condition B	16	4	3	14	9	27	6	24	16	17	4	18	16	27	29	16	22	20	45	22	31	34	37	36	21
TDAE Condition C	2	3	20	24	47	15	8	10	26	14	1	37	22	33	44	37	44	45	23	36	19	61	13	16	25

B. Secondary Alkylpyridinium: Aryl Bromide Scope

ArBr	1	2	3	4	5	6	7	8	9	10	11	12	13	14	15	16	17	18	19	20	21	22	23	24	Average
e-Chem Condition A	3	0	20	5	25	9	19	10	4	9	5	10	20	5	25	17	21	30	3	19	16	13	27	21	14
Mn Condition B	3	0	13	3	15	6	3	2	1	2	1	12	4	3	24	17	23	16	13	27	4	1	1	7	8

Scheme 2 Primary and secondary aryl bromide scope results. Percentages reflect product LCAPs that are the average of multiple runs (see ESI†). Formation of triphenylpyridine was not considered for LCAP determinations. (A) Primary substrate coupling conditions: (a) e-Chem = **1** (35 μmol, 1 equiv.), ArBr (52.5 μmol, 1.5 equiv.), NiBr₂(DME) (10 mol%), pyridine-2,6-bis(carboximidamide) dihydrochloride (12 mol%), NaI (1 equiv.), DMA [0.1 M], 0.9 V, 60 °C, 14 h. (b) Mn = **1** (10 μmol, 1 equiv.), ArBr (15 μmol, 1.5 equiv.), NiBr₂(DME) (10 mol%), pyridine-2,6-bis(carboximidamide) dihydrochloride (12 mol%), Mn⁰ (2 equiv.), TBAI (1 equiv.), DMA [0.1 M], 80 °C, 24 h. (c) TDAE = **1** (10 μmol, 1 equiv.), ArBr (15 μmol, 1.5 equiv.), NiBr₂(DME) (10 mol%), bipyridine (12 mol%), TDAE (2 equiv.), TBAI (1 equiv.), DMA [0.1 M], 80 °C, 24 h. (B) Secondary substrate coupling conditions: (a) e-Chem = **2** (35 μmol, 1 equiv.), ArBr (52.5 μmol, 1.5 equiv.), NiBr₂(DME) (10 mol%), 4,4′-di-tert-butyl-bipyridine (12 mol%), [Ni(terpy)₂]·2PF₆ (5 mol%), NaI (1 equiv.), DMA [0.1 M], 1 mA constant current, 30 V maximum 2.5 F mol⁻¹, rt. (b) Mn = **2** (10 μmol, 1 equiv.), ArBr (15 μmol, 1.5 equiv.), NiBr₂(DME) (10 mol%), 4,4′-di-methoxy-bipyridine (14 mol%), Mn⁰ (2 equiv.), NMP [0.1 M], 45 °C, 24 h.

Analogous to the results with primary alkyl pyridiniums (Scheme 2A), the electrochemical conditions compare favorably with the chemical reductant for reactions with the 2° pyridinium **2**, with an average LCAP of 12% *vs.* 9% for Mn⁰. Unlike the 1° pyridiniums, aryl bromides bearing protic functional groups afford low product LCAPs under both conditions (**ArBr-8**, **ArBr-10**, and **ArBr-11**). Interestingly, however, **ArBr-3** and **ArBr-5** afford higher product LCAPs than analogous reactions with 1° alkylpyridinium **1** using both the electrochemical and Mn⁰ conditions. We speculate that this change in scope between 1° and 2° alkylpyridiniums may be due to the faster and more irreversible formation of the 2° alkyl radical, which requires in turn a faster oxidative addition of the ArBr. Thus, electron-rich aryl bromides (*e.g.*, **ArBr-8**, **ArBr-10**, and **ArBr-11**) are less effective with alkylpyridinium **2**.

Investigations with informer halides

Encouraged by the scope with the 24 ArBr's detailed above, we also investigated the coupling of both 1° and 2° alkylpyridinium salts with informer halides to assess the efficacy of these methods in the context of complex, drug-like substrates (Fig. 2).[11] Because we focused this study on ArBr scope, the informer aryl iodides (**X14** and **X15**) and chlorides (**X16**, **X17**, and **X18**) were left out of this study. We investigated the scope across the remaining 13 informer halides with alkylpyridinium substrate **1** under the three sets of conditions (electrochemical, Mn⁰, and TDAE) and **2** under electrochemical and Mn⁰ conditions.

Results for the cross-coupling of 1° alkylpyridinium **1** with informer halides **X1–X13** are shown in Scheme 3A. The average LCAPs were 12%, 10%, and 15% for the electrochemical, Mn⁰, and TDAE conditions, respectively.

For all three sets of conditions, the scope was similar. An apparent limitation for the electrochemical conditions are pyridyl and fluorine containing partners **X3**, **X4**, **X7** and **X9–11**, although it should be noted that simpler 3-bromo-5-phenylpyridine was successfully employed in our previous report and that the complexity of the informers makes such generalizations difficult.[8a]

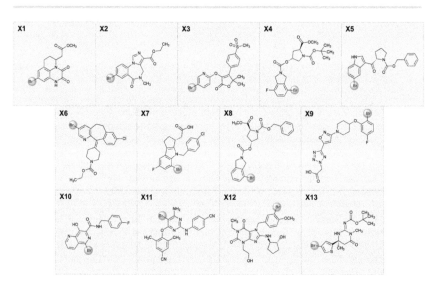

Fig. 2   Structure of informer halides.

A. Primary Informer Halide Scope

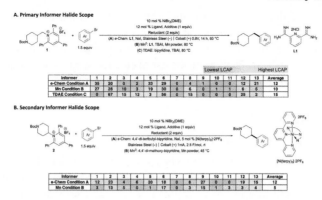

Informer	1	2	3	4	5	6	7	8	9	10	11	12	13	Average
e-Chem Condition A	35	20	0	2	33	29	0	4	1	0	0	12	21	12
Mn Condition B	27	28	10	3	19	30	0	6	0	1	1	6	5	10
TDAE Condition C	0	67	15	12	3	56	0	15	0	0	0	25	2	15

B. Secondary Informer Halide Scope

Informer	1	2	3	4	5	6	7	8	9	10	11	12	13	Average
e-Chem Condition A	12	23	4	6	20	18	0	8	27	0	0	19	16	12
Mn Condition B	3	13	5	0	1	17	0	3	15	1	3	3	4	5

Scheme 3 Primary and secondary informer halide scope results. Percentages reflect product LCAPs that are the average of multiple runs (see ESI†). Formation of triphenylpyridine was not considered for LCAP determinations. (A) Primary substrate coupling conditions: (a) e-Chem = 1 (35 μmol, 1 equiv.), ArBr (52.5 μmol, 1.5 equiv.), NiBr$_2$(DME) (10 mol%), pyridine-2,6-bis(carboximidamide) dihydrochloride (12 mol%), NaI (1 equiv.), DMA [0.1 M], 0.9 V, 60 °C, 14 h. (b) Mn = 1 (10 μmol, 1 equiv.), ArBr (15 μmol, 1.5 equiv.), NiBr$_2$(DME) (10 mol%), pyridine-2,6-bis(carboximidamide) dihydrochloride (12 mol%), Mn^0 (2 equiv.), TBAI (1 equiv.), DMA [0.1 M], 80 °C, 24 h. (c) TDAE = 1 (10 μmol, 1 equiv.), ArBr (15 μmol, 1.5 equiv.), NiBr$_2$·DME (10 mol%), bipyridine (12 mol%), TDAE (2 equiv.), TBAI (1 equiv.), DMA [0.1 M], 80 °C, 24 h. (B) Secondary substrate coupling conditions: (a) e-Chem = 2 (35 μmol, 1 equiv.), ArBr (52.5 μmol, 1.5 equiv.), NiBr$_2$(DME) (10 mol%), 4,4′-di-tert-butyl-bipyridine (12 mol%), [Ni(terpy)$_2$]·2PF$_6$ (5 mol%), NaI (1 equiv.), DMA [0.1 M], 1 mA constant current, 30 V maximum 2.5 F mol^{-1}, rt. (b) Mn = 2 (10 μmol, 1 equiv.), ArBr (15 μmol, 1.5 equiv.), NiBr$_2$(DME) (10 mol%), 4,4′-di-methoxy-bipyridine (14 mol%), Mn^0 (2 equiv.), NMP [0.1 M], 45 °C, 24 h.

As shown in Scheme 3B, the couplings with 2° alkylpyridinium 2 were more challenging with average LCAPs of 12% and 5% for electrochemical and Mn^0 conditions, respectively. Here, electrochemical conditions provided the broadest scope, with 6 informer halides leading to LCAPs ≥15%.

Conclusions

In summary, this report describes a systematic comparison of the cross-electrophile couplings of two, nearly identical 1° and 2° alkylpyridinium salts under three different modes of reduction. The use of HTE techniques facilitated the cross-couplings and analysis of reactions using 24 diverse aryl bromides and 13 complex informer halides using electrochemical (using a sacrificial cobalt anode and stainless-steel cathode), and non-electrochemical conditions (using Mn^0 and TDAE). Overall, the electrochemical conditions compare favorably with conditions using chemical reductants. In addition, and perhaps most importantly, the trends in ArBr scope were in general similar between the 1° and 2° alkylpyridinium salts, showing that one can extrapolate successes with one pyridinium class to the other with a reasonable level of confidence. This knowledge will be useful for chemists employing electrochemical deaminative couplings in pharmaceutical discovery and other synthetic endeavors. It strengthens the confidence that methods developed with chemical reductants can be translated to electroreductive approaches, and that the scope of newly developed electroreductive deaminative methods is likely to be similar to those developed with Mn^0. Perhaps even more importantly, this study provides future researchers with the knowledge that the scope of ArBr's is likely to be similar between 1° and 2° alkylpyridinium salts, facilitating the use of electrochemical deaminative cross-couplings broadly in parallel medicinal chemistry campaigns.

Experimental

General procedure for HTe-Chem experiments

In a nitrogen-filled glovebox, to 1 mL vials (secured in a 24-well aluminum block) equipped with 5 × 2 mm PTFE-covered magnetic stir bars was added pyridinium (175 μL, 0.2 M solution in DMA, 35 μM, 1 equiv.), aryl bromide (105 μL, 0.5 M solution in DMA, 52.5 μM, 1.5 equiv.), NaI (50 μL, 0.7 M solution in DMA, 35 μM, 1 equiv.), and pre-complexed NiBr$_2$·DME catalyst/ligand mixture (23.3 μL, 0.15 M solution in DMA, 3.5 μM, 10 mol%) sequentially. The electrodes were inserted, and the HTe-Chem reaction block was assembled inside the glovebox. The reactor was then connected to the external power supply inside the glovebox and heated to the appropriate temperature on an IKA stir plate. Upon reaching the desired temperature, the reactions were electrolyzed. For constant voltage mode ($V = 0.9$ V) for 14 h (overnight), and for constant current mode ($I = 1$ mA, 2.5 F mol^{-1}). After electrolysis, the HTe-Chem reactor was allowed to cool to rt (if applicable), taken outside of the glovebox, and disassembled. A 5 μL aliquot of the crude reaction mixture was taken and diluted in 400 μL of DMSO for UPLC-MS analysis.

Preparation of nickel stock solutions for HTe-Chem experiments

In a nitrogen-filled glovebox, a 4 mL vial (equipped with a stir bar) was charged with NiBr$_2$·DME catalyst (238.50 μmol) and appropriate ligand (1.2 equiv. regarding Ni) and DMA (1341.65 μL) was added to prepare the final 0.15 M stock solution. The mixtures were stirred for ~20 minutes (resulting in a slurry) before dosing into the reaction vials. The slurry was continually stirred at 1000 rpm while dosing.

Conflicts of interest

There are no conflicts to declare.

Acknowledgements

We thank NIH (R35 GM131816). At UD, data were acquired on instruments obtained with assistance of NSF and NIH funding (NSF CHE0421224, CHE1229234, CHE0840401, and CHE1048367; NIH P20 GM104316, P20 GM103541, and S10 OD016267). We thank Dr Jiantao Fu and Prof. Christo Sevov for helpful conversations.

References

1 (a) L.-C. Campeau and N. Hazari, *Organometallics*, 2019, **38**, 3–35; (b) J. Gu, X. Wang, W. Xue and H. Gong, *Org. Chem. Front.*, 2015, **2**, 1411–1421; (c) X. Hu, *Chem. Sci.*, 2011, **2**, 1867–1886.

2 A. W. Dombrowski, N. J. Gesmundo, A. L. Aguirre, K. A. Sarris, J. M. Young, A. R. Bogdan, M. C. Martin, S. Gedeon and Y. Wang, *ACS Med. Chem. Lett.*, 2020, **11**, 597–604.

3 (*a*) D. A. Everson, R. Shrestha and D. J. Weix, *J. Am. Chem. Soc.*, 2010, **132**, 920–921; (*b*) D. A. Everson, B. A. Jones and D. J. Weix, *J. Am. Chem. Soc.*, 2012, **134**, 6146–6159; (*c*) S. Wang, Q. Qian and H. Gong, *Org. Lett.*, 2012, **14**, 3352–3355; (*d*) S. Biswas and D. J. Weix, *J. Am. Chem. Soc.*, 2013, **135**, 16192–16197; (*e*) D. A. Everson and D. J. Weix, *J. Org. Chem.*, 2014, **79**, 4793–4798; (*f*) E. C. Hansen, C. Li, S. Yang, D. Pedro and D. J. Weix, *J. Org. Chem.*, 2017, **82**, 7085–7092; (*g*) D. J. Charboneau, H. Huang, E. L. Barth, C. C. Germe, N. Hazari, B. Q. Mercado, M. R. Uehling and S. L. Zultanski, *J. Am. Chem. Soc.*, 2021, **143**, 21024–21036.

4 (*a*) J. Cornella, J. T. Edwards, T. Qin, S. Kawamura, J. Wang, C.-M. Pan, R. Gianatassio, M. Schmidt, M. D. Eastgate and P. S. Baran, *J. Am. Chem. Soc.*, 2016, **138**, 2174–2177; (*b*) K. M. M. Huihui, J. A. Caputo, Z. Melchor, A. M. Olivares, A. M. Spiewak, K. A. Johnson, T. A. DiBenedetto, S. Kim, L. K. G. Ackerman and D. J. Weix, *J. Am. Chem. Soc.*, 2016, **138**, 5016–5019; (*c*) J. T. Edwards, R. R. Merchant, K. S. McClymont, K. W. Knouse, T. Qin, L. R. Malins, B. Vokits, S. A. Shaw, D.-H. Bao, F.-L. Wei, T. Zhou, M. D. Eastgate and P. S. Baran, *Nature*, 2017, **545**, 213–218; (*d*) D. C. Salgueiro, B. K. Chi, I. A. Guzei, P. García-Reynaga and D. J. Weix, *Angew. Chem., Int. Ed.*, 2022, **61**, e202205673.

5 (*a*) R. Martin-Montero, V. R. Yatham, H. Yin, J. Davies and R. Martin, *Org. Lett.*, 2019, **21**, 2947–2951; (*b*) H. Yue, C. Zhu, L. Shen, Q. Geng, K. J. Hock, T. Yuan, L. Cavallo and M. Rueping, *Chem. Sci.*, 2019, **10**, 4430–4435; (*c*) S. Ni, C.-X. Li, Y. Mao, J. Han, Y. Wang, H. Yan and Y. Pan, *Sci. Adv.*, 2019, **5**, eaaw9516; (*d*) J. Liao, C. H. Basch, M. E. Hoerrner, M. R. Talley, B. P. Boscoe, J. W. Tucker, M. R. Garnsey and M. P. Watson, *Org. Lett.*, 2019, **21**, 2941–2946; (*e*) J. Yi, S. O. Badir, L. M. Kammer, M. Ribagorda and G. A. Molander, *Org. Lett.*, 2019, **21**, 3346–3351; (*f*) J. C. Twitty, Y. Hong, B. Garcia, S. Tsang, J. Liao, D. M. Schultz, J. Hanisak, S. L. Zultanski, A. Dion, D. Kalyani and M. P. Watson, *J. Am. Chem. Soc.*, 2023, **145**, 5684–5695.

6 (*a*) K.-J. Jiao, D. Liu, H.-X. Ma, H. Qiu, P. Fang and T.-S. Mei, *Angew. Chem., Int. Ed.*, 2020, **59**, 6520–6524; (*b*) B. L. Truesdell, T. B. Hamby and C. S. Sevov, *J. Am. Chem. Soc.*, 2020, **142**, 5884–5893; (*c*) M. Yan, Y. Kawamata and P. S. Baran, *Chem. Rev.*, 2017, **117**, 13230–13319; (*d*) L. Yi, T. Ji, K.-Q. Chen, X.-Y. Chen and M. Rueping, *CCS Chem.*, 2022, **4**, 9–30; (*e*) J. L. S. Zackasee, S. Al Zubaydi, B. L. Truesdell and C. S. Sevov, *ACS Catal.*, 2022, **12**, 1161–1166; (*f*) W. Zhang, L. Lu, W. Zhang, Y. Wang, S. D. Ware, J. Mondragon, J. Rein, N. Strotman, D. Lehnherr, K. A. See and S. Lin, *Nature*, 2022, **604**, 292–297.

7 M. D. Palkowitz, G. Laudadio, S. Kolb, J. Choi, M. S. Oderinde, T. E.-H. Ewing, P. N. Bolduc, T. Chen, H. Zhang, P. T. W. Cheng, B. Zhang, M. D. Mandler, V. D. Blasczak, J. M. Richter, M. R. Collins, R. L. Schioldager, M. Bravo, T. G. M. Dhar, B. Vokits, Y. Zhu, P.-G. Echeverria, M. A. Poss, S. A. Shaw, S. Clementson, N. N. Petersen, P. K. Mykhailiuk and P. S. Baran, *J. Am. Chem. Soc.*, 2022, **144**, 17709–17720.

8 (*a*) J. Fu, W. Lundy, R. Chowdhury, J. C. Twitty, L. P. Dinh, J. Sampson, Y.-h. Lam, C. S. Sevov, M. P. Watson and D. Kalyani, *ACS Catal.*, 2023, **13**, 9336–9345; (*b*) Y. Liu, X. Tao, Y. Mao, X. Yuan, J. Qiu, L. Kong, S. Ni, K. Guo,

Y. Wang and Y. Pan, *Nat. Commun.*, 2021, **12**, 6745; (*c*) L. J. Wesenberg, A. Sivo, G. Vilé and T. Noël, *J. Org. Chem.*, 2023, DOI: 10.1021/acs.joc.3c00859.

9 S. Tcyrulnikov, Q. Cai, J. C. Twitty, J. Xu, A. Atifi, O. P. Bercher, G. P. A. Yap, J. Rosenthal, M. P. Watson and M. C. Kozlowski, *ACS Catal.*, 2021, **11**, 8456–8466.

10 J. Rein, J. R. Annand, M. K. Wismer, J. Fu, J. C. Siu, A. Klapars, N. A. Strotman, D. Kalyani, D. Lehnherr and S. Lin, *ACS Cent. Sci.*, 2021, **7**, 1347–1355.

11 P. S. Kutchukian, J. F. Dropinski, K. D. Dykstra, B. Li, D. A. DiRocco, E. C. Streckfuss, L.-C. Campeau, T. Cernak, P. Vachal, I. W. Davies, S. W. Krska and S. D. Dreher, *Chem. Sci.*, 2016, **7**, 2604–2613.

DISCUSSIONS

Flow cells and reactor design: general discussion

Anas Alkayal, Mickaël Avanthay, Belen Batanero, Pim Broersen, Richard C. D. Brown, Luke Chen, Ping-Chang Chuang, Toshio Fuchigami, Shinsuke Inagi, Dipannita Kalyani, Kevin Lam, Maya Landis, T. Leo Liu, Matthew J. Milner, Robert Price, Naoki Shida and Thomas Wirth

DOI: 10.1039/d3fd90042h

Anas Alkayal opened discussion of the paper by Thomas Wirth: Have you tried to run a control reaction in the absence of oxygen? Did this lead to undesired compounds such as dimers? I would like to take this occasion to mention that I am always interested in Professor Wirth's research. I am keen to attend his lectures at conferences. Furthermore, I usually share his latest papers with my colleagues in our group meetings/literature review weeks.

Thomas Wirth answered: There is no reaction in the absence of oxygen, no superoxide radical anion can be generated. The substrate might decompose in an uncontrolled way, depending on its oxidation potential.

Anas Alkayal asked: Since you have not realised a big yield of by-products, does that mean the flow chemistry makes the procedure more productive?

Thomas Wirth replied: Often flow chemistry does produce cleaner products due to better defined reaction conditions. Typically shorter reaction times compared to batch also can suppress the formation of side products.

Anas Alkayal questioned: When you start the reaction there are 3 phenyl rings around the double bond, but the corresponding carbonyl product has just 2 phenyl rings; why was one of the rings released?

Thomas Wirth answered: In the reaction the double bond is cleaved similar to an ozonisation. Therefore, two different carbonyl-containing products are being obtained. Any aldehyde obtained in the reaction is further oxidised and cannot be isolated.

Ping-Chang Chuang asked: In the paper (https://doi.org/10.1039/d3fd00050h), higher product yields are achieved when applying back-pressure in your flow system. What is the reason for this kind of set-up?

Thomas Wirth replied: A pressurised system probably allows a better mixing in the biphasic system. Back-pressure regulation has been applied to have a careful control over the residence time. Results without a back-pressure regulator indicate less defined reaction conditions.

Luke Chen requested: Please can you comment on the mass balance given the sub-optimal yields. Is this the result of poor conversion of starting material or are there other impurities being formed? If there is high starting material recovery, can the process be recycled to achieve full conversion?

Thomas Wirth responded: Usually there is no starting material left and only product is formed. We are currently investigating the stability of the product(s) towards the reaction conditions and if decomposition could be a reason for lower yields. Other side products have not been observed or identified.

Toshio Fuchigami asked: In order to generate super oxide ions efficiently, gas diffusion electrodes seem be more suitable than ordinary cathodes. Have you tried?

Thomas Wirth replied: This is an excellent suggestion. We have not tried this, but will certainly find out if the use of gas diffusion electrodes can enhance such reactions.

Shinsuke Inagi enquired: Where does the oxygen reduction occur, in the liquid or gas phase? Although the following reaction takes place in solution, superoxide ions can also be generated in the gas phase and dissolve readily in solution due to their improved solubility. If this works, the segmented flow approach would be very significant.

Thomas Wirth answered: In the batch reaction, the oxygen reduction will take place in the liquid phase. We assume the same happens in the flow reaction, but here also the gas is in direct contact with the electrode for a potential reduction. One would need to carefully think of a possible experimental setup for distinguishing between these reactions.

Richard C. D. Brown said: With reference to the mechanism proposed in Fig. 3 of your paper (https://doi.org/10.1039/d3fd00050h), radical cation and superoxide intermediates are generated at opposing electrodes. These species are then required to reach each other through mass transport in order for them to react to give intermediate **A**. Have you investigated the influence of the interelectrode gap on the reaction?

Thomas Wirth responded: In this reaction, we have not yet investigated the influence of the interelectrode gap. In other reactions with oxygen gas we have done this and did not find a meaningful relation between the outcome of the reaction and the interelectrode gap.

Mickaël Avanthay asked: Is it possible to saturate the solution for the flow reactor with oxygen (or to run a "batch solution" through the flow reactor), to avoid the need of a gas inlet and a back-pressure regulator?

Thomas Wirth replied: Using an oxygen saturated solution is possible, but will yield only trace amounts of product. Additional oxygen is necessary (also in the batch reaction, which is performed under an oxygen atmosphere).

Belen Batanero enquired: Have you tested another electrolyte or solvent, for instance lithium perchlorate or DMF? Flow is not high; have you tried reactions with faster flow? What should happen?

Thomas Wirth responded: We have not yet investigated other solvents or electrolytes as we used mainly the reported (batch) reaction conditions. Higher flow rates lead to lower conversions.

Belen Batanero commented: In this reaction the low yield can be justified by a further reaction of the obtained product (ketone) with superoxide anion.

Thomas Wirth replied: We are currently investigating the ketone stability under the reaction conditions.

Matthew J. Milner asked: Dioxetanes such as the proposed intermediate are in special cases stable enough for isolation, suggesting that the decomposition of **B** to form the product ketone may be slow enough for the dioxetane to persist in solution at least long enough to reach an electrode. What is the lifetime of such di- and tri-substituted dioxetanes? Is it possible that the dioxetane is being further oxidized or reduced at one of the electrodes, especially given the constant current conditions and the relatively high potential required to oxidize an alkene?

Thomas Wirth responded: We have never attempted to isolate the proposed dioxetane intermediates. They probably can be oxidised further if they have a sufficiently long lifetime under the reaction conditions.

Mickaël Avanthay opened discussion of the paper by Richard C. D. Brown: The selectivity increases as more mediator is used with very good selectivity at 1 equivalent. Would one observe an erosion of that selectivity with an excess of mediator due to reduction of the intermediate radicals by the mediator?

Richard C. D. Brown replied: We did not originally perform simulations using an excess of the mediator (phenanthrene) as it is desirable to use as little mediator as possible for practical reasons. However, the point raised is highly relevant as the concentration of the mediator is expected to affect the flux of mediator towards the electrode and, consequently, the flux of mediator radical anion away from the electrode. To address the question more fully we have performed additional simulations with 2.0 and 10 equivalents of mediator (Fig. 1 here). For the present reaction system, the effects are rather subtle, and can be summarised as follows:

• Both show almost identical selectivity profiles to the simulation using 1 equivalent of phenanthrene. Formation of the cyclised product is favoured (reaction layer detachment is achieved very rapidly).

• The mechanism switches from direct to mediated after a shorter time interval as the amount of mediator is increased.

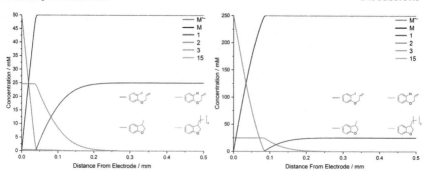

Fig. 1   Concentration profiles for species 1, 2, 3, 15, M and M$^{\bullet-}$ for simulated electrolysis under constant potential of -2.8 V $vs.$ SCE at $t = 1$ s. Left: mediator loading of 2.0 equivalents. Right: mediator loading of 10 equivalents.

• Increased flux of the mediator radical anion pushes the homogeneous reaction layer outwards from the electrode, and "detachment" from the electrode occurs sooner with higher concentrations of mediator.

Toshio Fuchigami commented: Cathodic reduction of halides usually generates radical and/or anionic species. Therefore, mediators are usually used in order to generate radicals selectively.

Richard C. D. Brown responded: Yes, that is correct. The purpose of our simulations is to better understand where the homogeneous coupled chemistry takes place relative to the electrode, and the factors controlling this. We also highlight that it is not necessary for the mediator to be more easily reduced than the substrate (with respect to reduction potentials), as is frequently stated in the literature.

Toshio Fuchigami remarked: Cathodically generated radical species are further reduced to generate anionic species. Sonication should be effective for avoiding the further reduction of radical species since the resulting radical species move away from the cathode. Furthermore, sonication causes local heating, which promotes radical cyclization.[1]

1 Y. Shen, M. Atobe and T. Fuchigami, *Org. Lett.*, 2004, **6**, 2441, DOI: 10.1021/ol049152f.

Richard C. D. Brown responded: This is an interesting point, and increasing temperature and sonication would indeed be anticipated to have a beneficial effect on mass transport and rates of cyclisation as is highlighted in Professor Fuchigami's elegant studies of oxidative cyclisation reactions using sonoelectrosynthesis. It would be interesting to examine the influence of sonication using our flow reactors, although there are some technical challenges that we need to overcome.

Pim Broersen asked: What happens to the system once you introduce turbulence to it, would the selectivity go down since more substrate would reach the electrode? Also, would you say that this is one of the few flow systems where laminar flow would be preferred over turbulent flow to increase selectivity?

Richard C. D. Brown answered: This is an important point, and any increase in mass transport will compress the diffusion layer towards the electrode. Turbulent flow would be expected to increase the flux of the substrate to the electrode, pushing the homogeneous reaction layer closer to the electrode. There would be conditions where turbulent flow would lead to decreased selectivity – as implied in the question – depending on the opposing fluxes of the mediator (**M**) radical anion and substrate. Therefore, in some situations laminar flow may be advantageous. Our feeling is, though, that even with turbulent flow, detachment of the reaction layer (albeit much closer to the electrode) may be achieved with sufficient flux of mediator radical anion and very fast electron transfer (ET) and coupled chemistry. We have yet to simulate turbulent flow conditions, although preliminary simulation using a rotating disc electrode did show the effect of compressing the diffusion layer so that it was closer to the electrode.

Naoki Shida enquired: The model study of flow electrolysis with the molecular mediator by COMSOL seems highly useful. Is it possible to use this type of simulation to predict the concentration of mediator for the bulk conversion of the substrate under flow electrochemical conditions?

Richard C. D. Brown responded: The simulations will prove useful in determining optimum conditions such as the concentration of the mediator required in preparative flow reactions. In the present case, the key criterium is to achieve detachment of the homogeneous ET from the cathode. We are currently developing models to identify the key parameters or groups of conditions where this can be achieved. It does, however, require some knowledge of kinetic and physical parameters for the specific process to input into the simulations.

Luke Chen opened discussion of the paper by Dipannita Kalyani: Do you see reproducibility in each of the wells across the plate and are duplicate wells required? How well can the reactions from the small, plate scale be transferred to more synthetically useful scales and have you been able to isolate products?

Dipannita Kalyani replied: We have done validation studies to show the reproducibility across wells. We have been able to scale up the reactions from HTe^-Chem scale to 0.3 mmol scale with a minor change in reaction conditions.

Kevin Lam asked: Is there a specific reason for choosing a cobalt electrode?

Dipannita Kalyani answered: Use of the cobalt electrode was important to suppress the background reaction in the absence of electricity.

T. Leo Liu requested: Kalyani, thanks for sharing the interesting cross-electrophile results. Can you comment on the difference with pyridinium *vs.* alkyl halide regarding reaction selectivity and scope?

Dipannita Kalyani responded: We have not done the systematic comparison yet.

Kevin Lam asked: Are pyridinium compounds of choice for the industry? They seem to lead to a lot of waste and not to be the most atom-economical options.

Dipannita Kalyani answered: Pyridinium chemistry is fine for medicinal chemistry.

Belen Batanero enquired: Have you used a flow cell or a batch conventional cell? Is your reaction performed under constant current conditions?

Dipannita Kalyani replied: Have not used a flow cell yet. We have used IKA ElectraSyn for scale ups. Constant current or constant voltage conditions work for our reactions.

Kevin Lam asked: What is the best way to start collaborating with Merck & Co., Inc. Rahway, NJ, USA?

Dipannita Kalyani answered: Best starting point is to connect with a scientist at Merck & Co., Inc. Rahway, NJ, USA.

Maya Landis communicated: Would it be possible to investigate the likelihood that a substrate would be suitable for your reactions using other electrochemical techniques, *e.g.*, cyclic voltammetry (CV), and do these techniques play a large role in R&D at companies like Merck & Co., Inc. Rahway, NJ, USA?

Dipannita Kalyani communicated in reply: Yes CV is appropriate to use as necessary.

Luke Chen opened a general discussion: My colleagues in medicinal chemistry are reluctant to try electrochemistry, let alone make use of flow and other reactor types. What advice and suggestions do you have to help bridge the gap in accessibility for the average synthetic organic chemist to make it easier to adopt new, developing technologies?

Richard C. D. Brown replied: It is a great time to try electrosynthesis, either in flow or batch reactors. There is a rapidly expanding array of transformations available including many that are well aligned with medicinal chemistry targets. There is also a good range of commercial equipment – budget permitting – to help lower the barrier and to provide a greater degree of reproducibility. Some caution should be applied, however, and a bad initial experience is more likely to deter future efforts. It is important to make sure that electrochemistry is appropriate for the planned synthesis, and can meet the objectives in terms of rates of production. This is relevant to the specific electrosynthesis and the equipment being used. Batch electrolysis cells are more familiar to the synthetic chemist, but flow offers an easier pathway to scale up. Providing time and freedom to employees to explore emerging technologies is important. This may be supported by collaborations with academia, and there are a growing number of groups focussing on electrosynthesis that would welcome interactions with industry. Great value can also be gained from collaborating with electrochemists and electrochemical engineers to avoid pitfalls, and disappointment. Visits to academic laboratories, or sponsoring a PhD student who can also spend some time engaged in technology transfer at the company, can also contribute to increase uptake of new technologies.

Dipannita Kalyani responded: Having training materials is helpful. Also showing the value of EChem in projects to access targets that are otherwise difficult to access can help folks to appreciate and use EChem.

Thomas Wirth answered: Nowadays with commercially available equipment the threshold to use this technique should be lower, but advice from a specialist might still be needed in order to understand the many reaction parameters which can play a role in electrochemical reactions.

Robert Price asked: In a lot of the work presented at this discussion, single flow cell modules are employed and, in some cases, full conversion of reactants is not achieved. Do you see the progression of these flow cells as remaining as single modules that could be linked in series to achieve full conversion, or is it likely that stack platforms/modules (like those developed in the fuel cell/electrolyser fields) could be employed to reduce flow cell footprint and increase conversion? Could this technology be borrowed (with modifications) from existing designs within the aforementioned communities?

Dipannita Kalyani answered: We have not used flow cells for the pyridinium chemistry.

Richard C. D. Brown responded: Achieving high conversions in a reasonable timeframe has been one of the challenges in preparative organic electrosynthesis. This can be addressed through increasing electrode surface area/volume ratios and by improving mass transport.[1] Both of these features are commonly designed into flow electrolysis cells to improve productivity. Turbulence promoters are commonly used to improve mass transport in simple parallel plate reactors.[2] We have developed narrow interelectrode gap flow cells with extended channels as one way to achieve high conversion in a single pass.[3] Linking two or more cells in series can be viewed as effectively extending the length of the flow channel, and the potential (current) can be optimised for each stage. Higher conversion may also be achieved by either slowing down the flow rate or by multiple passes through the reactor, which is commonly applied in industrial electrosynthesis. An excellent example is the stacked bipolar electrode cells used at BASF on a massive scale.[4] It is correct that cells used in other areas, such as fuel cells and water purification, can and have been adapted for organic electrosynthesis and knowledge from other areas should be taken advantage of. In some cases the materials will need to be adapted to allow for use with organic solvents.

1 D. Pletcher, R. A. Green and R. C. D. Brown, *Chem. Rev.*, 2018, **118**, 4573–4591, DOI: 10.1021/acs.chemrev.7b00360.
2 A. A. Folgueiras-Amador, A. E. Teuten, D. Pletcher and R. C. D. Brown, *React. Chem. Eng.*, 2020, **5**, 712–718, DOI: 10.1039/d0re00019a.
3 (a) R. A. Green, R. C. D. Brown and D. Pletcher, *J. Flow Chem.*, 2015, **5**, 31–36, DOI: 10.1556/jfc-d-14-00027; (b) R. C. D. Brown, *Chem. Rec.*, 2021, **21**, 2472–2487, DOI: 10.1002/tcr.202100163.

4 D. Hoormann, H. Pütter and J. Jörissen, *Chem. Ing. Tech.*, 2005, **77**, 1363–1376, DOI: 10.1002/cite.200500086.

Thomas Wirth replied: Stacked flow reactors are already reported and have been used to increase conversion and/or productivity. Industrial scale stacked reactors are also reported, often including cooling layers/liquids for thermal management.

Naoki Shida enquired: Automated electrosynthesis is highly desired to generate large amounts of data. Flow electrosynthesis and high-throughput electrosynthesis are critical enablers. Is there any idea or comment on implementing these technologies for automated electrosynthesis?

Richard C. D. Brown replied: Parallel batch electrosynthesis systems have been developed and some are commercially available. These formats are undoubtedly useful for reaction optimisation and high throughput synthesis. We don't have any direct experience of these, but technical points to consider would include ensuring that electrodes can be easily cleaned and electrolyte doesn't complicate the analysis or isolation. Flow systems with on-line real-time analysis seem attractive for optimisation as conditions such as flow rate, current density, concentrations *etc.* can all be varied using computer control. Complications include potential for electrodes fouling, and the need to change the cell to investigate different electrode materials. Together with experts in self-optimisation (Professor Mike George and co-workers), we explored optimisation of an alcohol oxidation in the Ammonite 8 flow electrolysis cell using a "self-optimisation" system developed by their team.[1]

1 J. Ke, C. Gao, A. A. Folgueiras-Amador, K. E. Jolley, O. de Frutos, C. Mateos, J. A. Rincón, R. C. D. Brown, M. Poliakoff and M. W. George, *Appl. Spectrosc.*, 2022, **76**, 38–50.

Thomas Wirth responded: We have already reported some results using automated electrochemistry, which are also taken on by several other groups working in this area as fully automated systems are now also commercially available. See ref. 1–3.

1 N. Amri and T. Wirth, *Synlett*, 2020, **31**, 1894, DOI: 10.1055/s-0040-1707141.

2 N. Amri and T. Wirth, *Synthesis*, 2020, **52**, 1751, DOI: 10.1055/s-0039-1690868.

3 N. Amri and T. Wirth, *J. Org. Chem.*, 2021, **86**, 15961, DOI: 10.1021/acs.joc.1c00860.

Pim Broersen asked: When you do some technoeconomic analysis on scaling electrocatalytic reactions such as large scale production of hydrogen, the material use quickly becomes prohibitive. A PEM electrolyzer is for example much more expensive to build than a nickel alkaline electrolysis system. When looking at the products made in your reactions, there is a bit more leeway as they are more expensive. Where do you think the limit is in material use for the electrodes for these more complex chemicals? Will it be possible to design these systems at scale when more expensive electrode materials such as gold or platinum need to be used, or is it similar to the electrocatalysis field and are we limited to cheap materials such as stainless steel and carbon?

Richard C. D. Brown responded: Cheaper materials such as graphite and steel will always be attractive for large scale applications, and fortunately, they are applicable to many electrosyntheses. New electrode materials – or applications of known materials – is certainly an area of interest. A good example is leaded bronze for use in cathodic reductions in place of lead, the latter being undesirable for production of pharmaceutical intermediates. Personally, I have never performed a technoeconomic analysis, but I believe that this will ultimately have to be justified by the value of the product and the lifetime of the electrodes. Innovations such as depositing costly electrode materials on cheaper substrates may also contribute to providing workable solutions where cheaper electrode materials are not suitable.

Thomas Wirth answered: For large-scale production cheaper electrode materials are certainly required, but depending on the process even noble metal electrodes can be a solution for processes adding a high value to a certain product.

Conflicts of interest

There are no conflicts to declare.

Faraday Discussions

ROYAL SOCIETY OF CHEMISTRY

PAPER

Concluding remarks: A summary of the *Faraday Discussion* on electrosynthesis

Kevin D. Moeller

Received 12th August 2023, Accepted 6th September 2023

DOI: 10.1039/d3fd00148b

A summary of the *Faraday Discussion* presented in this issue and a perspective on that discussion is presented. The work highlights the specific science contributions made and the key conclusions associated with those findings so that readers can identify papers that they would like to explore in more detail.

Introduction

Electrochemistry offers an opportunity to conduct oxidation and reduction reactions under neutral conditions, trigger interesting new umpolung reactions, and recycle chemical reagents so that they can be used in catalytic amounts.[1] Yet in spite of these opportunities and impressive advances in electrosynthetic techniques by individuals in the energy conversion community, for many years the larger synthetic community has mostly ignored electrochemistry and viewed it as a highly specialized, niche technique. During this time, broader applications of electrosynthesis were mainly studied by a small, dedicated group of scientists who worked to better understand the key elements of electrochemical experiments, to determine how electrosynthetic reactions could be optimized and developed, and to provide examples of how electrochemistry could be utilized in more complex settings. The hope was that this effort would illustrate, for the larger synthetic chemistry community, the potential of electrochemical methods so that they might be tempted to capitalize on the approach in their own work. This effort was not undertaken in vain, and today a rapidly growing number of synthetic chemists are exploring and taking advantage of electrochemistry. This influx of new people, talent, and ideas is driving the development of electrochemical methods in ways far beyond what the original practitioners of the field could have ever imagined. Today, it is an exciting time to take stock of what we have been learning and to think about what new opportunities lie ahead in the future. It was in this light that the *Faraday Discussion* on electrosynthesis was held.

The discussion opened with remarks by Professor Toshio Fuchigami (Professor Emeritis, Tokyo Institute of Technology, and Director, Sagami

Department of Chemistry, Washington University in St. Louis, St. Louis 63130, MO, USA. E-mail: moeller@ wustl.edu

Fig. 1 Professor Toshio Fuchigami with his Spiers Memorial award, received at the 2023 Electrosynthesis Faraday Discussion.

Chemical Research Center; Fig. 1) (https://doi.org/10.1039/D3FD00129F). Fuchigami outlined the history of electrochemistry over his career, highlighting direct and indirect electrochemical reactions, the development of new electrochemical mediators, and the use of electroauxiliaries, ionic liquids as solvents, modified electrodes, conducting polymer films, thin-layer flow cells, new electrode materials, recyclable hypervalent iodide oxidants, solid-phase electrodes, paired electrolyses, bipolar electrodes, *N*-centered radicals, and selective fluorination reactions. The review illustrated how these methods (with many contributions from the Fuchigami group) laid a foundation for much of the electrosynthetic chemistry being pioneered today. This historical perspective told a "story" of what electrochemistry was capable of and how we might achieve it.

This electrochemical "story" is still being developed, with much of the current push occurring along two main, very broad, themes. First, a large effort is underway to discover the new synthetic capabilities that electrochemical methods make available, capabilities that increase our ability to build (or detect) molecules in ever more efficient and selective ways. Second, significant effort is being put into the development and application of electrochemical approaches to meet societal needs in more sustainable ways. The *Faraday Discussion* addressed work that fit into both of these themes with an emphasis on providing the new mechanistic insights needed to drive innovation. What follows is a summary of those discussions that is meant to point readers to the original papers in the issue, papers that address topics they are interested in exploring in greater detail.

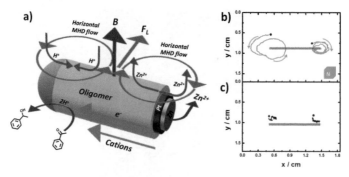

Scheme 1 The use of magnetic fields to propel Janus swimmers through solution. Image taken from https://doi.org/10.1039/D3FD00041A.

Main meeting

The main body of the meeting commenced with an intriguing paper by A. Kuhn and co-workers that describes the use of a magnetic field to propel designed particles through solution (https://doi.org/10.1039/D3FD00041A). The particles, comprising a Zn-core coated with a chiral oligomer, mediate the reduction of ketones to chiral alcohols with high enantioselectivity (Scheme 1). Movement of the particles through solution in the presence of the magnetic field aids convection on the surface of the particle in a way that raises the efficiency of the chiral reduction reactions by an order of magnitude. It is a method that offers a unique approach to minimizing the mass transport properties that can impede the use of heterogeneous catalysts. The group discussion of the paper focused on how the particles, or "swimmers", are propelled through solution, the nature of the oligomers on the surface of the particles, the scope of the reactions that might be possible, including potential oxidation reactions, and the chemistry that happens on the Zn-terminus of the particle. The discussion was driven by a desire to understand more deeply how the chemistry works, with an eye toward how the "swimmers" could be applied in future synthetic efforts.

The next paper in the session also focused on using technological advances to forward an electron transfer process. In this case, Long Luo and his group discussed how controlling the frequency of an alternating current reaction can alter the selectivity of H/D- (or H/T-) exchange reactions involving tertiary amine substrates (Scheme 2). When the reactions were performed using current state-of-the-art methods, they are typically non-selective and lead to total exchange of all

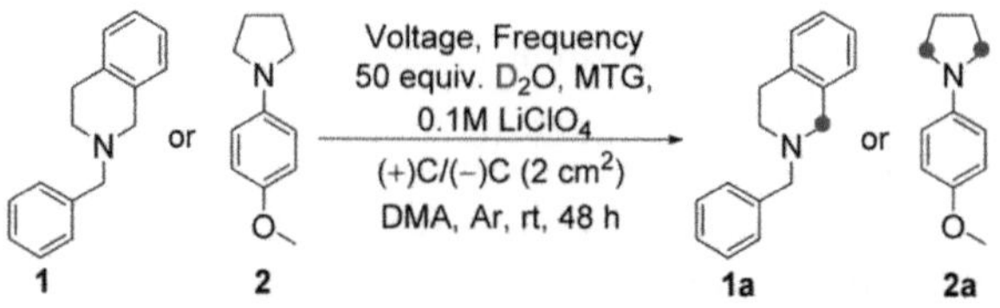

Scheme 2 H/D-exchange using direct- or alternating-current electrolyses. Image taken from https://doi.org/10.1039/D3FD00044C.

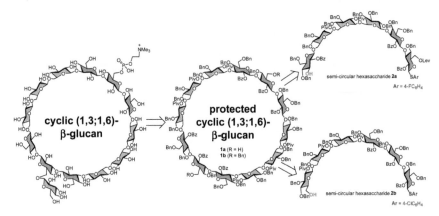

Scheme 3 An automated route to macrocyclic sugar natural products. Image taken from https://doi.org/10.1039/D3FD00045A.

relevant hydrogens (https://doi.org/10.1039/D3FD00044C). The talk detailed a mechanistic picture of the desired reaction pathway (oxidation of the amine followed by deprotonation to form a radical that then abstracts a deuterium from a deuterated thiol) and highlighted its rate relative to a pair of potential side reactions (dimerization of the initially formed radical intermediate and over-oxidation). The data presented then showed how that understanding could be used to select either the use of direct- or alternating-current electrolysis for the transformation. For the two substrates highlighted here, very different reaction conditions were needed to accomplish the desired outcome. For the functionalization of **1**, the use of a direct current proved to be the method of choice due to an underlying dimerization reaction that interfered with the alternating-current reactions. For the functionalization of **2**, the subsequent D-atom abstraction reaction was fast, leading to a reaction that benefited from the use of an alternating current. The discussion of the paper focused on how these conclusions were reached, how the selectivity of the reactions was obtained for the processes, and what other examples could be used to illustrate the true power of the method for inducing selectivity into chemical transformations.

T. Nokami and coworkers then presented their use of electrosynthesis for the automated synthesis of complex cyclic dodecasaccharides (Scheme 3) (https://doi.org/10.1039/D3FD00045A). Automated syntheses require a method that is consistent, reproducible, and flexible enough to handle changes in the structure of various substrates. In this case, a constant-current electrolysis of anomeric thioethers provided just such a method. The constant-current electrolysis allowed the potential at the anode to adjust to both changes in potential, caused by variations in the large substrates employed, and the use of a second thioether with a higher potential as a part of a dimerization–cyclization sequence. In the dimerization–cyclization sequence, oxidation of a thioether with a lower potential was used to generate the reactive species needed for the dimerization step, while oxidation of a thioether with the higher potential allowed the cyclization. The discussion of the paper following the presentation focused on the coupling steps, the intermediates involved, and the role that the electrolyte in the

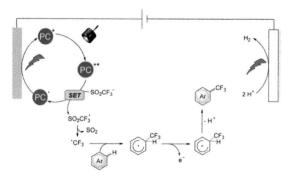

Scheme 4 A photoelectrocatalytic method for the trifluormethylation of arenes. Image taken from https://doi.org/10.1039/D3FD00076A.

reaction plays in that process. There was a general agreement that the ability to carry out the anodic oxidation on molecules of the size employed was truly impressive in terms of the mass-transport issues that frequently hinder heterogeneous reactions with larger substrates.

Next up was an intriguing paper by J. Stuwe and L. Ackerman that describes a systematic study of 21 different photocatalysts for use in the photo-electrocatalyzed CH-trifluoromethylation of arenes (Scheme 4) (https://doi.org/10.1039/D3FD00076A). As is often the case, the use of photoelectrochemistry was necessary for the reactions because it allowed access to high oxidation potentials without the need for high electrode potentials. In the science presented, it was found that two metal-free catalysts ([Mes-Acr]ClO_4 and [TAC] ClO_4) were optimal for accomplishing the desired transformations. The catalyst, light, and current were all required for generation of the trifluoromethyl radical. The study provides an excellent example of how constant-current electrolysis can allow a systematic study by adjusting to the various oxidation potentials of the catalysts without a need to change reaction conditions. Discussion of the paper by attendees at the meeting focused on mechanistic questions to do with the light source, and how the reactions led to improved yields relative to non-electrochemical methods while still producing the products with similar selectivities. For most cases, the two optimal catalysts behaved similarly, but for a more complex example associated with a natural product synthesis, [TAC]ClO_4 was a superior choice. At this time, it is not clear why one catalyst was better than the other one in these cases, an observation that highlights the need to screen catalysts for a given reaction.

The use of pyruvate as an electrolyte was shown by C. Bruggeman, K. Gregurash, and D. P. Hickey to afford significant advantages for the electrochemical reduction of NAD$^+$-biomimetics (Scheme 5) (https://doi.org/10.1039/D3FD00047H). Typically, the reduction of NAD$^+$-biomimetics is complicated by dimerization of the radical intermediate generated, a radical that is supposed to undergo a second reduction and then protonation step. The authors screened a number of additives/electrolytes to prevent the dimerization and found that one of them, sodium pyruvate, was effective, led to significant yields of the desired NADH mimic, and afforded far lower yields of the unwanted dimer. Methyl pyruvate was also compatible with suppression of dimer formation. The

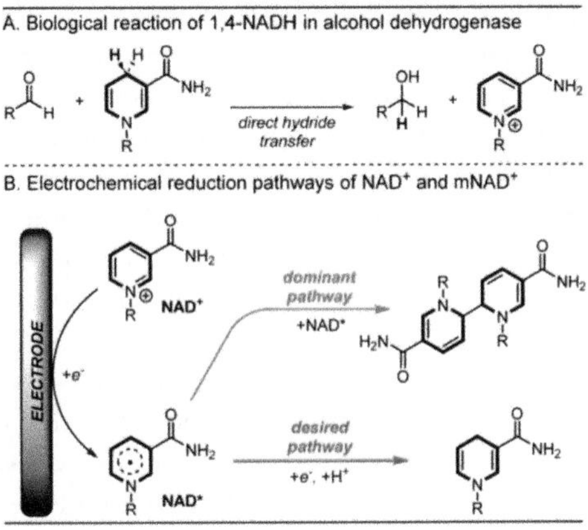

A. Biological reaction of 1,4-NADH in alcohol dehydrogenase

B. Electrochemical reduction pathways of NAD⁺ and mNAD⁺

Scheme 5 The use of pyruvate to channel NAD^+-mimic reduction toward the desired NADH-derivatives. Image taken from https://doi.org/10.1039/D3FD00047H.

discussion of the paper focused on the reason for this observation, and during that discussion a proposed mechanism where the NAD-radical is stabilized by a reversible electron transfer to the pyruvate was suggested. Alterations to the electrode material and other reaction conditions were discussed, and the authors pointed out that these changes did not influence the reaction to a great extent. The paper and the discussion that followed reflected a general theme of many of the subsequent discussions, a theme that focused on how electrolytes, counter-ions, and additives influence the reactive intermediates and products generated by the electrochemical transformation.

V. Flexer, along with W. R. Torres and N. C. Zeballos, then presented their efforts to demonstrate how lithium ions could be isolated as lithium carbonate from brines comprising mixed cationic species (Scheme 6) (https://doi.org/

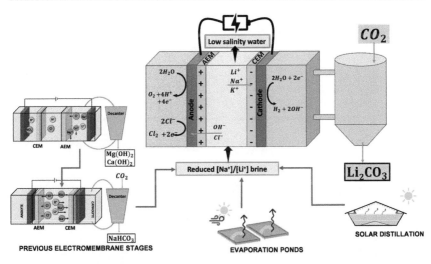

Scheme 6 A membrane electrolysis approach for the isolation of lithium carbonate from brine mixtures. Image taken from https://doi.org/10.1039/D3FD00051F.

10.1039/D3FD00051F). The project fits into a larger societal-need to isolate lithium without resorting to large evaporation pools that frequently throw away water in arid regions. The approach works by taking advantage of membrane electrolysis where cations in the brine solutions are transported from an anodic chamber to the cathodic chamber. Ca- and Mg-ions are transported across the membranes more rapidly than Li-ions, facilitating their removal from the original brine solution. The amount of sodium present in the resulting mixture can then be reduced by the precipitation of sodium bicarbonate at the correct pH. Following this pre-treatment, the remaining Li-, Na-, and K-cations are pushed across a new membrane to a cathodic chamber where carbon dioxide is being reduced to form carbonate. With proper pH control in this cathodic chamber, selective precipitation of $LiCO_3$ from the solution can be achieved. The discussion of this paper focused on technical details of the membranes used, the scale of the process that can be conducted, and how the process might be merged with other efforts to remove heavy metals in order to reduce its overall environmental impact.

At this point, T. L. Liu and his group discussed mechanistic efforts to understand how the biaryl byproducts that frequently complicate Ni-mediated cross-coupling reactions are formed (Scheme 7) (https://doi.org/10.1039/D3FD00069A). The work indicated that the rate-limiting step for the reaction is the oxidative addition of a cathode-derived Ni(I)-species to an aryl halide. This step is followed by the reductive generation of a Ni(II)-species by Ni(I) in solution, which in turn triggers a ligand transfer where an aryl-group on the Ni(II)-species is transferred to a second Ni(III)-species generated from the initial oxidative addition. The result is the formation of a bis-aryl Ni(III)-species that undergoes reductive elimination to form the undesired biaryl byproduct. The identification of the Ni(II)-species led to the development of a new method to trap it with a benzylic radical, leading to a different cross-coupling reaction, and the identification of approaches to avoid biaryl formation in the initially planned cross-

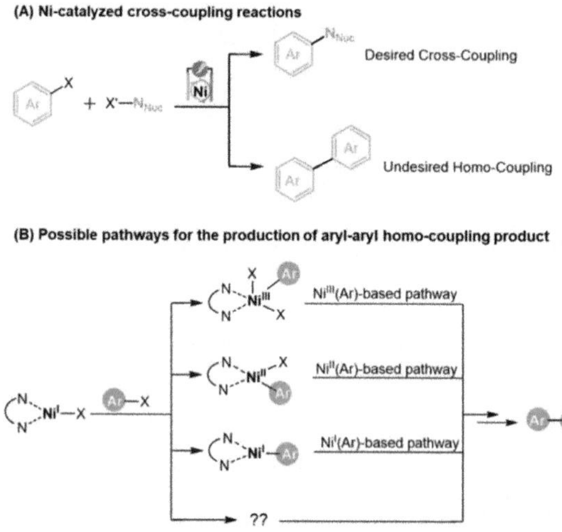

Scheme 7 A mechanistic study on the origin of biaryl products in Ni-mediated cross-coupling reactions. Image taken from https://doi.org/10.1039/D3FD00069A.

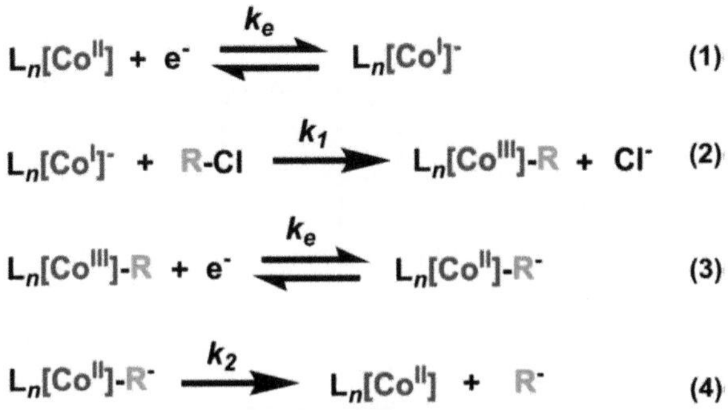

$$L_n[Co^{II}] + e^- \xrightleftharpoons{k_e} L_n[Co^I]^- \qquad (1)$$

$$L_n[Co^I]^- + R\text{-}Cl \xrightarrow{k_1} L_n[Co^{III}]\text{-}R + Cl^- \qquad (2)$$

$$L_n[Co^{III}]\text{-}R + e^- \xrightleftharpoons{k_e} L_n[Co^{II}]\text{-}R^- \qquad (3)$$

$$L_n[Co^{II}]\text{-}R^- \xrightarrow{k_2} L_n[Co^{II}] + R^- \qquad (4)$$

Scheme 8 The influence of electrolyte on Co- and Fe-mediated reduction reactions. Image taken from https://doi.org/10.1039/D3FD00054K.

coupling reactions. The discussion that followed the presentation focused on a number of technical questions about the generation of the Ni(II)-species, its possible generation at the anode, and the role electrolytes might play in the process. What was abundantly clear from the paper and the discussion that followed was how mechanistic insight about the process was being used to develop new synthetic protocols.

Work presented by D. G. Boucher, Z. A. Nguyen, and S. D. Minteer then showed that a similar mechanism operated in Co- and Fe-based systems and illustrated the role electrolyte can play in mediating such processes (https://doi.org/10.1039/D3FD00054K). Shown in Scheme 8 is the mechanistic paradigm presented for these reactions. In this case, an initial reduction afforded a Co(I)-species that then underwent the oxidative addition step to form a Co(III)-intermediate. The Co(III)-intermediate was reduced to a Co(II)-intermediate that then had the potential to transfer an anion. However, the anion was not automatically transferred. In this example, the authors found that the fate of the Co(II)-intermediate was determined by the electrolyte used. When electolytes with softer cations were used, the Co(II)-complex was stable and reaction (4) in Scheme 8 did not readily occur to produce the anion. When electrolytes with harder cations were used, dissociation of the Co(II)-complex to form the anion (R^- in Scheme 8) was favoured. Similar observations were made for the analogous iron-mediated chemistry. The discussion that followed focused on the role of ion pairing in the electrolyte solution and questions concerning the overall solvation of the anions and organometallic species present. The general consensus was that the systems defied a single simple explanation, and that work to gain a better understanding of electrolyte/intermediate interactions was essential for the design of future reactions.

A similar electrolyte dependence was observed by Bernardo Frontana-Uribe and his group (https://doi.org/10.1039/D3FD00064H). In their paper highlighting the utility of biomass-derived dihydrolevoglucosenone (Cyrene™) as a new environmentally compatible solvent, they observed that the reduction of diphenyl ketone derivatives afforded different products when tetrabutyl ammonium tetrafluoroborate was used as the electrolyte, relative to reactions where $LiClO_4$ was used as the electrolyte (Scheme 9). For starters, dihydrolevoglucosenone did prove to be an effective solvent for both reactions. It has

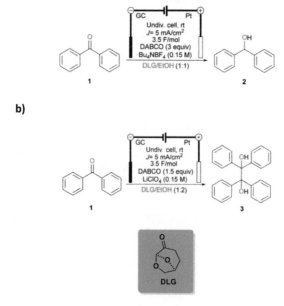

Scheme 9 The use of dihydrolevoglucosenone as an environmentally benign solvent and the role of electrolytes in product determination. Image taken from https://doi.org/10.1039/D3FD00064H.

a polarity similar to dimethylformamide or acetonitrile, although the reactions did require the use of ethanol as a cosolvent because of its viscosity. The discussion of the paper focused on the solvent in terms of its stability to acid or nucleophiles (not an issue to this point), the possibility that the use of dihydrolevogulcosenone as a solvent might induce enantioselectivity into a reaction (currently the solvent has influenced the diastereoselectivity of reactions, but there is no evidence of enantioselectivity), the overall conductivity of the system studied, and the how the electrolyte altered the selectivity in the reactions presented. With respect to the electrolyte, the discussion again focused on the ability of the cations to complex the alkoxides being formed, with the use of lithium leading to complexation of the alkoxide and optimization of the radical character of the intermediates generated. This favoured the dimerization pathway.

Of course, the electrodes used in electrolysis can also play a large role in determining the outcome of the reaction. Y.-J. Liao, S.-C. Huang, and C.-Y. Lin presented a paper that discussed the use of modified electrodes for the conversion of biomass into synthetic building blocks for materials science applications (Scheme 10) (https://doi.org/10.1039/D3FD00073G). The worked focused on ITO electrodes that were coated with a nano-NiOOH borate film. The films were applied to the ITO electrodes by electrodeposition. The authors showed how the quality of subsequent alcohol oxidation reactions at the electrodes depended on the method used for the synthesis of the nano-NiOOH borate film, and then illustrated for the oxidation of 1,6-hexane diol that either adipic acid could be formed at low pH and high potential in a flow system or 6-hydroxyhexanoic acid

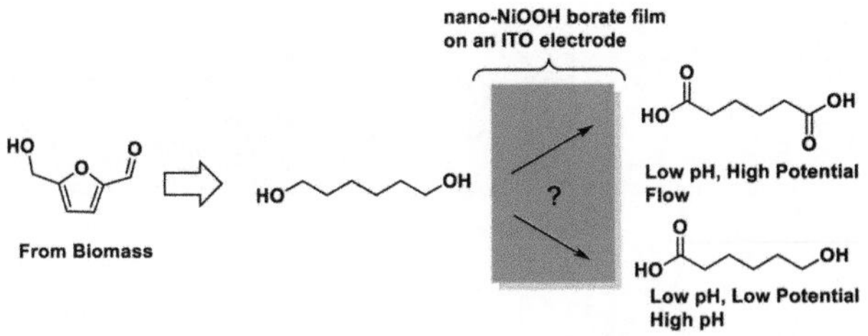

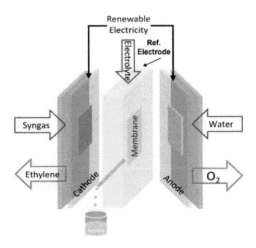

Scheme 10 The use of nano-NiOOH borate films on ITO for the oxidation of biomass derived alcohols.

could be formed, at either low pH at low potentials or at higher pH values. Reaction optimization afforded a chance to look more carefully at the mechanism of the oxidation pathways.

The focus on electrodes continued with a paper presented by Professor Jeannie Tan and her group (Scheme 11) (https://doi.org/10.1039/D3FD00058C). While the capture of carbon dioxide to form carbon monoxide has been optimized, the conversion of CO into value-added products remains problematic. In this paper, the authors point out that the conversion of CO to ethylene is highly dependent on the nature of the cathode material. They explored bimetallic catalysis comprising a Cu(core)-nanoparticle deposited with either Co, Pd, or Ag on a carbon-GDL base. The Ag–Cu nanoparticle provided the best performance. The discussion that followed focused on optimization of the system, how products were separated, the rationale for choosing the alloys studied, the source of the hydrogen atoms on the ethylene product (syn gas or a water/methanol solvent), and the possibility of labelling studies. Throughout, the discussion focused on how we can think about the process and the role of the electrode in accomplishing the desired transformation.

In another paper focused on the use of functionalized electrodes, T. Ashraf from the group of B. T. Mei (along with co-workers A. P. Rodriguez and Guido

Scheme 11 Optimizing the conversion of CO to ethylene. Image taken from https://doi.org/10.1039/D3FD00058C.

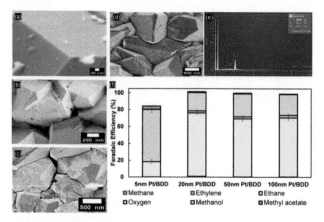

Scheme 12   New electrode materials for decarboxylation reactions. Image taken from https://doi.org/10.1039/D3FD00066D.

Mul) described their development of platinum-nanoparticle-coated boron-doped diamond electrodes and their use for the electrochemical decarboxylation of acetic acid (Scheme 12) (https://doi.org/10.1039/D3FD00066D). The long-range goal of the work is the removal of acids from pyrolysis oil, a step that is involved in converting the oil into value-added materials. Because of the importance of the reactions, the mechanism of the decarboxylation reaction was studied at a variety of electrodes. The Pt-nanoparticle-coated electrodes improved selectivity for the formation of Kolbe-type products. Boron-doped diamond turned out to be the best base electrode material for the experiments because of its high over-potential for oxygen evolution. The reactions led to the formation of

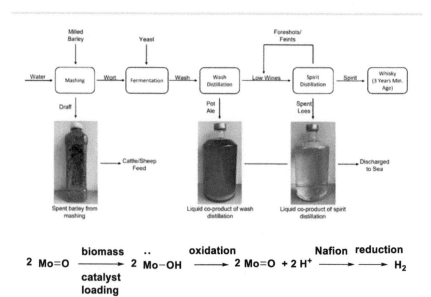

Scheme 13   Studying the utilization of whisky waste streams for the production of hydrogen. Image taken from https://doi.org/10.1039/D3FD00086A.

methanol as a by-product due to the generation of hydroxyl radicals at the electrode surface. The discussion that followed focused on what the electrodes looked like following the reaction and how stable the surfaces were with respect to changes during electrolysis, the procedure for how the nanoparticles are made and how the shape of those nanoparticles can be controlled, the spectroscopy of the surfaces, and the generality of the reaction for the production of more complex Kolbe products looking towards the future.

In a paper that built upon a theme of utilizing biomass waste to synthesize products of value, Robert Price and co-workers examined the use of PMA ($H_3[PMo_{12}O_{40}]$) as a M=O-based catalyst to utilize the organic waste being generated from whisky production as a source material for the production of hydrogen gas (Scheme 13) (https://doi.org/10.1039/D3FD00086A). The chemistry involves oxidation of the organic waste stream that in turn reduces the catalyst and loads it with hydrogen atoms. The catalyst is then regenerated by oxidation leading to the release of protons that then pass through a Nafion membrane and get reduced at a cathode to form hydrogen gas. The work examined the nature of the waste stream from whisky production in order to determine at which stage of the process the waste generated would lead to the most efficient electrochemical process for the production of hydrogen. In this study, it was determined that the use of Pot Ale from the second step of the whisky process was the best source material for catalyst loading (about 70% coating of the catalyst) leading to the largest amount of hydrogen production. However, it appeared that the use of the Draff waste from the first step could be optimized further for near complete efficiency in the catalyst loading step. The discussion of this paper revolved around the ability to scale the process, the energy efficiency of the overall approach, what could be done with the new waste generated from the loading step, why the choice of the PMA catalyst, and how one might accomplish the reaction catalytically given that the loading step in the process is currently a stoichiometric reaction.

In a manuscript describing the synthesis of catalytically active zirconium-doped manganese oxide nanoribbons, S. K. Chondath and M. M. Menamparambath presented their work on the synthesis of materials at a water/chloroform

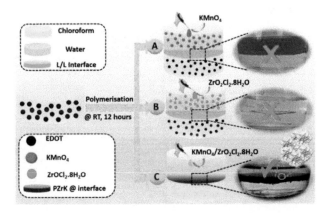

Scheme 14   The synthesis of functionalized nanoribbons on an interface. Image taken from https://doi.org/10.1039/D3FD00077J.

interface (Scheme 14) (https://doi.org/10.1039/D3FD00077J). The authors pointed out that while electrochemical methods are great for making uniform, pure conducting polymers, they are not always compatible with making those polymers with metal catalysts imbedded in them. For that reason, the authors have used a chemical oxidation to trigger polymerization from an EDOT-monomer at a water/chloroform interface. The interface was used to confine the synthesis to a two-dimensional plane. The result of the chemistry was the synthesis of nanoribbons imbedded with K-OMS-2 structures that are doped with Zr. The Mn-catalysts have the basic MnO_2 structure KMn_8O_{16}. During the discussion of this paper, questions focused on what was next, now that the nanoribbons are available (are there applications for the material in electrosyn-thesis), on what the role of the Zr-complex is during the synthesis since the reaction did not proceed without it, and how scalable the process is.

While many of the papers presented during the meeting discussed how mechanistic insights can be used to improve existing electrosynthetic reactions, the consideration of the electrochemical mechanisms can lead to entirely new approaches to synthesis. In their paper, S. Hosseini, J. A. Beeler, M. S. Sanford, and H. S. White presented a method for conducting oxidation reactions under reductive conditions and reduction reactions under oxidative conditions (Scheme 15) (https://doi.org/10.1039/D3FD00067B). The reactions work by generating

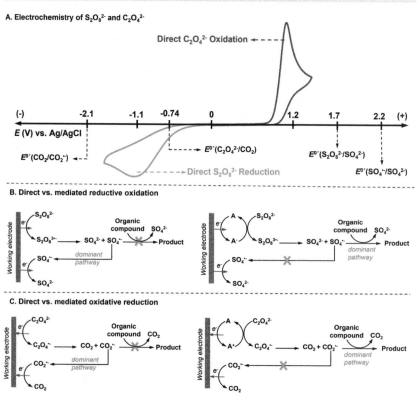

Scheme 15 Conducting reductive oxidations and oxidative reductions. Image taken from https://doi.org/10.1039/D3FD00067B.

reactive intermediates that fragment to afford reactive oxidants or reductants. For example, the mediated reduction of $S_2O_8^{2-}$ leads to a radical anion that fragments to form sulfate and sulfate radical anions ($SO_4^{\cdot-}$). The sulfate radical anion is a very strong oxidant. Hence, the original reduction generates a species capable of oxidizing a variety of organic substrates. Because the oxidant is generated at a cathode, it can be produced without worrying about the oxygen evolution reaction, a scenario that opens the door to new types of applications. In the opposite direction, the authors showed how the mediated oxidation of oxalate ($C_2O_4^{2-}$) affords a radical anion that then fragments to form the radical anion of carbon dioxide ($CO_2^{\cdot-}$). This radical anion is a strong reductant that will transfer an electron to organic molecules in order to generate carbon dioxide. In this way, the original oxidation reaction generates a powerful reductant, and again sets up the opportunity to carry out unique chemistry. The conversation following this presentation focused both on the details of the transformations and on the opportunities the reactions made available for new functional group compatibility. Professor Richard Brown asked an intriguing question about how one could take advantage of the reactive species migrating away from the electrodes and their interactions with groups that were attracted to the electrode surface, since the cathode attracts electron-poor groups. Could the reactions be used to conduct a selective oxidation of an electron poor species attracted to the cathode and in so doing reverse the normal selectivity for an oxidation reaction?

R. Ali, T. Patra and T. Wirth discussed how flow electrochemistry can be used to develop practical alternatives to the ozonolysis reaction (Scheme 16) (https://doi.org/10.1039/D3FD00050H). For this transformation, a convergent paired electrochemical strategy is used to combine the product generated by the cathodic reduction of oxygen with the product generated from anodic oxidation of an electron rich olefin.[19] The radical anion and radical cation generated undergo an addition reaction that affords a cyclic dioxetane structure that then fragments to a ketone. Key to this reaction is the biphasic (gas/liquid) reaction that illustrates how gases can be utilized in a flow reactor. The biphasic system (employing a segmented solvent approach) led to excellent mixing and facilitated the transformation. The discussion of this paper focused on issues associated with gas diffusion, where the oxygen was reduced, and suggestions for improving gas solubility with a focus on reaction optimization.

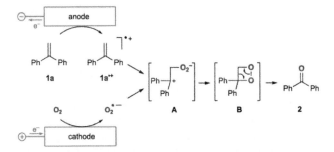

Scheme 16   A flow electrochemical method for the cleavage of olefins. Image taken from https://doi.org/10.1039/D3FD00050H.

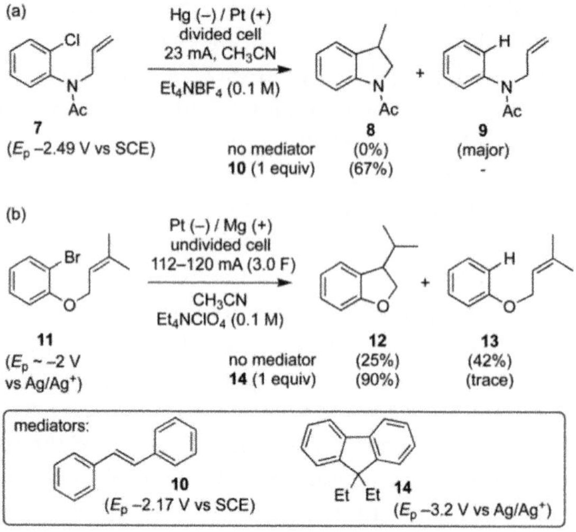

Scheme 17 Mass-transport phenomena in mediated electrolyses. Image taken from https://doi.org/10.1039/D3FD00089C.

A collaborative project between the groups of G. D. and R. C. D. Brown discussed the intriguing observation that mediated reduction reactions can work with a mediator that is reduced at a more negative potential than the substrate (Scheme 17) (https://doi.org/10.1039/D3FD00089C). The fact that the products generated were indeed a product of a mediated process and not a direct reduction was established by demonstrating that the mediated electrolysis reaction led to a different product than the direct (reaction (a) in Scheme 17), and then showing that the example with the more negative mediator led to the product from the indirect electrolysis (reaction (b) in Scheme 17). The presentation went on to make the point that the normal consideration of potentials ignores mass transport at the electrodes. Using a series of simulations, the authors illustrated how at the very start of the reaction both the substrate and the mediator were reduced, but following the initial reduction that consumed the starting material at the cation, reduction of the mediator occurred. As that mediator migrated away from the cathode, it encountered unreduced substrate and reduced it in a manner that prevented the substrate from reaching the electrode surface. This left only mediator at the electrode, which was reduced, migrated away from the cathode, and again encountered substrate that was migrating toward the electrode. The cycle was repeated, leading to formation of the product from indirect electrolysis. The result of the work is an opportunity to rethink how we design new mediator/reaction systems. The discussion of the manuscript focused on technical questions, rate constants, and an effort to gain a better understanding of how to think about the systems. Professor Fuchigami raised the idea of an alternate method for generation of the cyclization product (typically derived from the mediated process) from a direct reduction reaction using sonication.

Finally, B. Garcia, J. Sampson, M. P. Watson, and D. Kalyani presented their efforts to compare electrochemical and chemical cross electrophile coupling

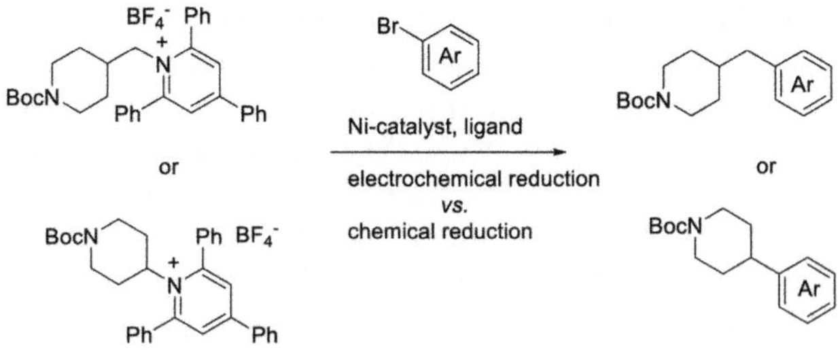

Scheme 18 A comparison of chemical *vs.* electrochemical Ni-mediated cross electrophile coupling reactions.

reactions utilizing primary and secondary alkylpyridinium salts (Scheme 18) (https://doi.org/10.1039/D3FD00120B). The talk presented details of a study that used high-throughput experiments to conduct a systematic comparison of the scope of the aryl bromides that could be utilized with both substrates. It was found that the trends observed for the different arylbromides (with 37 different arylbromides examined) were similar to the use of the two different alkylpyridinium salts. The study included 13 more complex aryl bromides in the library. For most cases, the electrochemical and chemical approaches were similar. The chemistry illustrated a general approach for evaluating the true potential of an electrochemical method in the context of its chemical alternative, a comparison that is essential if we hope to define, for a larger synthetic community, the true value of the electrochemical methods being developed.

Conclusions

New synthetic chemistry is often the product of new mechanistic insight. It is our understanding of how things work that drives innovation, and it was an effort to gain that understanding that drove the discussion during this meeting. In one form or another, the question asked after every single presentation was "how does this work?". Participants sought details that would enable them to best think about the methods discussed and better design their own experiments in the future.

Those discussions led to several themes that were consistently raised. In connection with efforts to develop new synthetic capabilities, questions were asked about how we can more carefully define the role of electrolytes in our reactions and then use those observations as design features in connection with future transformations, how we can capitalize on what we are learning about the design of electrode surfaces and the subsequent control of product selectivity, and how we can translate new, intriguing scientific discoveries like reductive oxidations into new methods that accomplish things that cannot be done any other way.

In connection with efforts to utilize electrochemistry to make more sustainable processes, the questions asked concerned how we can best balance fundamental studies with providing practical solutions to societal challenges, and whether the problems we are working on are truly solvable with the fundamental discoveries being made. It was generally understood that it is the answer to these questions that will help define the long-range utility of electrochemistry and how the larger community adopts and takes advantage of those methods in the future.

While the meeting impressively focused on mechanistic details of the methods being developed, at the end of the discussion the synthetic-organic-chemistry part of me was left with a feeling that we can still do better. On many occasions, we appropriately pointed out how synthetic chemists would benefit from paying more attention to the electrochemical details of the reactions they are running. Participants familiar with electrochemical methods for energy conversion and harnessing available sources of carbon, like carbon dioxide, know how to think about surface control of reactions, electrocatalysis, and cell optimization in ways that would enrich broader synthetic efforts. But, that information exchange is a two-way street.

When carrying out electrosynthesis, do we think enough about the physical organic chemistry part of our electrochemical reactions? We talk about the selectivity of kinetically controlled reactions, but how many transition state pictures are included in those discussions, even though it is the transition state that determines that selectivity? How many times do we as a community define if a reaction is under thermodynamic or kinetic control, or talk about electrolytes as counter ions for alkoxides in a broader context of what is known about enolates and/or alkoxides in general? How many times is a reaction discussed without illustrating a structure or potential mechanism that would provide a context for that discussion?

These comments are not meant to be a criticism, just an observation and a challenge. As a community, can we get more out of our question "how does this work?" if we think about both electrolytes, electrode surfaces, cell design, and the physical organic chemistry behind the reactions we are pursuing? The answer to that question is yes. The "story" of what electrochemistry offers for the larger community is being defined at impressive pace. By more completely considering both the electrochemistry and the physical organic chemistry elements of that story, we who spend our time at that interface can better define for others how to think about electrochemical events and the reactions they trigger. In so doing, we will make the chemistry accessible to an even greater number of scientists, and it is that expansion of the number of talented people working on electrochemical methods that will truly drive new innovation and a still more expansive future.

Conflicts of interest

There are no conflicts to declare.

Acknowledgements

Financial support for this work was provided by the National Science Foundation Center for Synthetic Organic Electrochemistry (CHE-2002158).

Notes and references

1 For selected reviews of electroorganic synthesis see: (*a*) J. B. Sperry and D. L. Wright, *Chem. Soc. Rev.*, 2006, **35**, 605–621, DOI: 10.1039/b512308a; (*b*) J. Yoshida, K. Kataoka, R. Horcajada and A. Nagaki, *Chem. Rev.*, 2008, **108**,

2265–2299, DOI: 10.1021/cr0680843; (c) B. A. Frontana-Uribe, R. D. Little, J. G. Ibanez, A. Palma and R. Vasquez-Medrano, *Green Chem.*, 2010, **12**, 2099–2119, DOI: 10.1039/c0gc00382d; (d) R. Francke and R. D. Little, *Chem. Soc. Rev.*, 2014, **43**, 2492–2521, DOI: 10.1039/c3cs60464k; (e) M. Yan, Y. Kawamata and P. S. Baran, *Chem. Rev.*, 2017, **117**, 13230–13319, DOI: 10.1021/acs.chemrev.7b00397; (f) J. E. Nutting, M. Rafiee and S. S. Stahl, *Chem. Rev.*, 2018, **118**(9), 4834–4885, DOI: 10.1021/acs.chemrev.7b00763; (g) K. Mitsudo, Y. Kurimoto, K. Yoshioka and S. Suga, *Chem. Rev.*, 2018, **118**(12), 5985–5999, DOI: 10.1021/acs.chemrev.7b00532; (h) N. Sauermann, T. H. Meyer and L. Ackermann, *ACS Catal.*, 2018, **8**, 7086–7103, DOI: 10.1021/acscatal.8b01682; (i) G. S. Sauer and S. Lin, *ACS Catal.*, 2018, **8**(6), 5175–5187, DOI: 10.1021/acscatal.8b01069; (j) S. Möhle, M. Zirbes, E. Rodrigo, T. Gieshoff, A. Wiebe and S. R. Waldvogel, *Angew. Chem., Int. Ed.*, 2018, **57**(21), 6018–6041, DOI: 10.1002/anie.201712732; (k) Y. Okada and K. Chiba, *Chem. Rev.*, 2018, **118**(9), 4592–4630, DOI: 10.1021/acs.chemrev.7b00400; (l) M. Elsherbini and T. Wirth, *Acc. Chem. Res.*, 2019, **52**(12), 3287–3296, DOI: 10.1021/acs.accounts.9b00497; (m) P. Xiong and H. –C. Xu, *Acc. Chem. Res.*, 2019, **52**(12), 3339–3350, DOI: 10.1021/acs.accounts.9b00472; (n) G. Hilt, *ChemElectroChem*, 2020, **7**(2), 395–405, DOI: 10.1002/celc.201901799; (o) H. Li, Y. –F. Xue, Q. Ge, M. Liu, H. Cong and Z. Tao, *Mol. Catal.*, 2021, **499**, 111296, DOI: 10.1016/j.mcat.2020.111296; (p) S. B. Beil, D. Pollok and S. R. Waldvogel, *Angew. Chem., Int. Ed.*, 2021, **60**(27), 14750–14759, DOI: 10.1002/anie.202014544; (q) F. Marken, A. J. Cresswell and S. D. Bull, *Chem. Rec.*, 2021, **21**(9), 2585–2600, DOI: 10.1002/tcr.202100047; (r) M. Dörr, M. M. Hielscher, J. Proppe and S. R. Waldvogel, *ChemElectroChem*, 2021, **8**(14), 2621–2629, DOI: 10.1002/celc.202100318.

Poster titles

Bromide-mediated silane oxidation: a convenient and practical counter-electrode process for undivided electrochemical reductions, **M. Avanthay, O. Goodrich, M. George and A. Lennox,** *University of Bristol, United Kingdom*

Legion: next-generation parallel, high-throughput electrochemical array with diverse analytical and synthetic applications, **B. Gerroll, M. Personick, K. Kulesa, E. Hirtzel, X. Yan and L. Baker,** *Texas Agricultural and Mechanical University, U.S.A.*

Electrosynthesis of homoquinone and oxygenated heterocycles in the reduction of 1,2-naphthoquinone in the presence of phenacyl bromide, **M. Batanero, L. Pastor, N. Salardón and S. Cembellin,** *University of Alcalá, Spain*

Reduction by oxidation: mediated oxalate oxidation in DMF-water mixtures for the reductive hydrodechlorination of 4-chlorobenzonitrile, **J. Beeler and H. White,** *University of Utah, U.S.A.*

The electroactive species of aliphatic aldehydes during their electrocatalytic oxidation at gold electrodes, **C. Bondue and K. Tschulik,** *Ruhr University Bochum, Germany*

A spectroscopic investigation of glycine production on a $CuPb_{1ML}$ catalyst, **P. Broersen and A. Garcia,** *University of Amsterdam, Netherlands*

Improved electrosynthesis of biomass derived furanic compounds *via* redox mediation design, **E. Carroll, D. Boucher and S. Minteer,** *University of Utah, U.S.A.*

A reaction with potential: an electrosynthesis of 1,3,4-oxadiazoles from N-acyl hydrazones, **L. Chen, J. Thompson and C. Jamieson,** *GlaxoSmithKline Medicines Research Centre* and *University of Strathclyde, United Kingdom*

Development of an electrochemical benzylic $C(sp^3)$-H amidation reaction, **A. Choi, O. Goodrich, M. Edwards, A. Atkins, M. George, D. Heard and A. Lennox,** *University of Bristol, United Kingdom*

(Photo)electrochemical formate production from biomass by using copper oxide as a selective electrocatalyst, **P.-C. Chuang and Y.-H. Lai,** *National Cheng Kung University, Taiwan*

Biomimetic mineralisation of conformal TiO_2 protection layers on nanostructured photoelectrodes for sustainable fuel production, **P.-C. Chuang and Y.-H. Lai,** *National Cheng Kung University, Taiwan*

Developing an electrochemical assay for the directed evolution of enzymes for electrosynthesis, **R. Gerulskis, D. Boucher and S. Minteer**, *University of Utah, U.S.A.*

Electropolymerization of aromatic monomers driven by a streaming potential, **S. Iwai, I. Tomita and S. Inagi**, *Tokyo Institute of Technology, Japan*

Progress in the development of an electrochemically-reversible hydride transfer mediator for organic synthesis, **D. Karr and D. Smith**, *San Diego State University, U.S.A.*

Non-native electrocatalysis using engineered non-heme iron enzymes, **L. Kays, J. Buchholz, D. Boucher and S. Minteer**, *University of Utah, U.S.A.*

Solution-based synthesis of Earth-abundant materials for photoelectrochemical solar fuel production, **Y.-H. Lai**, *National Cheng Kung University, Taiwan*

Tuning carbon nanomaterials for electrochemical and bioelectrochemical reduction of nitrocompounds, **M. Landis, T. Sudmeier, N. Grobert and K. Vincent**, *University of Oxford, United Kingdom*

Efficient electrosynthesis of CO from electrocatalytic CO_2 reduction using a polyaniline|gold nanoparticle core-shell nanofiber modified electrode, **T.-H. Wang and C.-Y. Lin**, *National Cheng Kung University, Taiwan*

Electrifying Friedel-Crafts intramolecular alkylation, **E. Lunghi, P. Ronco, F. Della Negra, B. Trucchi, M. Verzini, E. Casali, C. Kappe, D. Cantillo and G. Zanoni**, *University of Pavia, Italy*

Surveying the structural dependence of electrochemical properties and performance of verdazyl radicals in organic redox flow batteries, **M. Milner, S. Fricke, D. Diddins, E. Horst, A. Korshunov, A. Heuer, M. Grünebaum, M. Winter and A. Studer**, *University of Münster, Germany*

Electrolyte induced cage effects for enantioselective electrosynthesis, **Z. Nguyen, K. McFadden, D. Boucher and S. Minteer**, *University of Utah, U.S.A.*

Parametrization of catalytic system for organic electrosynthesis by real-time mass spectrometry, **P. Nikolaienko, A. Cuomo, M. Asfia, R. Cloth and K. Mayrhofer**, *Forschungszentrum Jülich, Germany*

Synthesis of N^+-doped triphenylene derivatives by electrochemical intramolecular pyridination and their optoelectronic properties, **Y. Ohno, S. Ando, D. Furusho, R. Hifumi, I. Tomita and S. Inagi**, *Tokyo Institute of Technology, Japan*

Dihydrolevoglucosenone-EtOH (Cyrene®-EtOH) mixture as solvent in electrochemical reduction of carbonyl compounds, **J. Ramos-Villaseñor, J. Sotelo-Gil, S. Rodil and B. Frontana-Uribe**, *Centro Conjunto de Investigación en Química Sustentable Universidad Autónoma del Estado de México-Universidad Nacional Autónoma de México, Mexico*

Tri-metallic electrocatalysts for prompting C-C bond formation during CO_2 reduction(CO_2R), **I. Brewis, A. Jedidi and S. Rasul**, *Northumbria University, United Kingdom*

Turning essential oil into a degradable polymer by catalytic oxidation, **T. Seko, N. Shida and M. Atobe**, *Yokohama National University, Japan*

Electrochemical C-N coupling reaction by π-extended haloarene mediator, **N. Shida, S. Yoshinaga and M. Atobe**, *Yokohama National University, Japan*

Photoelectrochemistry for sustainable hydrogen generation and isotope separation, **E. Sokalu, I. Sato, H. Matsushima and K. Brinkert**, *University of Warwick, United Kingdom*

Electrochemical polymer reaction of poly(3-hexylthiophene) *via* anodic C-H phosphonylation, **K. Taniguchi, I. Tomita and S. Inagi**, *Tokyo Institute of Technology, Japan*

Computational fluid dynamic modelling of electrochemical reactor for CO_2 conversion to ethylene, **A. Virdee, J. Tan, A. Jahanbakhsh, M. Van der Spek, M. Maroto-Valer and J. Andresen**, *Heriot-Watt University, United Kingdom*

Towards the electrification of metal-ligand cooperative catalysts, **S. Kasemthaveechok, D. Tocqueville, F. Crisanti and N. von Wolff**, *Paris Cité University, France*

Efficient and selective electrosynthesis of 4-aminophenol at circumneutral pH from the electrocatalytic reduction of nitrophenol over the nickel-iron phosphide modified electrode, **P. Ong, S. Huang, N. Lerkkasemsan and C. Lin**, *National Cheng Kung University, Taiwan*

Electrochemical reactor dictates site selectivity in *N*-heteroarene carboxylations, **G. Sun, P. Yu, W. Zhang, W. Zhang, Y. Wang, L. Liao, Z. Zhang, L. Li, Z. Lu, D. Yu and S. Lin**, *Eastern Institute of Technology, China*

The Green Chemistry Poster Prize for the best poster was awarded to Mr Mickaël Avanthay from the University of Bristol, United Kingdom, for their poster on "Bromide-mediated silane oxidation: a convenient and practical counter-electrode process for undivided electrochemical reductions".

The Reaction Chemistry & Engineering Poster Prize for the best poster was awarded to Mr Zach Nguyen from the University of Utah, U.S.A., for their poster on "Electrolyte induced cage effects for enantioselective electrosynthesis".

List of participants

Professor Dr Lutz Ackermann, *Georg-August-Universität Göttingen/Institut für Organische und Biomolekulare Chemie, Germany*
Dr Anas Alkayal, *University of Surrey, United Kingdom*
Mr Talal Ashraf, *University of Twente, Netherlands*
Mr Mickaël Avanthay, *University of Bristol, United Kingdom*
Professor Lane Baker, *Texas Agricultural and Mechanical University, U.S.A.*
Dr William Barton, *Lubrizol Limited, United Kingdom*
Professor Dr M. Belén Batanero Hernán, *Universidad de Alcalá, Spain*
Mr Joshua Beeler, *University of Utah, U.S.A.*
Miss Rita Blasi Alsina, *E-watts Technologies Sociedad Limitada, Spain*
Dr Christoph Bondue, *Ruhr-University Bochum, Germany*
Dr Dylan Boucher, *University of Utah, U.S.A.*
Mr Pim Broersen, *University of Amsterdam, Netherlands*
Professor Richard Brown, *University of Southampton, United Kingdom*
Ms Emily Carroll, *University of Utah, U.S.A.*
Mr Luke Chen, *University of Strathclyde/GlaxoSmithKline, United Kingdom*
Dr Anthony Choi, *University of Bristol, United Kingdom*
Mr Ping-Chang Chuang, *National Cheng Kung University, Chinese Taipei*
Miss Molly Colgate, *Royal Society of Chemistry, United Kingdom*
Dr Robert Davidson, *Pfizer, Ireland*
Miss Tracey Dean, *Royal Society of Chemistry, United Kingdom*
Dr Mark Dowsett, *Alvatek, United Kingdom*
Professor Dr Victoria Flexer, *National Scientific and Technical Research Council-National University of Jujuy, Argentina*
Mr Ching Wai Andy Fong, *University of Leeds, United Kingdom*
Professor Robert Francke, *Leibniz Institute for Catalysis, Germany*
Professor Dr Bernardo Antonio Frontana-Uribe, *Centro Conjunto de Investigación en Química Sustentable Universidad Autónoma del Estado de México-Universidad Nacional Autónoma de México, Mexico*
Emeritus Professor Toshio Fuchigami, *Tokyo Institute of Technology, Japan*
Mr Rokas Gerulskis, *University of Utah, U.S.A.*
Professor David Hickey, *Michigan State University, U.S.A.*
Dr Bee Hockin, *Royal Society of Chemistry, United Kingdom*
Dr Seyyedamirhossein Hosseini, *University of Utah, U.S.A.*
Miss Shih-Ching Huang, *National Cheng Kung University, Chinese Taipei*
Professor Dr Shinsuke Inagi, *Tokyo Institute of Technology, Japan*
Mr Suguru Iwai, *Tokyo Institute of Technology, Japan*
Dr Dipannita Kalyani, *Merck & Company Incorporated, U.S.A.*
Mr Dylan Karr, *San Diego State University, U.S.A.*
Mr Luke Kays, *University of Utah, U.S.A.*
Dr Chris King, *United Kingdom*
Professor Dr Alexander Kuhn, *University of Bordeaux, France*

Professor Yi-Hsuan Lai, *Department of Materials Science and Engineering, National Cheng Kung University, Chinese Taipei*

Dr Kevin Lam, *University of Greenwich, United Kingdom*

Ms Maya Landis, *University of Oxford, United Kingdom*

Ms Yun Ju Liao, *National Cheng Kung University, Chinese Taipei*

Professor Song Lin, *Cornell University, U.S.A.*

Professor Chia-Yu Lin, *Department of Chemical Engineering, National Cheng Kung University, Chinese Taipei*

Professor Dr Tianbiao Leo Liu, *Utah State University, U.S.A.*

Mr Enrico Lunghi, *Università degli Studi di Pavia, Italy*

Professor Long Luo, *Wayne State University, U.S.A.*

Miss Kirsty McRoberts, *Royal Society of Chemistry, United Kingdom*

Dr Mini Mol Menamparambath, *National Institute of Technology Calicut, India*

Mr Matthew Milner, *University of Münster, Germany*

Professor Shelley Minteer, *University of Utah, U.S.A.*

Professor Dr Kevin Moeller, *Washington University in St. Louis, U.S.A.*

Professor Andrew Mount, *School of Chemistry, University of Edinburgh, United Kingdom*

Dr Jacob Murray, *Lubrizol Corporation, United Kingdom*

Mr Zachary Nguyen, *University of Utah, U.S.A.*

Dr Pavlo Nikolaienko, *Helmholtz Institute Erlangen-Nürnberg for Renewable Energy, Germany*

Professor Dr Toshiki Nokami, *Tottori University, Japan*

Mr Yushi Ono, *Tokyo Institute of Technology, Japan*

Dr Wade Petersen, *University of Cape Town, South Africa*

Dr Robert Price, *University of Strathclyde, United Kingdom*

Mr José Manuel Ramos Villaseñor, *Universidad Nacional Autónoma de México, Mexico*

Dr Shahid Rasul, *Northumbria University, United Kingdom*

Mr Colin Robertson, *Hiden Analytical, United Kingdom*

Mr Pietro Ronco, *Università di Pavia, Italy*

Dr Hugh Ryan, *Royal Society of Chemistry, United Kingdom*

Mr James Scanlon, *University of Surrey, United Kingdom*

Mr Leon Schneider, *Julius-Maximilians-Universität Würzburg, Germany*

Mrs Rosalind Searle, *Royal Society of Chemistry, United Kingdom*

Mr Tatsuya Seko, *Yokohama National University, Japan*

Professor Dr Naoki Shida, *Yokohama National University, Japan*

Miss Eniola Sokalu, *University of Warwick, United Kingdom*

Miss Jessica Sotelo Gil, *Universidad Autónoma del Estado de México, Mexico*

Dr Jeannie Tan, *Heriot-Watt University, United Kingdom*

Mr Kohei Taniguchi, *Tokyo Institute of Technology, Japan*

Miss Ashween Virdee, *Heriot-Watt University, United Kingdom*

Dr Niklas von Wolff, *Centre National de la Recherche Scientifique, France*

Dr Susan Weatherby, *Royal Society of Chemistry, United Kingdom*

Dr Nipunika Weliwatte, *Lubrizol Corporation, U.S.A.*

Professor Dr Thomas Wirth, *Cardiff University, United Kingdom*

Dr Toshiaki Yamaguchi, *National Institute of Advanced Industrial Science and Technology, Japan*

Dr Peng Yu, *Eastern Institute for Advanced Study, China*